A Study of Microscopic Polish on Flint Implements

Irene Levi Sala

BAR International Series 629
1996

Published in 2019 by
BAR Publishing, Oxford

BAR International Series 629

A Study of Microscopic Polish on Flint Implements

ISBN 9780860548102 paperback
ISBN 9781407349350 e-book

DOI https://doi.org/10.30861/9780860548102

A catalogue record for this book is available from the British Library

This book is available at www.barpublishing.com

BAR Publishing is the trading name of British Archaeological Reports (Oxford) Ltd. British Archaeological Reports was first incorporated in 1974 to publish the BAR Series, International and British. In 1992 Hadrian Books Ltd became part of the BAR group. This volume was originally published by Tempvs Reparatvm in conjunction with British Archaeological Reports (Oxford) Ltd / Hadrian Books Ltd, the Series principal publisher, in 1996. This present volume is published by BAR Publishing, 2019.

BAR
PUBLISHING

BAR titles are available from:

BAR Publishing
122 Banbury Rd, Oxford, OX2 7BP, UK
EMAIL info@barpublishing.com
PHONE +44 (0)1865 310431
FAX +44 (0)1865 316916
www.barpublishing.com

To Nissim 1936-1989

Abstract

This is a report on research carried out to investigate the microscopic polish which appears on the edges and surfaces of prehistoric flint tools after they are used on various materials. A natural polish also occurs on the surface of flint under some depositional conditions and the project investigated both this natural polish and its possible effects on polish produced by use. Interest in microscopic polish was pioneered by L.H. Keeley who in the late 1970s published several long papers and a book which examined polish on flint. Keeley suggested very forcefully that the polish produced on flint tool edges by use has a distinctive texture and appearance determined by and characteristic of the material on which the tool was used. Keeley and his followers have since published studies of many archaeological assemblages where the interpretation of function has been based mainly on the distinctiveness of polish appearance. The present research was prompted by "microwear analysis" of implements from several sites of different ages which suggested that post-depositional processes had negatively affected the surface of the flint implements and altered or degraded the polish making it difficult to characterize and casting doubt on the distinctive nature of polish on flint.

An experimental programme was then devised to produce polish on flint surfaces under a variety of controlled conditions using different materials.

The experiments were designed to investigate the processes involved in polish formation rather than to replicate tasks which prehistoric people are likely to have carried out.

The results suggest that:

a) flint is polished by the removal of asperities from its surface during use which then act as abrasives between the flint and the worked material;
b) consequently use produces one polish on the flint microtopography in several stages of development;
c) the rate of polish development depends on length of use, hardness of used material, lubrication, pressure and flint texture;
d) striations and other linear features are caused mainly by the flint particles from asperities and edge damage but that added abrasives do affect the quantity and size of the striations;
e) polish distribution on the microtopography is related to the hardness of the material used, external abrasives and the coarseness of the flint;
f) distribution of polish on the tool edge is related to the task performed;
g) in order to interpret function correctly; polish appearance and distribution are not sufficient. Other clues such as edge morphology and macrowear have to be taken into consideration.

Two reports are included regarding the attempted functional reconstruction of archaeological artefacts from two sites: Hengistbury Head in Dorset and Kebara Cave in the Carmel Hills. Due to the altered condition of the flint surfaces only a limited level of interpretation was possible.

Acknowledgements

This research has benefitted from the help of many to whom I am grateful. I would like to thank first of all my supervisor Dr. Mark H. Newcomer for all the help and advice during the many years of my research and for getting me interested in stone tools in the first place.

During the work described in the second part of this project my supervisor was Dr. Dafydd R. Griffiths. To him go my thanks not only for his guidance and encouragement throughout the period but particularly his patience in imparting to me some of his scientific knowledge to carry out this work. I am particularly grateful to him for the time spent and the valuable comments given on reading the earlier drafts of this thesis.

Prof. Ofer Bar Yosef of Harvard University (previously of the Institute of Archaeology Hebrew University Jerusalem) has very kindly allowed me to study both the material from Kebara and from many of the other sites sampled for this work. I am grateful also for his encouragement to continue along this path in moments of doubt.

Dr. Ruth Amiran, Dr. R.N.E. Barton, Dr. Desmond Collins, Mrs. Joan Crowfoot Payne, Dr. Roger Jacobi, Dr. I. Gilead, Dr. Avi Gopher, Dr. Nigel Goring-Morris, Dr. Tamar Noy, and Prof. Abraham Ronen have all put artefacts from their sites at my disposal for study and I am deeply grateful for their kindness and cooperation.

Nick Barton has also read earlier drafts of the Hengistbury Head section and I thank him for making my English more readable.

My appreciation to Itzik Gilead for the lively and useful discussions and to Liliane Meignen for enlightening me on the Mousterian.

I would also like to thank my friends and colleagues at the Institute for lively discussions and suggestions especially Kathy Ataman, Roger Grace, Margaret Maher, and Norah Moloney. Appreciation and thanks are offered for the special cooperation received from Chris Bergman in my experimental work and much moral support in the early days of my work. My gratitude goes to Stuart Laidlaw without whose help the quality of the pictures in this thesis would not have been the same.

The drawings have been made by Jeff Wallis, Christine Wilson, Tessa Rickards in the U.K., and Mariana Grinblat in Jerusalem.

A special *thank you* to Corine Cadranel, Tracy Morris and Barbara Scott who have lovingly and patiently helped with the intricacies of word processing and other problems...

And finally to Davide, Monica and Marina, whose patience, encouragement and understanding supported me throughout and specially to Vicky without whom this would not even have begun.

Preface

The work described in this thesis falls into two or more or less distinct parts although the second has been inspired by the first. Part I starts with Chapter I which is an historical review of recent microwear research. Chapter 2 describes briefly what the aims of my project were and explains how they have had to be modified in Part II.

The contents of Chapter 3 are intended as a description of all the various steps and techniques generally followed and/or recommended for use in functional analysis by incident light microscopy at magnifications of x50 to x400 and by scanning electron microscopy. Comments and observations have been offered which are the result of the experience acquired while carrying out this research.

In Chapter 4 artefacts from two sites (microdenticulates from a Mesolithic site in Dorset and scrapers from a Mousterian cave site in Israel) have been studied. An attempt was made at a reconstruction of possible tool uses although the post-depositional alterations affecting their surfaces allowed only a more limited level of interpretation than would have been possible had use-wear polish survived on their edges. A list of all the sites sampled whose arrtefacts were found to be in a similar state as far as surface alterations were concerned is also supplied. Appendix 1 lists the experiments carried out to investigate the probable uses of microdenticulates in great detail, while Appendix 2 reviews the edge wear and breakage pattern of the archaeological artefacts. Some cautionary remarks regarding examples of a higher, but not necessarily more accurate, level of interpretations are given in Chapter 5.

Part II deals with the new research focus resulting from the problems encountered in Part I. Details of the revised aims are given in the first part of Chapter 6. The same chapter also contains description of all the experiments carried out. They have been divided into three categories:

a) experiments on use-wear traces,
b) experiments on the processes of polish formation,
c) experiments on post-depositional processes.

The division is not rigorous due to overlap, particularly since some of the artefacts used in a) and b) were then subjected to post-depositional processes in c). All the artefacts in a) were used by hand and most though not all the artefacts in b) were subjected to mechanical polishing. Chapter 6 is concluded by enumerating the limitations of this project.

In conclusion, Chapter 7 starts by presenting the model of polish formation as it was proposed by Anderson-Gerfaud in 1980, and goes on to propose a new model based on the results of the various experiments and on the observed flaws in the previous model. The results of the various aspects of this research are reviewed together and suggestions for future research are given.

Irene Levi Sala 1929–1991

Irene Levi Sala's successful academic career as a highly motivated and innovative archaeologist ended with her untimely death in an air crash over Southern Chile on February 20, 1991.

Irene was a latecomer into the world of archaeology. After raising her daughters and pursuing a successful career in business with her husband Davide, she decided to follow her interest in archaeology. In 1979, she became a student at the Institute of Archaeology, London, where she earned a B.A. honours degree in 1982.

Her enthusiasm for microwear research began while she was still working on her undergraduate thesis, and later developed more fully into the Ph.D dissertation presented in this volume. In the early 1980s microwear was a relatively new branch of archaeology. It was not surprising, given her meticulous approach, that Irene soon discovered that several of the accepted methods of research were not without their problems. She began her own research by studying the visible alterations on the surfaces of flint implements in order to discover the processes involved in 'polish' formation and surface modification caused by post-depositional processes. Following a series of meticulous experiments and a thorough analysis of their results, Irene reached the conclusion that polish formation was the result of abrasion of the flint surface and not an additive process, as had been widely accepted. These experiments also led her to conclude that post-depositional surface modifications affected the visible wear traces on the surface of an implement. Such natural modifications, if ignored, could lead to the misinterpretation of use-wear patterns by relating them to human activity. Her Ph.D dissertation was essentially dedicated to this important line of inquiry, but motivated by the concern that archaeologists should be aware of these natural effects on stone artefacts, Irene also lectured and published the results of her studies. Upon completion of her Ph.D in 1989 she continued her research as an honorary research fellow of the Institute of Archaeology in London.

The personal knowledge of the sites and their excavators Irene gained from field experience made it possible for her to use artefacts from excavations in England, Israel and Italy in her laboratory studies. During her career, Irene participated in various excavations in England and Israel, including several seasons at the excavations of Kebara cave. Her involvement with the Kebara project prompted the excavators to dedicate the first volume of the site's publication in her memory. Indeed, all those who took part in the same excavations as she, admired both her ability to accommodate herself to poor living conditions and physically strenuous work and the charm and intelligence with which she conversed with her fellow workers.

Irene regularly attended meetings of the Lithic Studies Society and the Palaeolithic and Mesolithic day conferences in England. Her last research results were presented in December 1990 at 'Les Gestes Retrouvés', an international microwear conference held at the University of Liège, where she found that her results had been corroborated by the work of Japanese colleagues. Both organizers and participants held her work in high regard, and the two volumes that resulted from the conference, published in 1993, were dedicated to her memory.

Irene was also an active member of the Lithic Studies Society for ten years, as well as a member of the Israel Prehistoric Society. Her deep interest in the prehistory of the Holy Land led her to endow the foundation originally called the Care-Archaeological Foundation, an organization which annually allocates grants to scholars world-wide for prehistoric research in Israel or for research on collections from Israel. This foundation, now called the Irene Levi Sala Care Archaeological Foundation, bears her name in tribute to her unflagging dedication to the field of prehistoric archaeology.

Finally, Irene Levi Sala was not just a respected colleague, but a well loved friend. She was kind and always good company, and made time for everyone, always willing to advise or help with any problem. Her untimely death left a gap among family members and numerous friends and colleagues. Her loss is still felt deeply; her contributions, both personal and professional, will always remain indelible.

Ofer Bar-Yosef
Harvard University, Department of Anthropology

Note from the Editors

Irene was a friend and colleague. At the time of her death, she was in the process of preparing her Ph.D dissertation for publication. We felt that Irene's important contribution to microwear studies should not go unpublished and therefore undertook to finish the task she had started. Although Irene may well have made changes to the original text, we decided to leave it virtually as she had written it. Our changes have been primarily those of presentation to suit the published format of the book.

We would have been unable to complete this task without the help of Dr. M. Czwarno who came to our aid and prepared the computerised copy. We are deeply grateful to him. Thanks too, to F. Swartz for the initial unravelling of computer problems. Without Barbara Scott's unfailing help and encouragement, we would have been lost. And to Irene's family, a sincere thank you.

Margaret Maher and Norah Moloney

Glossary/Definitions

Abrasion "The result of rubbing or wearing off or away" (Concise Oxford Dictionary).

Adhere As used in this thesis particles from the worked material are said to "adhere" to the surface of flint when they are not removed by cleaning with detergent, nylon brush and/or by ultrasonic cleaning. They can be removed by biological detergents and/or alkaline solutions. It is **not** intended to suggest that they are fused in one way or another to the surface (see also DEPOSIT).

Beilby Layer A term loosely used after Beilby's (1921:81) theory regarding wear on metal surfaces. He theorized that through high temperatures the surface was softened or liquified and smoothed material solidified as a layer on the surface. When this theory was transferred to glass polishing it was hypothesized that the silica particles were being removed and hydrolysed—in the presence of water and high frictional temperatures—and were being redeposited as amorphous silica gel on other parts of the surface as a layer. Although this "Beilby layer", or a "silica gel layer" is often invoked in the literature dealing with polish on flint, there seems to be no conclusive evidence for its presence even on glass.

Built up layer This term has been used in the literature dealing with microwear. In chemistry it is used to indicate particle growth (Iler 1979:220), whereas in physics and tribology it is used to indicate the accumulation of particles moved by plastic deformation from their original location on the surface. It is in this latter sense that it is used in the text.

Deposit As used here particles from the worked material are "deposited" on the surface of flint when they are visible microscopically and sometimes macroscopically before cleaning with warm water and detergent (see also Adhere).

Edge damage Edge scarring produced by natural processes and covers both microscopic and macroscopic observations. It is by definition not intentional.

Edge wear Macroscopic and microscopic edge scarring (microflaking) on the edge as well as edge rounding.

Elastic deformation Is said to occur when the tensile strength produces an "instantaneous and completely reversible" deformation and when "stress and strain are linearly proportional to one another" (Anderson *et al.* 1985:182–183).

Embedded As used by Anderson Gerfaud (1981) is referred to here in the sense that during a hydrolysed phase of the flint surface external particles (such as plant phytoliths) are trapped within the hydrolysed layer to become a permanent part of the flint structure (see also ADHERE & DEPOSIT). No embedded particles have been observed in this research.

Flow "To move with a continual change of place among the particles or parts" (Shorter Oxford English Dictionary).

Friction "...the resistance to motion which exists when a solid object is moved tangentially with respect to the surface of another which it touches, or when an attempt is made to produce such motion." (Rabinowicz 1966).

Hydrocarbon oil "...is an organic chemical compound of hydrogen and carbon" (Shorter Oxford English Dictionary). It is also a non-polar liquid. Hydrocarbons and non-polar liquids are unable to hydrolyse a flint surface.

Hydrolysis "Decomposition by chemical reaction with water" (Concise Oxford Dictionary).

Inland Used here to describe the area of a flint implement away from the edge towards the centre of the piece. Under the microscope the edge of a flint artefact is reminiscent of a shoreline and this term seemed appropriate since topographic terms are sometimes used to describe the flint surface.

Invasiveness Relative to the distance from the edge that flint seems polished.

Linear features Used as a generic term as in the literal sense meaning all indicators of motion whether grooves in the polish or streaks of polish on the unpolished surface.

Longitudinal motions Like sawing, cutting, reaping etc. where the movement is parallel to the working edge.

Macroscopic Visible to the naked eye (Shorter Oxford English Dictionary).

Motion The direction a tool moves in relation to the worked material during use.

Plastic deformation Is said to occur when you "increase the tensile strength until there is a permanent shape change". (Anderson *et al.* 1985:155–6).

Polish When flint surfaces are observed optically at x200 and certain altered areas of the flint exhibit, after thorough cleaning, a higher reflectivity than that of the adjoining surface, those reflective areas are said to be polished and the reflectivity is referred to as "polish".

Polished by Used here to mean a mechanical polishing of the flint by a worked material.

Polished on Used here to mean that the polish was produced by using the flint on a worked material by hand.

SILICA:

–Amorphous "Consists of regions of atomic order or crystals ofextremely small size, which by careful X-ray diffraction studies appear to have the cristobalite structure" (Iler 1979:27). "Any method which rearranges the surface molecules will give an amorphous surface" (Cornish 1961:17; Adam 1927:162).

–GEL Is a "coherent ,rigid three dimensional network of contiguous particles of colloidal silica". (Iler 1979:462) It can be formed in different ways but always in the presence of "a liquid medium, usually aqueous". (*ibid.*) which results in a chemical reaction between silica and water through hydrolysis yielding SiOH. It can be said to be hydrolysed amorphous silica.

STRIATIONS Linear features appearing on the polished flint surface whether as furrows or grooves. They are useful in reconstructing the direction of the tool's movement against the worked material.

SURFACE FLOW The movement of material from one part of a surface to another usually by friction. It is often, but not always, accompanied by high temperatures hence the distinction THERMIC SURFACE FLOW and ATHERMIC SURFACE FLOW.

TRANSVERSAL MOTIONS Such as scraping, whittling, planing, where the movement of the tool is perpendicular to the working edge.

TRIBOLOGY "It originates from the Greek *tribos* meaning rubbing. Hence it is the 'study of rubbing' or more correctly 'the study of the science and technology of interacting surfaces in relative motion'" (Quinn 1971:1).

USE-WEAR Used in this thesis to mean all traces left on the edge of a flint tool after use. It is used primarily to refer to use-induced polish but also to striations, edge scarring and rounding.

WEAR "The progressive loss of substance from the operating surface of a body occurring as a result of relative motion at the surface". (OECD 1969; Dowson 1979:499)

PART I
CHAPTER 1
INTRODUCTION

1.1 Development of Microwear Research

Given the very exhaustive surveys of the history of Functional Analysis which have appeared in recent years, only a brief outline of important earlier work will be presented here. This review will discuss in more detail the period following the advent of "microwear polishes" inaugurated by the publication of L.H. Keeley's work at Oxford between 1974 and 1977.

Attempts to reconstruct the function of stone tools have been made since the 19th Century (Frre 1800; Nilsson 1838–43; Evans 1897; Pfeiffer 1912; Vayson 1922) and have relied heavily on ethnographic parallels, speculation and occasional experimental replication of tasks. Wear traces on actual ancient tools were deduced either through macroscopic observation of the edges or through examination with a hand lens.

A comprehensive review of the literature from 1839 to 1978 was published by Deborah Seitzer Olausson; it did not appear until 1980 although it was compiled before the appearance of the proceedings of the First Conference on Lithic Use-wear held in 1977 in Canada (Hayden 1979). Olausson discussed the various achievements both chronologically and under headings relating to each specific breakthrough both in Europe and the U.S.A. Other reviews were compiled by Keeley (1977), Hayden and Kamminga (1979), Vaughan (1981), Odell (1982), Moss (1983b:9–22), Unger-Hamilton (1984), Cook and Dumont (1987), Juel-Jensen (1988a) and Levi Sala (in press).

With the appearance in 1964 of the English translation of Semenov's book *Prehistoric Technology*, interest in the function of stone tools was given new impetus. Semenov concentrated on the location of polish and striations for the interpretation of tool movement. Binocular microscopes with magnifications of x 20 to x 180 were used to observe not only edge scar, but striations and "polish" which at low magnifications usually appear only as a dull lustre. Semenov reported the results of experiments aimed at understanding kinematics, examining the tool's motion in sawing, scraping, whittling, boring etc., although he did not consider such experiments as an essential aspect of tool use reconstruction (1964:121).

Semenov suggested that in certain cases magnifications from x 300 to x 500 (*ibid.*:22 may be required for observation of characteristic features. Nonetheless, very few attempts were made by Semenov or others to identify the worked material except by ethnographic analogy or logical reasoning (Semenov 1964:83-142).

Most work done in the decade following 1964 concentrated on examination of tools at lower magnifications with a view to reconstructing tool movement and orientation. Tringham *et al.* (1974), Broadbent and Knutsson (1975) and Hayden and Kamminga (1973) tried to observe patterns of edge scarring from which to infer the relative hardness of materials.

The biggest qualitative jump attempted by use-wear analysis was first seen in a paper by Keeley in 1976. He reported that microscopic polish on flint edges produced by different worked materials was identifiable under the incident light microscope at magnifications of around x 200. While Vayson in 1920 was probably the first to suggest that polish could indicate the worked material, he did not follow up his claims with experiments (1920:470–1).

Keeley however, based all his work on thorough experimentation and the subsequent blind test (Keeley and Newcomer 1977) gave the impression that there is a strict correlation between distinct microwear polishes and different worked materials. Fifteen tools from retouched and unretouched flakes of Brandon flint were made and used by M.H. Newcomer in "many different ways on a wide variety of materials" (Newcomer and Keeley 1979:195). They were used to scrape, chop, drill, saw and whittle wood, scrape bone, butcher fresh meat, cut frozen meat and cut hide and bracken. Keeley examined the tools by his newly developed method and obtained the following published results:

- Identification of area of tool used 14 of 16 correct;
- Reconstruction of tool movement 12 of 16 correct;
- Identification of material worked 10 of 16 correct;

The count is out of a total of 16 because one tool was used on two different edges. The errors were partly attributed to Newcomer using a wooden chopping board in several experiments. Nonetheless this was considered a very "impressive" result.

The method was later supported by the Ph.D. theses of Anderson-Gerfaud and Vaughan in 1981 and the blind test carried out by Gendel and Pirnay in 1982 (see comments on both blind tests below). This one-to-one relationship of polish to used material has since been so widely accepted that the literature is full of expressions such as "bone polish", "wood polish", "hide polish" and even "meat polish".

Some criticism did follow Keeley's first results (Holley and Del Bene 1981) based mainly on the way the results were assessed and on the conviction that criteria other than microwear polishes were used implicitly in the identification of the worked materials. However, many researchers were only too happy to be able to read more into stone tools than just their morphology and technology of manufacture and accepted Keeley's work rather uncritically.

1.1.2 The Low Power Approach

Inspired by Semenov's work, Tringham and some of her students (1974) carried out extensive experimentation to observe whether edge scarring could be used to identify precise function. They concluded that there was a correlation between certain patterns of microflaking on flint tool edges, the motion in which the tool was used and the hardness of materials against which the tool was used.

Keeley (1980:7) criticised their experimental programme as being too mechanical and not replicating human activity. He also felt that in their effort to isolate too many variables they have produced "results which are unlikely to be comparable to the wear traces found on archaeological specimens and which are irrelevant to the inference of function on ancient implements."

The low power method was further developed by Odell, who was a member of the original Tringham team, in an attempt to establish it using magnifications of x 10 to x 100 "as a reliable analytical tool for deriving functional information from lithic artifacts" (Odell and Odell Vereecken 1980). In the Odells' work the results of a blind test of the method were published. The Odells, like Tringham, felt that they were able to distinguish intentional retouch and post-depositional scars from those caused by use. They have however attributed certain errors in their blind test to the fact that some retouch scars were mistaken for utilisation damage (*op.cit.*:117–118).

Their correct identification of precise worked material (as opposed to material hardness on a 4-part relative hardness scale) was 38.7%, while the Keeley-Newcomer blind test claimed 62.5%. Recent work suggests, as illustrated below, that the latter may have been an over-optimistic assessment of the accuracy of the Keeley method. Keeley (1980:83), Vaughan (1985:45), Moss (1983b:76) and Unger-Hamilton *et al.* (1987:274) concluded however, that not only is edge scarring due to use not diagnostic of worked material but, more importantly, that it is not always present. It is my observation that edge wear is, when present, a helpful element for the interpretation of tool use but cannot on its own be used as the indicator of precise function for the reasons mentioned above and as discussed in Chapter 3.

In contradiction to the assertion in Odell and Odell Vereecken (1980:90) that "in their criticism of the low power approach Keeley and Newcomer are correct in asserting that the most ubiquitous and frequently used index is scarring along the edges", Odell (1986:117) recently pointed out that "scholars employing low power techniques **do not rely solely on use fracturing** (Odell's emphasis) but on an entire complex of dislocatory and abrasive variables. Edge rounding and polish are crucial indicators of the existence and location of wear". He further feels that critics of the low power method have always employed non-stereoscopic metallurgical microscopes thus losing "salient visual images".

I have not personally looked at edge-wear under a *stereoscopic* microscope but experience in the experimental use of flint tools has not contributed to convince me of the reliability of information derived at such low magnifications. Only independent testing will confirm whether this is a robust technique and what levels of interpretation can be reached through it.

Another paper has recently been published by Shea (1987), a student of Odell's, who describes a series of blind tests. The experiments were carried out by undergraduate and graduate students of lithic technology workshops in Harvard and Boston Universities (one set of test tools were used by Odell) and the functional analysis based on the Odell method was carried out by Shea. The tests were, as Shea says, as much to test his own ability as to test the method. He divides the categories of worked material into animal, vegetal and inorganic as well as each being subdivided into soft, medium and hard. In view of this difference with the Odell scoring and the Keeley and Newcomer test it is difficult to make precise comparisons. There are no drawings or detailed description of the tools in Shea's test, neither is there an explanation of the way they were used or of the material on which each tool was used. No formal tools were involved but rather flakes were picked up by the experimenters from a knapping floor (Shea pers. comm.)

There is no doubt that IF the Odell method were reliably tested and shown to be valid, its greatest advantage would be the speed of analysis "ca. 5 min per tool" (*ibid.*:120). When using higher magnifications Moss points out (1983b:21) that it takes "up to an hour for any one piece". In the Tuebingen blind test the analysts found that they spent "on the average one and a half hours to analyse" each tool without the "exploratory experimentation to solve specific problems" which was denied to them as they were to complete the test on nineteen tools in only one week (Unrath *et al.* 1986:165). It is my experience that it can take much longer than one or two hours for complex pieces.

If a five minute per tool analysis could give accurate answers on the relative hardness of the material used as well as the motion, it would be worthwhile either to lower the threshold of expectations or to consider this as a first level of analysis.

1.2 Situation at present

Since 1979, many microwear analysts have been working in five continents interpreting the functions of tools in prehistoric assemblages, the majority probably using some form of the Keeley method (see the Registers of Microwear Analysts and their Research No. 1 and No. 2 Owen *et al.* 1984; Owen 1986). The activities of prehistoric people are thus supposedly "reconstructed" in great detail down to the traces of finger prehension even when the results fly in the face of archaeological knowledge. The literature abounds with descriptions of wear attributed to hafts, estimates of duration of work, faint traces on 1.5 million year-old tools interpreted as butchery (Keeley and Toth 1981) and distinctions between working bone and antler based on polish appearance alone (see Chapter 5 for examples of this type of work).

The Tuebingen multi-analyst blind test has shown that the most experienced microwear analysts are not especially successful in reconstructing such details on implements manufactured and used only months if not weeks before (Unrath *et al.* 1986:171).

1.2.1 Processes of Polish Formation

There is as yet no generally accepted explanation of the phenomenon of polish formation. It has been intriguing to many that a material as hard as flint can be polished by materials so much softer than itself. The relatively recent greater access to electron microscopy has produced several observations which may assist in understanding polish formation. Various theories have been put forward mainly in regard to the high gloss observed on Neolithic sickle blades. The models can be summed up in their broadest lines as follows:

A use wear polish is additive caused, in the case of sickles, by a deposition of plant silica on the flint surface (Witthoft 1967).

B polish is a subtractive process, the result of high attrition removing a microscopic part of the flint surface with or without any concomitant chemical process (Diamond 1979; Masson 1981).

C polish is a layer build-up due to increased temperature (caused by friction) resulting in the formation of amorphous silica gel trapping phytoliths and other elements from the plant cells which contribute to the build-up (Anderson-Gerfaud 1981).

D That polish is the result of some combination of the above processes (Del Bene 1979; Kamminga 1979; Meeks *et al.* 1982; Unger Hamilton 1984).

In 1980 Patricia Anderson reported the observation by SEM (Scanning Electron Microscopy) of what she interpreted as "phytoliths" embedded in the flint surfaces. The theory with which she explained the phenomenon was that "the flint surface changes locally to an amorphous gel during the working of plant materials and that the amorphous gel layer imprisons plant residues" (1980:186). She felt they could be diagnostic of the plant family being harvested.

Anderson's work was questioned by Masson who criticised these claims and, using the SEM found no phytoliths, amorphous silica or plant opal on a sample of sickles from Mureybet (Masson *et al.* 1981).

The weakness of Masson *et al.*'s conclusions seems due to the fact that the Mureybet sickles were burnt (*ibid.*:47) and that their conclusions were based on the amorphous silica not being observable on the sickles at present while Anderson claims the edges had passed through an amorphous silica stage during use and then recrystallized.

Subsequently Masson studied experimental sickles used to cut dry hay and concluded (1982:36) "No structural change of the flint or of other rock types was observed by SEM on the experimental pieces. This confirms the same observations carried out on the Neolithic sickles from Syria" (my translation). It is my experience that polish will develop from cutting dry hay only after very long hours of work especially when using Near Eastern flint.

Anderson's work was, on the other hand, supported by Vaughan, who had also attempted (1981:15; 1985:34) to identify reed families from the optical appearance of the polish. Vaughan suggested (1985:36) that the identification of "phytolith bodies fused within the plant polish" by SEM analysis is a more reliable way to recognise plant families.

Mansur-Franchomme (1983b) observed by SEM what she considers to be hide particles embedded in experimental flint tools. Hurcombe (1985:54) has observed by SEM "phytoliths incorporated into the obsidian surface after working grasses with experimental tools". The claims for residues of hide and meat being embedded in the surfaces of stone tools are even more difficult to explain. A process such as envisaged by Anderson could only occur when very high temperatures are reached during work. When flint is used to cut silica-rich grasses one might expect relatively high temperatures to be reached because of the contact of two very hard substances the silica in grass and flint.

There is however no evidence for these high temperatures and some experimental evidence against (see section 6.5). Other materials such as meat and hide are much more yielding than silica and, as less friction is produced, one cannot even *expect* very high temperatures to develop when they are worked.

Recent work by some authors has cast doubt on the theory of polish trapping organic particles, either because such a layer would be too thin to trap such particles or because there is as yet no concrete evidence that such a layer actually forms during use. Unger-Hamilton (1984:94 Fig 3b) has photographed a feature visually similar to one of Anderson's "phytoliths" (1980:Plate 8d) on flint used only "to polish other flint" suggesting that some features of the flint structure may be mistaken for phytoliths. Yamada (1986:123) has also observed the process of plant polish formation on siliceous shale under the SEM, concluding "it was possible that residues adhere to the surface, but impossible that they are embedded into the surface" because he observed that "hydration in the order of less than a micron may be occurring on the surface". I have observed the same features under the SEM on flint polished with diamond paste. This paste is made of pure carbon and hydrocarbon oil. It is therefore certain that the flint did not come in contact with any phytoliths. Furthermore figure 220 suggests that the thickness of the polished layer is well under a micron.

Andersen and Whitlow (1983) have used ion beam analysis and hydrogen profiling to study the nature of "glossy wear traces" (*ibid.*:470). They reported a "good correlation between changes in hydrogen profiles and the material worked" (*ibid.*:470). It seems however that the correlation is with the water content of the worked material on experimental implements. Hence working dry materials does not increase the flint's water uptake and working wet or moist materials does. More importantly this effect on the flint surface is transient since no such differences have been found on archaeological implements. Andersen and Whitlow suggest (*ibid.*:468) that the increase in water uptake indicates the formation of a silica gel layer of the order of 100 to 500 nm (nanometres equivalent to a fraction of a micron). They have actually only demonstrated that the hydrogen on the surface of the flint had been enhanced by working materials with a high water content and NOT that a silica gel layer had formed.

My own work regarding processes of polish formation is described in Part Two and suggests that polish is a smoothing of the flint surface produced by the removal of surficial asperities which are sometimes dragged across the surface flattening it. The smoother and flatter surface reflects light more uniformly and appears as a "polished" surface under the optical microscope.

The increase in the number of microwear practitioners mentioned earlier has led to an increase in the identification of problems with the Keeley method. A great overlap in the diagnostic features of "polishes" produced by the various worked materials was noted even by the most ardent supporters of the method (Vaughan 1985:30 and 46). It was furthermore observed that the same material produces different "polishes" under different circumstances and that "polishes" from two very different materials can look very similar (Moss 1983:86–87; Unger-Hamilton 1988:107; Unrath *et al.* 1986: 168).

Keeley's (1980) characterisation of the various "polishes" already hints at the problem, and Vaughan's very detailed descriptions (1985:27–46) highlight the cause of confusion. It is implied in the work of these researchers that in spite of this overlap they can distinguish between one polish and another.

Furthermore, the rather vague nature of polish descriptions such as "bright", "dull", "matte", "smooth" and "rough" makes identification of polishes a very subjective matter. In a survey of "bone polish" characteristics as described by Keeley (1980:43), Anderson-Gerfaud (1981(1):58), Moss (1983b:92) and Vaughan (1985:32) reported by Grace *et al.* (1988:220), it was noted that "bone polish can be rough or smooth, matte or bright with micropits or depressions or both and there seems to be no real agreement about the precise appearance of a distinctive bone polish".

1.2.2 Distinctiveness of Polish

It is necessary to bear in mind on reading these pages that when this research was undertaken the Keeley method of functional analysis was considered very reliable and that the doubts discussed in the preceding pages are relatively recent. The most serious criticism was the one which Holley and Del Bene (1981) proffered to the Keeley and Newcomer (1977) blind test following its presentation at the Vancouver Conference and its publication in 1979 (Hayden ed.). Keeley rightfully replied that instead of judging his method through experimental replication and microscopic observation, his critics chose to employ a binomial test to verify what they thought he saw.

The experimental replication and microscopic observation of microwear polish practised by me for the last five/six years have led me to conclude that polish is not distinctive according to the worked material. The arguments in favour of this conclusion are illustrated in Chapter 6.

In the blind test just mentioned the rate of success in the identification of the used material was sufficient to convince some archaeologists that polishes are distinctive and that consequently this was a promising method of functional analysis.

There is much controversy regarding the way blind tests are scored (Holley and Del Bene 1981; Bamforth 1988; Grace 1988) therefore another rescoring of other people's results will not be attempted here.

What one can learn from this controversy, however, is that in the case of any future blind tests there is a need for:

a) a unified scoring system to allow comparability;
b) a much larger number of tools per test to allow the application of statistical techniques;
c) tests to be carried out as in Keeley and Newcomer (1977) by independent experimenters and analysts and not by members of a team who are likely to be familiar with each other's experimental habits and thought processes.

In the meantime the following are some comments about Keeley's apparent high rate of success in the blind test :

1) Had polishes been clearly distinctive no one, especially not Keeley, would confuse bone polish with hide polish (tool N.15) or confuse bone and antler with wood (tools N. 11,6).
2) Much information allowing tool use to be reconstructed can be gleaned from edge morphology, edge angle and edge wear. The errors made by Keeley suggest that he may have incorporated this kind of information in his observation of the polishes. In fact his success rate is attainable even assuming polish is not material specific as the blind test carried out by Grace *et al.* (1988:228) has shown.
3) Polish distribution is highly correlated with friction and/or rigidity of the worked material which can be inferred from edge wear, striations, and the type of edge used. The experiments in the Keeley and Newcomer blind test were heavily biased towards rigid worked materials which develop polish faster than softer materials.

Another blind test whose success is often quoted in support of the Keeley theory of distinctive polishes is the one carried out by Gendel and Pirnay (1982:251–260).

The apparent high success rate is due to the fact that Pirnay was limited to using only three materials: wood, hide and bone a very unlikely situation when analysing an archaeological assemblage.

The three materials were at the same time quite different in their relative hardness which could have made it easier to choose between them. Furthermore a tolerant scoring method allowed points for only partly correct answers.

There is no doubt that polish is a very sensitive indicator of wear. When it is well developed and there is a patterning on the edges of several tools, polish can indicate that these edges have been used and strongly suggest the motion by the distribution and linear features. What polish cannot do is **by its appearance alone** (e.g. smoothness, brightness, greasiness) suggest the material used.

Many other clues need to be taken into account in order to interpret the material used, and these include edge angle and morphology, edge wear and flint characteristics.

Keeley has never stressed the importance of considering all these clues together. He and his adepts insist that the various worked materials produce very different and distinguishable polishes (Keeley and Newcomer 1977:37; Keeley 1980:83; Vaughan 1985:45–46; Bamforth 1988). Keeley does describe polish in terms that form a continuum of variation and to some extent overlap yet he leaves the clear impression that once the art is mastered wood polish would never be mistaken for hide polish, for example (*ibid.*: 83). In fact in 1981 he says:

> By the standards of experimental science, the existence of distinctive microwear polishes is amply demonstrated, and it is hypothetical only to the degree that all such empirical generalizations or correlations are (e.g. "the sun will rise tomorrow")(1981:349).

In the controversy that followed the publication of Newcomer *et al.* 1987 (Moss 1987; Hurcombe 1988; Bamforth 1988) many analysts have indicated that they have always been using other data such as tool morphology, edge morphology, raw material, edge wear etc., in order to reconstruct tool use.

In evaluating the Keeley method one has to bear in mind that while Keeley in his meticulous work did record edge angle, edge wear, striations and polishes independently, he concluded that "utilization damage was found to be of limited utility in distinguishing different activities and worked materials" (1980:174).

He did not find edge angle to affect the appearance of microwear polishes (*ibid.*:42–59) and did not stress the correlation of edge angle and edge morphology to certain uses or the information they can give the functional analyst. He furthermore noted that "the type of flint involved had no effect on the formation or appearance of any of the microwear features" (1980:16).

He was clearly implying that polish appearance was the only important clue. Consequently those analysts who have been complementing the Keeley method with other observations in addition to polish appearance have presumably done so as a result of their own experience acquired through experimentation in stone tool use. They may thus have effectively developed their own method, but their success does not validate the Keeley method per se.

In 1985 at the Tuebingen Conference on "Technical Aspects of Microwear Studies on Stone Tools", the widespread doubts concerning the reliability of the distinctiveness of microwear polishes for the identification of the used material were made public.

Two multi-analyst blind tests Newcomer *et al.* (1986) and Unrath *et al.* (1986), discussed at the Conference, revealed great problems with the method. The Newcomer *et al.* (1986:215) paper described the ease with which polish can be confused. In fact, in the blind tests polish produced by antler was identified as "plant polish" and polish from hide was taken for wood or plant. This paper has subsequently aroused much controversy but the critics seem to have overlooked that the Tuebingen blind test (Unrath *et al.* 1986:15) was just as disappointing with only 26% success rate in giving "specific answers" to worked material. These latest results produced by four analysts, some with several years of experience, come therefore, as a confirmation that there is a flaw in the method itself.

A new multi-variant method has recently been developed by Grace *et al.* (1988; Grace 1989) which takes into account explicitly and in a standardised manner, all the variables thought to be diagnostic of function. The variables are: raw material characteristics, edge morphology, macroscopic and microscopic edge wear, polish development and distribution. Observations are made macroscopically and microscopically at up to x200. Experiments to test interpretations are an integral part of the method.

Systematic recording and standard vocabulary will make it possible to replicate and compare observations. The results of the blind tests carried out to verify the method show "a far greater level of accuracy" than in previous tests (Newcomer *et al.* 1986:Table 1). Work was continued to improve the method which has since formed the basis of a Ph.D. thesis published in B.A.R. (Grace *et al.* 1988; Grace 1988, 1989). It is my opinion that this new approach will make functional analysis much less time consuming, less subjective and more realistic in its aims. It has done away with the very subjective information about the appearance of the polish itself relied on by Keeley (1980) such as its "smoothness, bumpiness, greasiness and melted snowbanks". Consequently the most obvious advantage of the low power approach (Odell *et al.* 1980), its speed, now seems far less important especially when compared to the degree of accuracy obtained by the new method.

When Keeley started his research, the accepted method of functional analysis in the West was by low power observation of edge scarring and striations as advocated in the U.S.S.R. by Semenov (1964) and continued in the U.S.A. by Tringham *et al.* (1974).
Keeley's work was directed mainly to show, correctly, that edge wear alone is a poor indicator of function and often cannot be observed because edges are retouched, because post-depositional edge scarring cannot be confidently isolated and lastly because certain tasks and certain edge angles lead to almost no scarring on the edge. Keeley's valuable contribution was to point out the importance of micropolish as observed at high magnifications under incident light microscopes.

Odell (1986:117) now maintains that users of his low power method do observe polish at x50 with the stereoscopic microscopes. However at such low magnifications one can barely ascertain the presence of polish let alone understand its stage of development. Furthermore, in order to study its distribution on the edge and on the microtopography of the flint, incident light microscopy at x100 and x200 is necessary. Functional analysis, including interpretation of the worked material, can be carried out on experimental implements if a careful approach is adopted taking into account edge characteristics, edge wear and **much experimental verification** of the conclusions.

It is by implicitly taking these factors into account that Keeley was so successful in his 1977 blind test.

Keeley's emphasis on the visual description of polish has made analysts following his method believe that they can learn to memorise the appearance of polish. Many including Keeley (1980) and Vaughan (1985) discuss polish types in terms of a reference collections implying that one could have some sort of Munsell chart to recognise polish. Many of Keeley's followers have either not experimented enough, or not supported their conclusions by sufficient experimental evidence, or worse still relied on photographic evidence to identify used material (see section 5.5). Unfortunately polish appearance alone is insufficient to obtain the complex and sophisticated reconstructions that the analysts using the high power method have been proposing since the publication of Keeley's work. In conclusion while Keeley was right to point out that relying on edge wear alone is insufficient to reconstruct function, one must stress that relying only on the appearance of polish is also insufficient to obtain the fine details archaeologists have now come to expect from the high power method.

1.2.3 Quantification

Attempts have been made to overcome the qualitative and subjective nature of polish description. This was done in the hope that quantitative methods will make the identification of contact material through polish appearance more reliable. Keeley (1980) himself attempted to measure the brightness of polishes with the light meter attached to the photographic apparatus of the microscope. He discounted its capacity to classify polishes. In 1982 John Dumont used interferometry with the same aim, but he acknowledges that the method has difficulties. As Dumont illustrates in his Plate 12 (1982:214), interference patterns seem to be a contour map of the polish. It does not seem, however, to measure criteria which will replace polish description.

In other attempts to attain objectivity, the profilometer was used in assessing use-wear traces by Akoshima (1981) and Kjel Knutsson (pers. comm.). They have both since abandoned the method as they did not find it could measure the fine nuances of the flint topography.

At the Valbonne Conference in 1986, Beyries *et al.* presented their work with the profilometer (Rugosimetre). A flint blade which was polished to a high gloss was examined and the profilometer perceived the polished areas as totally flat while some of the undulations are apparent to the naked eye and are certainly visible at x100 (figures 119 and 226).

The use of image processing techniques (Grace *et al.* 1985) borrowed from satellite photography have also been applied to use-wear traces. They were used to analyse the textures of polishes produced by contact with various materials. Although they may partly overcome the problem of describing polish objectively, the study emphasised that "polish textures do not cluster simply according to contact material" (Newcomer *et al.* 1986:212; Newcomer *et al.* 1988).

It may well be that the biggest stumbling block to quantitative methods is that different worked materials do not produce distinctly different "polishes" that can lend themselves to quantitative assessment. A further obstacle is that most techniques used to detect fine irregularities on relatively flat surfaces cannot be used on flint on which the original microtopography is irregular. It is consequently difficult to distinguish between those irregular surfaces and the surfaces produced by use-wear.

1.3 Analysing Archaeological Implements

All the problems described above have been encountered while analysing **experimental** flint tools which are carefully handled after use and stored singly in plastic bags. It is obvious therefore that archaeological implements present greater complications.

After discard in prehistoric times, archaeological artefacts may have been exposed to the elements for some time, trampled, swept in a rubbish dump, and buried for thousands of years during which time they were attacked by soil chemicals and abraded by surrounding sediments.

It has become widely recognised that artefacts can move within the sediment even without there being clear stratigraphic disturbance. Such vertical and horizontal displacement of artefacts is much more common than previously imagined. It has been highlighted by the conjoining of artefacts from several stratigraphic levels on a variety of sites (Cahen and Moyersons, 1977; Barton and Bergman, 1982; Villa, 1982; Rowlett and Robins, 1982; and others).

1.3.1 Post-depositional Surface Modification (PDSM)

The problems discussed above concern experimental investigation of the distinctiveness of polishes on experimental implements carefully handled after use. As mentioned earlier, analysing implements unearthed after thousands of years is bound to be more problematic.

Unfortunately while Keeley warned researchers to analyse only fresh implements, he did not define freshness as far as microwear is concerned. There are many pieces which will be considered fresh by archaeologists in view of the absence of abrasions on ridges and edges and yet are either patinated or affected by soil sheen. Keeley did not clarify that such post-depositional processes can affect and obliterate the polish itself and are obstacles to microscopic analysis. He covered the subject of naturally produced traces at length but only from the perspective of scratches and edge scarring and concluded:

> Most natural processes, then, leave traces which are unlikely to cause much confusion to the microwear analyst, once he is familiar with them and with true

microwear features, and provided he studies them at high enough magnifications. (1980:28-35).

Scratches and other wear traces which this author has observed both on archaeological artefacts and on those affected by trampling and movement in the sediment (see experiments described in section 6.5.1 and 6.5.2) are easily recognisable and need not be mistaken for use-wear. Keeley's statement is confusing however, since a more serious surface modification can be produced by natural processes.

The pioneers in the investigation of natural surface modifications on flint were Rottlaender (1975;1976) and Stapert (1976) working independently and on different aspects of the problem. Rottlaender attempted to produce certain commonly observed types of patina in the laboratory using chemicals. He was investigating this problem to ascertain whether it is possible to infer the age of a flint implement from the degree of patination on its surface. His work, though not directly aimed at the microwear traces, is nonetheless often quoted as it is the most recent on the subject in English (see Curwen 1940; Schmalz 1960).

Stapert on the other hand described a number of flint surface phenomena observed on Dutch artefacts and tried to "suggest possible modes of genesis" (1976:1). His reasons for carrying out the project were:

a) "some phenomena described below could easily be interpreted as traces of use";
b) "traces of use formerly present may be entirely or partly obliterated".

As Stapert pointed out, although the problem is generally known to exist, often work is done without sufficient account being taken of it, as in the case of Semenov's important work (1964).

Seitzer-Olausson (1980) points out that, "while Stapert's paper should be basic to all edge-wear studies, it has received relatively little notice" (*ibid.*:54). Since 1980, many functional analyses using the Keeley method have been carried out without a discussion of how it was ascertained that the observed "use-wear" traces were actually the result of use.

In a paper presented to the fourth International Flint Symposium at Brighton in 1983, Plisson and Mauger reported the results of experiments designed to test the resistance of use-wear polishes to soil chemicals (1988:3–16). This was prompted by the fact that they noticed "it is rare to observe microwear on archaeological pieces which are completely characteristic of polishes produced by experimental tool use" (*ibid.*; see also Gysels and Cahen 1981).

Degradation of the microwear polishes by soil chemicals was observed, in differing degrees according to the material worked, the quality of flint and the chemical solution employed.

Objections were raised by certain participants at the Brighton Symposium that the chemicals in the experiments, sodium hydroxide (NaOH), calcium oxide (CaO) and sodium carbonate (Na_2CO_3) were too concentrated or unlikely to occur in natural conditions.

Plisson subsequently presented the results of a further set of experiments to the Tuebingen Conference in 1985. In the new experiments, agents likely to occur in nature were used such as urine, ammonia, peat, fertilisers, etc. (Plisson 1986:111–116). Although polish degradation did not occur to the same extent as in the first instance, the experiments confirmed that polishes are degraded when exposed to such natural agents. As in all experiments attempting to replicate a process which occurs archaeologically over thousands of years, it is often necessary to increase one element (in this case chemical concentration) to obtain results within manageable time spans.

Plisson's contribution in pointing out the degradability of microwear polishes is very valuable. It is difficult to agree with him, however, when he concludes that it is possible to analyse archaeological tools after polish degradation has occurred: only patches of polish remain and much functional information is lost. The paper presented by this author at the 1985 Tuebingen Conference once more raised and emphasised the issue of complications caused by post-depositional factors affecting polishes and striations (Levi Sala 1986a).

Experimental replication of certain post-depositional phenomena on flint surfaces, purely through friction with sediment, showed the extent to which they affect use-wear polish and striations (even without the undoubtedly important added effect of chemical action). Polish was obliterated and replaced by more developed polish which is indistinguishable from that produced by use. Polish from contact with sediment was seen to round the edge (although definitely not use-induced) and to present a variety of characteristics usually considered indicative that the polish is use-wear. (Levi Sala 1986b Figure 3c, d, e; figure 5 e, g).

These results were thought-provoking, especially as they followed three years of work, analysing artefacts from 16 different sites using the Keeley method and getting negative results. In all cases no polish could be clearly attributed to use, due to alterations which had affected the flint surfaces microscopically on artefacts which seemed macroscopically "fresh". There are other factors besides use, friction in the sediment and soil chemicals which can affect flint surfaces. Post excavation handling including packing, refitting and laboratory cleaning are all factors which were investigated and are discussed in Chapter 3. Following the presentation and publication of the above-mentioned paper, post-depositional problems became more widely and openly discussed (Moss 1986a; Prost 1988 etc.).

1.3.2 Sample Size and Levels of Interpretation

Sampling is a problem faced by many branches of archaeology. It is even more of a problem in microwear studies due to the time-consuming nature of the work.
It is unfortunately complicated by the elimination of pieces which are unsuitable for microwear either due to their post-depositional surface modification or due to post-excavation damage. This creates a substantial bias in the sample population. Keeley (1980:120–121) gives figures which are very revealing regarding flint artefacts from the Lower Loam at Swanscombe (Kent).

Of the 267 flint artefacts available to him for sampling, only 66 were suitable for examination and of these only four showed use wear traces. If we consider his accuracy to be 62.5% as in the Keeley and Newcomer blind test (1977), only the 2 to 3 pieces out of the total 267 pieces which were correctly interpreted served as the basis for the interpretation of the site. This may seem an extreme case but it is more or less typical. With Clacton-on-Sea the ratio is only marginally better (*ibid*:86–87).

Unfortunately, not all analysts discuss the proportion of items studied against those eliminated and even fewer mention the number of pieces bearing wear traces on which their site interpretations are based.

The danger of over-interpreting the micropolish present on the stone tools lies mainly with the use to which the conclusions of a microwear analyst are put. Sometimes, though couched in cautious terms, the results of microwear analysis are used too optimistically by those responsible for the overall interpretation of the site.

On other occasions the functional analysts themselves are over-interpreting their observations. In both cases this could lead to erroneous reconstructions of prehistoric behaviour. Some examples which may help to illustrate this point are discussed in Chapter 5.

Microwear reports can have a profound influence on archaeological reconstruction and yet they may well be based on insufficient numbers of artefacts and/or lack experimental verification; many also lack a published documentation of the traces on whose basis the conclusions were drawn.

Section 4.3 lists the sites from which artefacts have been sampled, at the start of this work, for microwear analysis. All the artefacts sampled seemed fresh and unaffected by patina or rolling. It is worth pointing out that the ages of these sites range from the Middle Palaeolithic to the Chalcolithic and that their environmental and geographical settings were also very varied. Yet the number of pieces on which the microscopic traces observed on the edge could confidently be attributed to use was very small. It is unlikely that any archaeological significance could be attached to results from such small samples.

The work on archaeological implements described in Chapter 4 illustrates the problems encountered in interpreting function and highlights the causes of the scepticism which prompted the subsequent research described on Part Two. This research has aimed at trying to understand the processes involved in the formation of polish whether resulting from use or from post-depositional circumstances. Experiments (described in detail in Chapter 6) were carried out under controlled laboratory conditions, not to replicate human tasks, but to attempt to understand what is physically occurring on the flint surface by controlling the variables involved.

CHAPTER 2
AIMS AT THE OUTSET

It is clear from the literature review in Chapter 1 that the Keeley method of microwear analysis was considered a great breakthrough in the functional interpretation of stone tools when the project for my undergraduate dissertation was started. It had been consecrated by a blind test (Keeley and Newcomer 1977) which gave everyone great hopes about the possibilities of the method. Subsequent dramatic "results" through the reconstruction of both the occupation and population of a disturbed Belgian Mesolithic site Meer (Cahen and Keeley 1980) added even more to the already high hopes.

The method was self-taught (as it was not being taught anywhere at the time) by experimenting on various worked materials and attempting to recognise the visual differences between the polish produced by each. Any confusing similarity between the "polishes" was understandably attributed to my own inexperience. The project was to analyse a set of flint implements from the site of Ksar Akil, Lebanon and determine whether they were cores or scrapers in view of the difficulty in distinguishing between them typologically or technologically. It was not easy to carry out the task in the allotted time but most of the difficulties in interpreting the observed traces were attributed to the fact that the archaeological implements had been travelling around the world in cardboard boxes all packed together and had lain in wooden drawers opened and closed often for thirty years (Levi Sala 1982). By the time the postgraduate work was started not only had Keeley's thesis been published (1980), but it was followed by the doctoral dissertations of Anderson-Gerfaud (1981), Vaughan (1981) and Moss (1983), none of whom voiced any doubts about the method as it then stood. Only brief mention was made of the obstacles which they encountered in recognising polishes which too often resemble each other and the problems posed by post-depositional processes. When these problems and obstacles were discussed, however, they were dismissed, implying that once an analyst had acquired experience of many years' standing these stumbling blocks would not be in the way. In 1982 Gendel and Pirnay published their own blind test seemingly confirming the validity of the method. Flaws in this second blind test were not immediately noticed by me. They are discussed in Chapter 1.

Subsequently problems were encountered in isolating use-wear on archaeological implements from a site recently and carefully dug where each implement was bagged separately in a plastic bag. At first, understandably, all problems were attributed to inexperience. I continued to experiment in order to observe a large number of implements and to collect a wider catalogue of use-wear "polishes". Suspicions started to creep in, however, that some of the problems were connected with the vagueness of polish characterisation and some were related to post-depositional processes and these too were investigated through experiments.

It is only when we, as a group of researchers at the Institute of Archaeology, carried out the blind tests described in Newcomer *et al.* (1986) that my serious doubts about the theory behind the method itself were confirmed. We noticed how often even the most experienced among us mistook one supposedly distinct polish for another.

The experimental work involved in functional analysis of archaeological collections is described in Chapter 4. A discussion of how experimentation led to the new direction which my work has now taken is given in Chapter 6.

Investigating processes of polish formation has led to experiments which though not exhaustive are all the ones that could have been tackled within the period of a Ph.D. project. To understand in even more depth the process by which flint tool edges are polished by use and how such a polish is affected by post-depositional processes, is probably, a project for materials scientists to carry out.

My work does, at least, suggest very strongly that previous theories on polish formation are flawed and need serious revision.

Since the Tuebingen Conference of February 1985 and following the publication of Newcomer *et al.* (1986), many analysts are clearly admitting that polish from different worked material does look alike and that other criteria, besides the appearance of the polish, are needed to reconstruct the function of a tool.

It might therefore seem that my research project has been overtaken by events. In actual fact however most of the followers of the high power approach are still convinced that once well developed, polish from different used materials, is dissimilar enough to be distinguished (Bamforth 1988; Hurcombe 1988; Juel-Jensen 1988a).

There is, however, no logical and physical reason for this diversity independently of the model one adopts to explain polish formation. It is hoped that the experiments described in Chapter 6 will convincingly illustrate this point.

Chapter 3
Method

This Chapter deals with the methods employed by the writer during the various attempts at functional analysis of archaeological tools. The same methods, where applicable, were used in the study of the experimental tools and in the work concerned with the second part of this project. A description of the techniques involved in the various stages of work are also discussed.

3.1 Microscopy

3.1.1 Optical Microscopy

The observations described in this work were originally carried out at magnifications from x50 to x400 using a binocular Olympus Vanox incident light microscope. Magnifications of x100 and x200 were used most frequently as they are the most useful for observing the flint structure and the polish. Polish development and patterning on the microtopography were recorded at x200. At the highest magnification (x 400) the depth of field is so limited that the observation of irregular surfaces such as those on flint is very difficult. On the other hand, the lowest magnification (x 50) was useful for observing coarse or altered flint and the distribution of scars along the edge.

3.1.2 Scanning Electron Microscopy

The work described in PART II has sometimes been checked using a Scanning Electron Microscope (SEM) which can magnify images several thousand times. The use of the SEM for use-wear has many advantages, not least of which is the possibility of viewing tool edges at such magnifications without the problems of poor depth of field encountered with the optical microscope.

There are, however, disadvantages in attempting to analyse all the pieces by SEM. In the first place, not all microwear analysts can have access to this expensive equipment. Sample preparation can be time consuming as may be inferred from the writings of those advocating the use of the SEM (d'Errico 1985:446–447). Furthermore, although used edges appear smoothed when observed by SEM, images at x200 to x500 do not show "polish" as it is seen optically. This has also been remarked upon by Anderson-Gerfaud (1981 [i]: 130). Distribution of polish on the edge and on the microtopography cannot therefore be recorded.

Last but by no means least, when using very high magnifications we are only observing a minute pinpoint on the edge and not the edge of a tool. This is also the case with optical microscopy but more so with the SEM where much larger magnifications are possible.

It is necessary however to use the SEM for observing features related to the flint structure, or particular problems related to the polish as was done in this project. The magnifications most useful for observing polish by SEM were found to be from x700 to x6000 while magnifications of up to x10.000 were used to examine the flint structure.

Due to the irreversibility of gold coating on flint artefacts, attempts were made to observe flint surfaces without this coating. Unfortunately the surface of the flint was charging and did not allow good observation and/or good photography. Carbon coating was then used with reasonably satisfactory results. This coating has been cited by d'Errico and Varetto (1985:136–137) as reversible. It is my experience, however, that washing in warm water and detergent and subsequent ultra-sonic cleaning removes most but not all of the carbon coating. The question of coating is therefore to be added to the obstacles for the use of the SEM in experimental work requiring observation and photography during successive stages of a process or a work sequence.

3.2 Photography

The Olympus Vanox is fitted with a C 35 AD camera back and an automatic exposure unit. Ilford FP4 35mm film was used almost exclusively once it was observed that it strikes a good balance between speed and resolution. The problems encountered in obtaining good pictures do not, it would seem, depend so much on the film as on the flatness of the flint implement (given the limited depth of field of the microscope optics) and the good eyesight of the worker in choosing the correct orientation of the piece and the correct focus of the image. It is important for photography that the microscope be used on a sturdy bench relatively free from vibration.

Because of their irregular shape it is necessary to mount the flint implements on a bed of putty. The putty's consistency is very important as a very soft type will allow the specimen to move imperceptibly and blur the micrographs while too hard a type will not hold the implement at all.

These may be very elementary hints for an experienced laboratory microscopist but as most microwear/functional analysts may not have had any previous familiarity with microscopes it seems useful to mention some of the pitfalls.

All photomicrographs have been enlarged to 4.3 x 6.5 cm which is approximately 1.75 times the negative. It was decided always to use the same enlargement for the sake of comparison; even though different lighting conditions of the piece under the microscope can make an edge look very different at the same magnification. Figures 1a, 1b, 1c show a stage scale photographed with each set of microscope optics used in this research (x 50, x100, x200). These scales may be used to measure the size of the features on micrographs.

3.3 Data recording

On starting functional analysis, colour-coded cards were used on which experimental and archaeological artefacts were drawn to scale in outline only and the observed traces marked on the drawings as suggested by Keeley (1980:26–27). The areas photographed were pinpointed on the line drawing and assigned alphabetical references. In my published papers and in this thesis the location of the published or submitted photograph is highlighted with a square. The information to be recorded varied according to the purpose of the analysis. In view of the time involved in high power microwear studies, some analysts no longer examine large numbers of artefacts from a particular site but choose rather to study one particularly unusual or characteristic tool type or a type common to a several sites of different ages. Whatever the approach taken, the recording system must be adapted to the task. Dedicating some time at the start of every project to ascertain which information one is likely to require can save time later on.

Appendix 1 illustrates the data recording used when studying the Hengistbury Head microdenticulates. After the use-wear analysis phase of this thesis had been completed, a team of Institute researchers evolved a new data recording sheet (Grace *et al.* 1988:206 figure 1) which seems to be a very useful backbone on which to add new flesh according to the project at hand.

Photomicrographs, in ample quantities, were used to complement record cards, particularly in the work described in Part II. They are obviously indispensable in post-depositional experiments which are checked at regular intervals or at particular stages of the process. They were useful to document what was being observed microscopically and to illustrate the comments suggested by these observations.

3.4 Functional analysis

The experiments described in Part II to investigate polish formation suggest that there is no polish which is exclusively produced by one material alone. The various stages in the polishing process are discussed below in 3.4.4 under the heading of polish development. Even before the conclusions of this project were in hand, it was obvious that polish appearance was not generally diagnostic of the worked material. It became necessary, therefore, to adopt a method for functional analysis on archaeological implements which can yield the best results in each situation. The method relies on using a combination of all the variables likely to lead to a successful functional reconstruction. These variables are:

- edge morphology, edge wear, polish development and distribution.

The use of these variables is summarised in the following steps which, after cleaning of the artefacts (see 3.4.6 and 3.6.2 below) are to be followed in this order:

1. Artefacts are briefly scanned microscopically on all their surfaces, to gain a general impression of the condition of the flint, the presence of any traces which could be attributed to use and the presence of any obvious PDSM (see discussion below). When a pattern of polish distribution is found on the particular edges of a tool type, it is reasonable to infer that the polish is probably due to use since it is highly unlikely for an accident to cause a patterning of traces on the same edge of several similar artefacts.
2. When such a regular pattern has been observed and even when, as often happens, there is no clear patterning on the edges, it is a good idea to hold the tool in the hand to get an impression of how the tool might have been used. Information from the site can help with the formulation of one or more hypotheses.
3. Edge morphology, including the measurement of the edge angle and thickness, is recorded in order to narrow down the choices for the most probable hypothesis regarding used edge, motion, and worked material.
4. The flint microtopography is observed in order to ascertain to what extent the polish distribution may be affected by it.
5. The distribution of the polish on the microtopography is observed and recorded as this is affected by the hardness of the used material.
6. Edge wear is observed macroscopically and microscopically to decide whether, in conjunction with polish distribution in step 5, it is compatible with the hardness of the probable worked material and the use motion.
7. An experiment is carried out using an implement with the same characteristics (flint type and edge morphology) as the studied artefact, to replicate the wear traces actually produced by the hypothesised motion and worked material.
8. The experiment will reveal whether the tool is suitable for the task and whether the traces produced experimentally are the same as those observed on the archaeological artefact. The term **traces** is used here to include both polish development (its distribution on the edge and on the microtopography) and edge wear (its type and frequency).

It is not automatically assumed that artefacts sharing a common tool morphology will have been used for the same action and/or on the same material. The above procedure does not claim to give categorical answers on the function of the tool which, it is worth remembering, is always a matter of interpretation of the observed traces. The most likely function is the one that best fits the evidence.

The great unknown variable is of course the duration of use since it is possible for a harder material to produce the same polish as a softer material used for a longer period of time. It is unlikely, however, that the edge which is useful to cut hide for example, would be used to whittle wood. Therefore edge morphology and edge wear, together with the distribution of the polish on the edge and the microtopography, will suggest both the relative hardness of material and the motion in use. In the case of extremely hard materials such as shell or stone, there is a noticeable crushing of the flint crystals (figure 2) occurring only with use on such hard materials and which I have never encountered in working antler, wood or bone.

The preceding paragraphs describe the eight step procedure followed when no problems occur, but in practice problems do intervene at any or every step described above. In such cases it is important to answer only such questions as can be answered. When no polish has survived, it is not possible confidently to give answers regarding the worked material. It may, however, be possible to interpret the motion and the relative hardness of the material by observing the edge wear, the edge morphology and experimenting to test the hypothesis (Levi Sala 1987).

The actual implementation of this approach is illustrated in Chapter 4 in a discussion of the functional analysis of archaeological implements. It is however appropriate to make some comments on each of the variables discussed above.

3.4.1 Edge Morphology

To make an hypothesis regarding the possible use of an artefact, it is necessary to assess the functional capabilities of the working edge. The identification of the working edge is based on the patterning both of polish and edge scarring, or the presence of edge rounding.
Edge angle and thickness are then measured and recorded.

3.4.1.1 Edge Angle

Edge angles mentioned in this work have been measured with a goniometer placed on the angle formed by the ventral surface of a flake and the plane of the dorsal surface which is nearest to the utilised or retouched edge (Wilmsen 1968; Grace 1988). There are usually slight variations of the angle along the edge. Several points along the edge were measured and an average was taken. As no plastic goniometer was found, the areas which contact the artefact have been covered with adhesive tape to protect the flint surface from being damaged by a metal goniometer.

3.4.1.2 Edge Thickness

Edge thickness is measured with plastic callipers also at the mid-point of the edge. Roger Grace found that edge angle and thickness together with edge length, edge profile, and the shape of the tool were those "which seemed to provide reliable functional information" (1988:74). Although edge length, profile and tool shape were not systematically recorded in this work, they were kept in mind as a result of the discussions with Grace during the evolution of the method described in his Ph.D. thesis. The experimental work described in section 6.4 confirmed Grace's view on the importance of these variables.

3.4.2 Polish Distribution on the Edges.

Polish distribution on the edges of flint tools is indicative of the use motion and this has been discussed in great detail by Keeley (1980) and later by Vaughan (1981), Moss (1983b), Unger-Hamilton (1984) and others. In general there is agreement on the diagnostic elements of each motion. It is therefore not necessary to cover this ground again. It is only worth reiterating that the polish distribution is affected by the direction of use, the hardness of the worked material, the angle of attack, the angle of the edge and other aspects of its morphology, the flint microtopography and the pressure and expertise of the worker.

3.4.3 Polish Distribution away from the edges and "Hafting Traces"

Sometimes traces occur either on the dorsal ridges or on the ventral aspect of a flint artefact and which "make little sense as traces of utilisation but [do] conform to what is known or expected of wear traces from minor movements of a tool against its haft" (Keeley in Cahen *et al.* 1979:681). Following this observation microwear analysts have attributed to hafting many traces they could not account for. No experimentation seems to have been carried out to verify whether hafts do leave such traces (Vaughan 1987: 135–144 figure 2 and other papers in Stordeur ed. 1987). Experiments by Moss and Newcomer (1982) and by Unger-Hamilton *et al.* (1987) have shown that when a haft is efficient it does not leave **microwear** traces on the tool. Several hafting experiments were carried out by me during the work on Hengistbury microdenticulates to see whether or not they had been hafted. Hafts were made of wooden shafts in which a heated mixture of natural wax and resin was poured. Once the artefact was inserted, the mastic was allowed to cool and harden. No microwear traces were left by such hafts. Other sorts of hafts may leave microwear traces. It is advisable, however, to experiment before attributing to hafting all traces which "do not make sense" as suggested by Keeley (*ibid.*).

What is discussed here is the survival of microwear traces and does not refer to the survival of resin, tar, or other materials used as hafting mastics in prehistory and which are known to survive under certain conditions. What was **NOT** found after use on hafted artefacts are scratches, streaks of polish and other microscopic traces. Many post-depositional events can cause scratches, streaks of polish and striations to occur on flint implements see section 6.5. It is advisable, therefore, to attribute to hafting only such traces as have been carefully replicated experimentally (see also Grace 1989:49 ff.).

3.4.4 Linear Indicators

Certain linear indicators of direction and motion in use are observed on used artefacts. These have been given various names such as striations, striae, linear features, scratches, sleeks etc.

Kamminga (1979:192) defined a **sleek** as "a linear plastic deformation" and a **scratch or furrow** as "a tear in the surface material". Figures 64 and 65 illustrate the two features which appear quite different. The fact that the difference between "sleeks" and "scratches" would, however, only be visible by SEM makes the information gained by this distinction inaccessible to the majority of microwear analysts who are users of optical microscopes. The same is true of the painstaking work of Mansur-Franchomme in classifying several types of "striations", correlating their length and depth to various used materials and abrasive sizes. Until it was possible for me to use an SEM, Mansur-Franchomme's observations could not be applied and were not well understood.

For example "comet tail pits" appear in the SEM to be the effect of a larger crystal (or a mass of crystals) being removed and dragged against the surface, causing a comet shaped groove to occur (figure 66). Fern-like striations described by Mansur Franchomme (1983a:225 figure 5) can also be observed only by SEM and are very useful for understanding the polishing process as well as the direction of motion (figure 67).

In this thesis **Linear Features** is used as a generic term to encompass any indicator of the direction of motion as observed optically. The term **striations** is used to describe furrows or grooves in the polished surface which can be very fine or wider and deeper. Their correlation with used material and particle size was tested in the experiment described in section 6.4.7 but despite the use of the SEM no such correlation was observed. It seems rather that a variety of linear features can occur on the same edge. The term **streak** is used here to describe a line of polish such as that caused by retouch (figure 5).

3.4.5 Polish Development and Distribution on the Flint Microtopography

The different polish appearances which microwear analysts have observed so far are most probably due to differing stages in the development of the polish on the microtopography. The experiments described in Part II indicate that polish formation is a progressive and complex process depending on several variables: hardness of worked material, edge morphology, fluids, abrasives, lithic raw material, and pressure.

Although all these factors influence development and distribution of polish, duration of work is the most important variable and this remains unknown for archaeological implements. On experimental artefacts, in the early stages of work polish does not develop but rather a general brightness occurs without observable localisation on the microtopography. This brightness develops both when working soft and relatively hard materials (figure 3 and 4). The traces initially appear as a generic brightness rather than a polish; there is no observable localisation on the microtopography.

Soft or pliable materials, such as fresh hide, come in intimate contact with the flint topography and affect both the high and lower points simultaneously. In view of the extreme lubrication offered by fresh hide, attrition is minimal and "polish" is very slow to develop. The interpretation of this phenomenon may be that a smoothing of the surface is taking place but not a removal of asperities to the point where a flat surface is produced as is presumably the case when polish develops.

In the later stages of polish development material hardness affects the distribution of the polish on the microtopography while abrasives, liquids and backing can alter its speed of development and its invasiveness. The tasks performed govern the distribution of the polish along the edge, determining the extent of contact, penetration and interaction between the flint edge and the used material.

In the second part of this work the schematic representation advocated by Grace *et al.* (1988:208–210; figure 63 this volume) to classify polish has been adopted as it is based on the principle that polish categories "represent a continuum of polish development, from poor to well developed, almost any category can be produced by almost any material" (*ibid.*:220), an observation confirmed by my experimental work.

The stages of polish development are summarised as categories below:

A polish occurring as individual elements on the unpolished microtopography and not linked;

B polish linked in a pattern;

C polish very highly linked, covering most of the observed microtopography.

A+ and B+ are used to indicate stages of polish development intermediate between the above stages. I have used C+ and C++ to describe almost blanket polish covering most of the microtopography with rare empty spaces as in the case of "sickle gloss" (see section 6.4.2).

3.4.5.1 Material Hardness

The experimental work described and discussed in Chapter 6 suggests that polish forms on flint surfaces essentially as a result of attrition causing the removal and/or the redistribution of deformation material. The hardness of the material on which flint implements are used is consequently of great importance. It is necessary, therefore, to clarify what is meant by hardness. In materials science hardness is defined as follows:

> The **hardness** of a material is determined by pressing an indenter into its surface and measuring the size of the impression. The bigger the impression the softer the material (Anderson *et al.* 1985:211).

This test has not, to my knowledge, been applied to materials of so obviously different resistance as stone, bone, antler, wood and hide. It is probably due to the fact that it is not easily measurable particularly on materials such as hide which have significant **elastic** deformation. Hardness will be considered here in the sense of pliability rather than as defined in materials science. Many functional analysts have suggested a hierarchical relationship of relative hardness between the various materials mentioned above (Tringham *et al.* 1974:189–191; Odell and Odell Vereecken 1980:101; Grace 1989:113). It is suggested here that meat is the most pliable followed by hide, wood, antler, bone, shell and horn. This is not to be intended as a rigid classification. Some of these materials present a different hardness when they are fresh or dry (hide and wood) and whether or not they have been soaked (hide and antler).

In the wood grooving experiment section 6.4.6 moistening dry wood with distilled water caused the polish to become more invasive that is developing farther from the edge. This suggests that the wood had become more pliable since all the other variables had been kept constant. All harder materials (in the absence of a fluid such as water or oil) will remove the asperities protruding from the flint topography. When the material is extremely hard such as stone or shell, the crystals may be crushed into the surrounding interstices figure 2, 68 and 69). On the other hand, when the material is more pliable, crystals are dragged along the topography smoothing it (figures 65 and 70). Soft materials are unlikely to produce a crushing of crystals as illustrated above even if very hard abrasives have been added as the pliability of the worked material will prevent a load to be applied that will cause crushing. Very hard materials, however, if used for a prolonged period, may eventually flatten the topography and smooth the surface. The last two possibilities have, however, not been tested experimentally. Materials less hard than stone and shell (such as bone, antler, wood and hide) certainly do smooth the flint topography with prolonged use (figure 71).

The most difficult material to insert in this scheme of relative hardness is what is referred to in the microwear literature by the generic term "plants". Green grasses and reeds with a high silica content in the form of phytoliths and a high fluid content in the form of juices, usually produce a very well developed polish covering almost all the microtopography (figure 6). This only occurs after prolonged use however. Reaping rushes of a variety with a very high silica content (*Juncus effusus* G. Hillman pers. comm.) for up to sixty minutes produced no polish. The silica in the plants is probably responsible for the blanket polish development seen microscopically with prolonged use. The pliability of these plants accounts for the invasiveness seen on Neolithic sickle blades which are macroscopically glossy for several centimetres away from the edge.

3.4.5.2 Fluids

The importance of fluids in the development of polish on flint is discussed in more detail in Chapter 6. Experimental work described in sections 6.4.8 and 6.4.9 suggests that when working thin soft materials with a hard backing the hardness of the backing can alter the distribution of the polish making it more similar to that obtained when working hard materials. When using hard materials or soft materials on a hard backing, the high points in the microtopography are polished first. The later stages consist of a more or less gradual linking up of the polish depending on the amount of water or other liquid present. Experiments described in section 6.4 show the important role that an adequate quantity of a liquid medium plays in the speed of polish linkage. This explains the observation by microwear analysts that moist hide develops more visible polish than dry hide and that green plants cause "sickle gloss" to occur faster than dry plants (Unger-Hamilton 1983; Vaughan 1985:35–36; Mansur-Franchomme 1983b:100). In my experimental work it was observed that moist hide which is starting to dry produces the best developed polish in a relatively short time probably due to the right combination of moisture and attrition.

3.4.5.3. Abrasives

Added abrasives also play a crucial role in the development of polish. Experiments carried out scraping hide with ochre produced polish which was macroscopic after only 20 minutes of work.

Inversely with work on hard materials in the presence of a soft abrasive (see grooving wood with talc section 6.4.6), the high and low points of the microtopography are both affected. This makes it advisable not to infer hardness of worked material from polish development without taking this variable into consideration. When the evidence from polish development and distribution is suggesting a soft material and the edge damage is more indicative of a hard material, it is worthwhile verifying experimentally whether a soft abrasive might have been used. Components of environmental dust might well be rather harder than talc and would not act as a soft abrasive.

3.4.5.4 Lithic Raw Material and Polish Development

Flint microtopography is another very important variable in the development of polish. This was already noted by Vaughan (1985:27–29). In my work the experiments carried out with the coarser variety of Israeli flints have shown that polish may not develop even after flakes have, for example, been used to reap reeds for 60 minutes (EXP/82). In another instance very coarse Israeli flint scrapers used to scrape moist hide with ochre for 20 minutes developed very little polish while Brandon flint scrapers used for the same length of time developed a brightly macroscopic polish (see section 6.4.7).

When polish does develop on coarser flint the linkage is much slower than in more tightly packed varieties. On a ridged topography (figure 23) only the ridges are polished and the polish distribution may appear similar to linear features perpendicular to the edge even if the motion was perpendicular to the line of the ridges as in cutting. Furthermore, in the case of coarse flint, some surface protrusions can determine the degree to which an edge is coming in contact with the used material, hindering good adherence and affecting both polish distribution and polish development.

3.4.6 Residues

As mentioned in the Introduction, there has been much concern in the microwear literature about the residues from the worked material remaining on the flint surface. The subject is dealt with both in this Chapter under 3.6 Post-Excavation Handling and in Chapter 6 in the Experiments. It is useful, however, to describe the conditions in which residues have been observed experimentally and to suggest a hierarchical terminology:

Deposited Residues

After use and before any cleaning **all** used materials leave some residue such as sap, blood, dirt or grease which is **"deposited"** on the experimental artefact. This is usually removable by any mild detergent in warm water. It has no connection with polish whatsoever. Very often, as in cutting meat, de-hairing hide or reaping rushes in spite of extensive deposition of macroscopic residues, no polish is observed after washing the implement (figures 17 and 18 see also Van Gijn 1986:13–28). Loy (1983) has suggested that blood plasma can survive on archaeological tools but much controversy surrounds his research. It is highly unlikely that "deposited" residues defined here as removable by warm water and mild detergent will survive archaeologically.

Adhering Residues

After washing a used implement in warm water and mild detergent some used materials such as green wood will leave a more stubborn residue still **"adhering"** to the flint surface. It was shown that this is removed by movement in sand (figures 11, 12, 13 EXP/9 Section 6.5.1). As it is also removed by soaking in a biological detergent, it is therefore transient in spite of appearing to "adhere" to the surface after washing. It could, theoretically, be preserved on sites where organic materials have survived and where no movement in the sediment has scoured the surface of the flint artefacts. On sites like Nahal Heimar where organic material has survived, I have observed tar on some implements but **not** recognisable organic residues. It is unlikely to be present on most other sites.

Embedded Residues

As discussed in Chapter 1, Anderson-Gerfaud, Mansur-Franchomme, and Hurcombe have reported observing, by SEM, residues such as phytoliths, meat and hide particles from the used material on experimental and/or archaeological artefacts "embedded" in the flint surface. None of the researchers in our Institute: Moss (1983b); Unger-Hamilton (1984); Ataman (1989); Grace (1988 and 1989) has observed such embedded residues, even under the SEM. The researchers who have reported the presence of these residues have explained their survival by suggesting that they were "trapped" under a hydrolysed amorphous silica layer which developed during use. No in-depth investigations on the subject have been carried out so far by the advocates of this hypothesis to determine whether such a layer does in fact exist. It is here proposed that the term "embedded residues" is used only in those cases where the presence of a hydrolysed layer trapping and embedding such large sized residues has been ascertained. As mentioned in Chapter 1 and in Chapter 6 this project has yielded no evidence for a thick hydrolysed layer that could trap items the size of phytoliths.

3.4.7 Edge Wear

In view of the controversy, between the followers of the low power approach and those of the Keeley method or high power approach (see Chapter 1), the following comments on the importance of edge wear may be appropriate. It is necessary first of all to define the various expressions used to indicate wear on the edges of flint tools.

Although polish is doubtless a kind of wear, it appears to be generally assumed by many researchers that expressions of wear or damage incorporating the word **edge** relate to scarring intended only as microflaking on the edge. In this work the term **edge wear** is used to indicate macroscopic and microscopic **edge scarring** together with **edge rounding. Edge damage** is used to refer to **edge scarring** produced by natural processes and covers both macroscopic and microscopic observations. As **edge damage** is by implication not intentional it excludes deliberate retouch (see also Grace 1988:84).

3.4.7.1 Edge Scarring

The various types of fracture or scarring which occur on flint edges have been characterised by Crabtree (1972) and his terminology is largely in use today as codified by the Ho Ho Classification and Nomenclature Committee Report of the Vancouver Conference (Hayden ed. 1979:133–135). Cotterell and Kamminga (1979:97–112) end their careful investigation of the correlation of edge scarring and use with the following observation:

> Use fracture patterns are a valuable **aid** in determining the functions of certain types of prehistoric tools (my emphasis).

While there is no way to **prove** without an independent blind test that edge scarring alone is not diagnostic of used material, there is quite a consensus now among a large number of analysts that it cannot be used alone as a diagnostic indicator. As mentioned earlier, the reasons are as follows:

a it is not always present after use;
b it can be caused by accident in prehistory, after deposition, or during and after excavation
c it cannot be observed on retouched edges.

Furthermore Knutsson in Broadbent (1979) suggests that tools used for a long time on soft materials could produce the same edge scarring as short use on hard materials. That is not to say, however, that edge wear (including edge rounding) is not an important variable. But even the users of the low power method are now using other information (Odell 1986). When present and identifiable, edge scarring can be used **in conjunction with the information obtained from all the other variables,** to support or call into question the diagnosed hardness of the material and to suggest the direction of motion. Determining whether or not edge scarring is due to use or damage is generally very difficult particularly on retouched edges (see also Keeley 1980:27).

It is therefore suggested that only such scars as occur on the unretouched aspect be considered as probably use-induced.

3.4.7.2 Edge Rounding

Edge rounding has often been directly associated with scraping hide (Keeley 1980). It develops, however, when scraping either soft materials with coarse abrasives or hard naturally abrasive materials. In transversal motions such as scraping, edge rounding is visible on the aspect of the edge in contact with the scraped material. It rarely occurs with longitudinal motions such as sawing and, if present, will be light and observable on both aspects of the edge. High angled robust edges do not round as easily as acutely angled ones. With certain coarse raw materials edge rounding does not occur and/or is not observable. Scraping with a low angled and unretouched edge causes scarring rather than rounding to develop in the initial stages, especially when scraping harder materials. After such an edge has been stabilised edge rounding will start to occur. It is evident therefore that if work is interrupted before the edge is stabilised there will be no edge rounding. It is unlikely, however, that such edges will be used on the harder materials since an acute edge which is constantly breaking down would be inefficient.

There are cases when it is only possible to study edge wear and in this case low magnifications are adequate. This is particularly true where artefacts are affected by post-depositional surface modifications and where implements made from materials coarser than flint such as quartzite do not appear to yield useful information when observed optically at high magnifications. In these cases it is necessary to use either low magnifications or the SEM (Sussman 1988; Knutsson 1986).

Speed and the relatively low cost of the equipment required make analysis at low magnifications very attractive, particularly in the case of projects involving a large number of artefacts. A perfected and improved study of edge wear (scarring and rounding) could answer questions such as used v. unused, hard v. soft materials or cutting v. chopping. The initial overview might then be complemented by a subsequent in-depth analysis of a smaller number of artefacts such as particular tool types or interpretation of spatial distribution etc. The detailed analyses involving the more time consuming methods, at higher magnifications, would include all the variables described above.

3.5 Flint structure

Flint is composed of 98% pure silica in a crypto-crystalline form. It has a "framework of globular lepispheres with interstitial chalcedony" (Griffiths *et al.* 1987:49 figure 6.7A) with water being present both "as hydroxyl groups on grain and subgrain boundaries and along lattice defects, and as molecular water in microscopic and sub-microscopic fluid inclusions and adsorbed on surfaces" (Griffiths *et al.* 1987:48–49; see also Micheelsen 1966).

Most writers on microwear have noticed that the "grain size" or microstructure influenced the speed of polish development on flint artefacts (Vaughan 1985:28; Holmes 1987:92; Masson 1982:31-49).

Flint was however described in the literature as, for instance, "fine grained English chalk flint", suggesting that the "grain size" was homogeneous within English chalk flint, whereas there are in fact wide variations in flint even within the same nodule.

Bradley and Clayton (1987:81-89) have discussed the influence of flint microstructure on the pattern and rate of development of polish during use through experimentation and SEM observation. Griffiths *et al.* (1987:43–52) have studied the effect of heat treatment on flint microstructure and the effect of the change in structure on flaking characteristics.

These two studies have shown that:

- typical black flint consists of silicified skeletal fragments, spherical quartz "lepispheres" (which together make up the main framework structure of flint), and an interstitial "cement" of structurally disordered microfibrous quartz chalcedony;
- The fracture surface of fresh flint tends to go through the chalcedony phase and around the lepispheres giving the uneven microtopography of the flint surface;
- during the life of a nodule recrystallisation can occur in response to chemical changes in the surrounding pore waters, or heating, intentional or otherwise;
- recrystallisation can affect the structurally disordered silica, resulting in local redistribution of the structural and surface-bound water into microscopic and sub-microscopic fluid inclusions;
- in heated flint the water migrates to fluid inclusions and the chalcedony becomes similar to lepispheres in its fracture properties;
- whether its coarseness is determined by its original structure or caused by recrystallisation, coarse flint shows a greater resistance to microwear polish formation;
- flint containing a lower proportion of well crystallised quartz and a higher amount of chalcedony than recrystallised flint is more susceptible to both mechanical and chemical damage;
- the density of packing of the lepispheres within the flint structure also influences the form of the traces resulting from use;
- the presence of microfluid inclusions probably causes the waxy sheen seen on fractures made after heat treatment. The inclusions tend to partially reflect incident light rather than simply refract it as does normal flint; certain white patches in flint may well be due to reflections from crystals , not fluid inclusions and can cause intense glare leading to problems in observation.

The Bradley and Clayton observations referred to Upper Cretaceous chalk flint (probably Brandon) and two nodules of "beach pebble flint collected from the north coast of Ireland". The Griffiths *et al.* work was carried out with the "visually homogeneous" black flint from Brandon and the mottled grey flint from Potters Bar "containing lighter apparently coarser grained regions".

My experimental work replicating use-wear and post-depositional effects at Hengistbury Head was done mainly with Brandon and occasionally with Potters Bar flint. A variety of Israeli flint types were used to replicate wear traces and surface alterations when analysing artefacts from Israeli sites.

Some post-depositional and polish formation experiments were carried out simultaneously both on Israeli and English chalk flints for comparison.

A characterisation of the various Israeli flints used in the experiments has been attempted (section 6.4.1). Most of the Israeli flint types used are derived from Eocene formations. The flint was collected from the highland region of the Negev desert which was deeply dissected by active river valleys and erosional depressions. In the area "Flint generally occurs as small nodules, thin, slab-like bands, or as very large cobbles. These can be found either on flat surfaces which have been exposed due to erosion, or bedded in cliffs, hill slopes, and, in the case of the small nodules, in conglomerates" (Munday 1976:20). The present climate in the Negev is desertic and continental, witnessing temperature variations from 40 to 45° C in the daytime to -2 to -5° C at night. Climatic variation and erosion may affect the flint microstructure.

I have also used flint collected from the Carmel hills where the climatic conditions are much more temperate. This is much coarser to the touch than the Negev flint. Not much information is available in the literature regarding flint sources in Israel or elsewhere in the Levant, their age, weathering and characteristics.

It is not clear under what conditions the Israeli flints studied here have been formed. Some of these flints, however, are not only highly resistant to polish formation but are very reflective even when freshly knapped. They often cause glare at higher magnifications and the surface is sometimes "pedestalled" at x200. Figure 7 illustrates the surface seemingly consisting of individual crystals mounted on pedestals and separated by large hollow spaces. Diane Holmes (1987:92) has noted that Egyptian flint, similar to some of the Negev varieties, is very slow to develop use-wear polish. She had to abandon the planned functional analysis of these implements due to this problem and to post-depositional alterations to the flint surface (Holmes 1989).

3.5.1 Post-Depositional Surface Modification (PDSM)

PDSM is the scourge of functional analysis at magnifications above x100. Patinated flint is known to be unsuitable for use-wear studies based on polishes (Keeley 1980:29).

Unfortunately, some studies are still being carried out on patinated flint without any discussion of how the difficulties of observing polish on patinated flint have been surmounted.

Now white patination, a common phenomenon associated with natural weathering, is not the only process capable of obliterating use-wear polishes. When pronounced "soil sheen" or "gloss patina" also renders flint surfaces unsuitable for microwear analysis (Stapert 1976; Plisson and Mauger 1988; Levi Sala 1986 a and b).

Soil sheen does in fact "polish" flint and affect use-wear polish by degrading it and by making it impossible to separate the area polished by use from the surrounding area "polished" by taphonomic processes; this is confirmed by the work described in this volume (see experiments in section 6.5). Consequently all artefacts strongly affected by soil sheen cannot be considered reliable for microwear analysis based on examination at high magnifications. (Levi Sala 1986 a and b; Stapert 1976; Plisson and Mauger 1988:4).

When Israeli flint has been affected by soil sheen and other types of PDSM the glare, which is present in some varieties even on freshly knapped implements, increases and impedes observation at high magnifications. As with other types of flint, it is not possible to distinguish use-wear from post-depositional polishing.

3.6 Post-excavation handling

It is often mentioned in the microwear literature that post-excavation handling of archaeological artefacts can cause 'wear' traces which might constitute "background noise" when analysing edges (Plisson 1983; 1985). The various stages of handling after excavation have been investigated and experiments done to study the preservation of flint surfaces.

3.6.1 Washing Packing and Storage Procedures

Washing on site with a nylon brush may remove organic deposits if they have survived. It is best avoided, therefore, when there is the possibility of having such residues analysed. Contrary to Plisson's observations referred to above, on-site washing does not seem to affect the condition of the flint or the polish. This was deduced from an experiment in which Brandon flint was brushed with sediment for five minutes without affecting either the polish or the surrounding unpolished surface. This experiment is described in section 4.2.2 (Levi Sala 1987:143–150).

On the other hand, several pieces of flint packed together for transport and final storage are liable to develop edge damage. Such edge damage may remove the area bearing the use polish, if it has survived, and add other abrasion traces.

The use of metal instruments to remove encrustations is also to be discouraged; if absolutely necessary plastic pointed implements should be used instead. Metal implements leave a metallic deposit on the flint surface which is not easily removed. This "polish" is easily distinguishable from non-metallic polish but can obscure an area where use polish had survived.

3.6.2 Cleaning in the Laboratory

Keeley (1980:10–12) suggested the cleaning of flint implements prior to microwear analysis with white or methylated spirit, hot water and an ammonia-based detergent; he recommended rubbing with the fingers and soaking. This is followed by "immersion of the piece in warm HCl (10% solution)" followed by "NaOH (20 to 30% solution)".

For the NaOH solution, 20 to 30 minutes was considered safe; the aim being to remove transient organic residues from experimental tools while the HCl solution was used to remove carbonate deposits from archaeological specimens, but "both archaeological and experimental implements were treated with both solutions, in order to maintain strict comparability" (*ibid.*:10–12). Whilst this cleaning method is still used by some researchers, it was criticised by his colleagues at the Vancouver Conference (Hayden 1979:190–191). I for one have abandoned it, and have noted that a growing number of colleagues have had second thoughts though not all have entirely abandoned chemicals other than detergents (Cahen and Caspar 1984:277–308; Unger-Hamilton 1988:64–66; Grace 1989:63–64; Anderson-Gerfaud 1986:74; Plisson 1986:114; Moss 1983b:105).

As a result of careful experimentation on flint artefacts to study the process of patination, Schmalz writes "Alkaline solutions, even when very dilute, attack the [flint] grains exposed to their action, etching the grain surfaces and in extreme cases perceptibly reducing the grain size" (Schmalz 1960:49). See also the experiments described in sections 6.5.9 to 6.5.12. in this volume regarding the patination of flint.

When using flint tools experimentally some organic deposits from the worked material seem to remain on the surface even after washing in a household detergent, especially after use on green wood and certain plants. Figure 14 shows features still adhering to the flint surface after cleaning with detergent and dabbing with methylated spirit and acetone. They have been produced by whittling reeds and are similar in appearance to "striations" and polish. Figures 15 and 16 illustrate their disappearance after contact with sand in a rotating drum although the polish that had modified the surface of the flint itself had survived.

Linear features caused by the deposition of organic matter from the worked material which are removable by low contact forces such as movement in the sediment are not likely to survive archaeologically. The question of organic residues has also been discussed under section 3.4.6 above.

I have noted, however, that soaking or thorough but gentle washing with **biological** ammonia-free detergent in warm water, followed by immersion in an ultrasonic tank, in a plastic container filled with distilled water, will remove all such organic deposits from experimental implements without the need for NaOH as suggested by Keeley. Subsequent cleaning by NaOH, tried experimentally in several cases and documented photographically before and after, did not appear to improve the condition of the polished surface. This suggests that the biological cleaning had removed all the removable deposits and that there was nothing more to be gained by using NaOH.

The situation regarding archaeological artefacts is however a different matter. Organic residues do not often survive and are in any case not readily recognisable by the microwear analysts. Some work has been reported in arid areas (Broderick 1979, 1988; Shafer and Holloway 1979) but this is in my view a very specialised field. If this expertise is available to the excavators, cleaning of the artefacts should be completely avoided. Carbonate encrustations and soil deposits however do occur more frequently on the flint surfaces. When flint is encrusted its structure is masked and any surviving polish would not be observable. Moss had also advocated abandoning the Keeley cleaning method because "I began to compare chemically cleaned and uncleaned implements and found absolutely no difference in the appearance of the wear traces" (1983b:105).

She has now slightly reversed her opinion because she has "encountered on artifacts…soil deposits which needed chemical removal as suggested by Keeley" (Moss 1986c:94).

My experience has been that a brief soaking in a biological detergent followed by an ultrasonic cleaning as described above will eliminate most unwanted post-depositional deposits. In cases where there were carbonate encrustations, a bath in a 10% HCl solution for up to 30 minutes at room temperature has been sufficient.

The artefacts from Nahal Heimar, a cave site in the Arava desert, were covered by some deposit from the soil. They were subjected to the Keeley cleaning method before I had started using biological detergents. Figures 19 and 20 show that the cleaning was very thorough but no traces whatsoever were left on the tools. It is quite possible that there were no well developed use-wear traces on these tools in the first place either because they were a "hoard" or because many were "arrow heads", but useful traces may have been removed by the drastic cleaning.

Some obstinate encrustations of non-carbonate material such as the ones which occurred on implements from Kebara cave, will not be removed by the cleaning described earlier including the 10% HCl bath. There is evidence of bone dissolution on site and the cave is extensively used by bats. It is therefore possible that the encrustations are phosphates from the dissolution of bones and/or from the bat droppings.

Sodium hydroxide (NaOH), an alkaline solution, can therefore patinate flint and hydrochloric acid (HCl) although less likely to have an adverse effect on the flint, seems to be superfluous when there are no encrustations.

Cleaning during microscopy is carried out with cotton buds dipped in methylated spirit to remove grease from contact of fingers with the flint. Occasionally acetone is used though very sparingly as it tends to cause false traces mimicking polish and striations. When putty comes in contact with the flint it is necessary to remove the artefact from the slide, wipe off the putty with a dry soft tissue and wash again with detergent and warm water. This procedure is necessary in order to remove the false traces which grease from the putty produces.

3.6.3 Finger Prehension

Contrary to some suggestions, it seems unlikely that prehension with the fingers even for prolonged periods can leave lasting traces on flint implements, especially since use on meat for 90 minutes does not produce developed polish (Juel-Jensen 1988a:55).

Her observation is confirmed by an experiment in which two flint implements, **EXP/71** and **EXP/72**, were rubbed against moistened raw hide for one hour each without any polish developing. What is being observed by others (Plisson 1985:100) is probably not true polish in the sense of a deformation of the flint to give a smooth surface. It is suggested therefore that traces, if any, left on flint by fingers even after extensive handling are removable by a biological detergent and are unlikely to survive archaeologically.

3.6.4 Refitting

This procedure is very useful in reconstructing flaking sequences in the manufacture of flint implements. It is, however, liable to cause polish, edge scarring and striations of varying importance (in quantity and intensity) to develop on flint surfaces (see also Plisson 1985:99). Although such traces are likely to be random, refitted pieces are best avoided as candidates for functional analysis.
Refitting could be done after microwear analysis has taken place, at a later stage of the site analysis.

Careful handling of artefacts destined for functional analysis is therefore recommended although use-wear polishes have been shown to resist cleaning with a nylon brush.

CHAPTER 4

FUNCTIONAL ANALYSIS OF ARTEFACTS FROM ARCHAEOLOGICAL SITES

4.1 A TOOL TYPE FROM HENGISTBURY HEAD

The microdenticulates from the Powell Mesolithic site at Hengistbury Head, Dorset are "a highly characteristic group of artefacts" in the assemblage (Barton 1981:15). They are made on bladelets well under 6 cm. in length with a thickness of between 3 and 5 mm. They are either curved or curved and twisted in profile (figure 24). Of the 34 pieces examined 26 are denticulated on a concave edge, while the remainder are too broken to be characterised (see Appendix 2). The edge angle of the microdenticulated edge varies between 30–45° and rarely exceeds the upper limit. The retouched edge normally consists of a series of small contiguous notches which are extremely regular and equally spaced, giving the tool its fine "saw tooth" profile. The manufacturing technique suggested by Bocquet (1980) involves the removal of tiny flakes along one edge of a blank using another artefact. The resulting notch scars are highly characteristic (figure 24) and occur on the ventral surface when the retouch is produced from the dorsal side (which is mostly the case here) or vice versa. Although use-wear can affect the appearance of the teeth there is no question that the retouch is the result of use alone. Experiments with unretouched blanks on a range of materials failed to produce a single example with the kind of regular serration found on the Hengistbury artefacts (see Appendix 1, No.12).

4.1.1 Method

The functional analysis was originally undertaken using the method elaborated by Keeley (1980). The flint surfaces were observed by optical microscopy using an Olympus Vanox microscope as described in the Chapter on Methodology. The Keeley method relies on the observation of polishes and striations which are then used to infer the kind of materials used and the tasks performed. The problems being encountered in the application of this method of identifying polish and determining the materials used have been already discussed in Chapters 1. Among other things, there were problems due to "post-depositional surface alteration" (PDSM) described below and in Chapter 6.

A more realistic approach was therefore adopted which relies not only on polish identification but on recording of macro edge morphology and edge scars, all microscopic edge wear traces including polish when present, (Grace 1981 , Grace et al. 1988) and on experimental replication as described in Moss and Newcomer (1982).

Prior to examination, both experimental and archaeological artefacts were cleaned with warm water and detergent and then immersed in distilled water in an ultrasonic tank. Occasionally spirit and/or acetone was used to remove fingerprints. Only two archaeological pieces (I 6/2Q/I and E9/16/4) were immersed in 5% HCl to try and remove some spots of what looked like varnish which, in the event, were not removed. The reasons for abandoning the Keeley method of cleaning is discussed in detail in Chapter 3.

4.1.2 Experiments to Determine Tool Function

Experiments were carried out on a variety of materials in an attempt to match the rather perplexing traces observed on the archaeological pieces (see Microscopic Analysis below). Apart from Experiment 3 below in which an unretouched artefact was used, all experimental pieces were microdenticulates. The examples were used both in the hand and hafted to test their efficiency. They were all of East Anglian flint except the artefact in Experiment Seven which was made from a local beach cobble. A detailed description of the individual experiments appears in Appendix 1, but they are summarised here as follows:

1. Cutting hide with microdenticulates.
2. De-hairing fresh hide.
3. De-hairing and cutting hide with unretouched edges.
4. Cutting meat.
5. Cutting cartilage and sinew.
6. Cutting and whittling dry reed (*Phragmites*).
7. Reaping *Phragmites*.
8. Reaping green bracken (*Pteridium*) stems.
9. Whittling seasoned wood.
10. Incising green birch (*Betula*).
11. Sawing shell.
12. Sawing bone.
13. Cutting rushes (*Juncus*).

4.1.3 Observations Resulting from the Experiments

The small size of the tool, its thin cross section and low edge angle appeared to rule out a variety of heavy duty tasks and this observation was subsequently confirmed by the experimental work. On 24 out of 34 archaeological specimens (Appendix 2), the individual teeth making up the serrated edge were worn down in a relatively even manner. Experiments again showed that this would exclude use on very hard materials such as stone, shell or bone.

4.1.3.1 Hafting

1. Experiments confirmed what was already suspected from the size and morphology of these tools that they were difficult to use in the hand for long periods and in heavy tasks. The tools were too small and sharp for comfortable handling for more than short periods. Despite the obvious advantages of using a handle, no diagnostic hafting traces were observed on any of the experimental implements. This is not surprising since an efficient haft would not allow the tool to move.
2. The likelihood that several implements were inserted in a row in a haft was rejected because :

a) the concave edges of the microdenticulates would have made them impractical to use in a row, even in a crescentic haft.

b) the curved and sometimes twisted profile would likewise make them unsuitable for inserting in a handle.

c) there was no evidence for consistent modification of the tools in order to facilitate hafting. This might have been achieved by snapping both ends of the implement. In fact only four were broken in this way, the majority (24) being more or less complete.

3. The experimentally hafted tools did not develop microscopic traces similar to those observed on the archaeological specimens and tentatively supposed to be hafting traces (Appendix 1).
4. At least one of the microdenticulates (K8/2b/1) was modified in such a way as to suggest handholding. The distal end of the tool was abruptly truncated which would have provided a comfortable finger grip for an unhafted tool (figure 24 item 10).

4.1.3.2 Functional Efficiency

Tixier (1963:124) has suggested that fine denticulated saws from the Caspian have the same edge morphology as modern steak knives, and were probably used in much the same way. Considerable differences exist, however, in the overall shape of the North African and British Mesolithic artefacts. For example, whereas the Maghreb examples are long and straight-edged, the Hengistbury tools are much shorter, have concave edges, and are also curved and sometimes slightly twisted.

Juel Jensen (1988b) has studied similar microdenticulates found from the Danish Late Mesolithic to the Early and Middle Neolithic. The preservation of the flint surfaces appears to be markedly better than in Hengistbury Head. Juel Jensen was able to conclude in spite of the preservation of polish on the potential used edges that " the tools were used for *processing* rather than for cutting movements oriented perpendicular to the edge. The worked material remains an open question." (*ibid.*:231).

Despite the differences, it is clear that the most obvious function of these tools is cutting or sawing. This has been amply demonstrated in the experiments which showed that other actions requiring a transversal motion (e.g. whittling, de-hairing) are far less efficiently carried out than, say, by using a high-angle edge.

A slightly surprising discovery, given the highly standardised nature of the tool, was the fact that denticulation did **not** seem to greatly improve performance in any of the experiments involved. In fact, a long straight edge performed much more efficiently, especially in the cutting of hide thongs and meat and detaching cartilage from bone. Likewise, experiments in reaping plants such as bracken proved slow and cumbersome and in any case these plants could be gathered using much simpler methods like uprooting. Although comments of this kind are inevitably subjective, they give some indication of how we might begin to evaluate the function of the microdenticulates.

4.1.4. Materials used in the Experiments

4.1.4.1 Hard Materials

The experiments showed that in the case of use on dry reeds, dry wood and shell, the teeth were worn down in a manner which is not observed on the archaeological pieces. The central part of the edge completely lost its denticulation while the adjacent teeth were still very sharp. (See figure 25).

Even the 10 archaeological microdenticulates described as differentially worn in Appendix 2 did not present this pattern of wear. In those, a few teeth were occasionally worn down more than the rest. It was therefore concluded that the material(s) worked could not have been as hard as dry reed, seasoned wood or shell.

4.1.4.2 Soft Materials

The experiments suggest that the idea of use on hide, meat, cartilage and sinew should be discarded, on the grounds that work on these materials could be carried out just as easily using an unretouched edge. Furthermore, it was not possible to match the polish produced by these tasks and what was observed on the archaeological implements (compare figures 21, 22 to 26, 27).

Dumont (1983:143) put forward meat cutting as a possible use for microdenticulates from Star Carr. There were only two artefacts involved in his interim report and due to the lack of polish in the scars, he concluded "that the denticulation occurred after or shortly before the tool ceased being used to cut meat or fresh hide". It is my observation that cutting meat and fresh hide does not leave a well-developed polish (cf. Dumont 1983:figure 4) but rather a "shine" which can be produced by many soft materials as well as by soil (figure 28). Many of the experimental tools used on *both* soft and hard materials have no polish in scars (figure 29). It would seem that the function of meat/fresh hide cutting does not fit the evidence from Hengistbury where the edge is concave and the wear on the teeth of the serration was greater than that produced experimentally by cutting meat. In my opinion the material(s) used, was not as hard as dry reed, but not as soft as meat.

The presence of bracken (*Pteridium*) in the Early Flandrian (Scaife pers. comm.), together with its potential use as bedding material made it a reasonable candidate for study. The bracken was reaped from the vicinity of the site in early and late Summer. In June, after 10 minutes of use (figure 30), the resulting polish showed some similarity with the polish observed on a few of the archaeological pieces. The slight resemblance might be due to the polish being not too well-developed. By September bracken is no longer very green and the cutting was carried on for 20 minutes (figure 31). The polish was better developed but the similarity with the "polish" on the archaeological pieces was still doubtful.

The use of birch (*Betula*) bark vessels and mats is well known in the North European Early Mesolithic (Clark 1954, Bokelmann 1986). Experiments were carried out involving the cutting and incising of the outer bark prior to removing it from the tree. Although these tasks produced polish (figure 32), it was not similar either in appearance or in distribution to the polish on the Mesolithic artefacts; the minimal edge damage plus the wear on the denticulations, however, were broadly comparable.

Microdenticulates were also used experimentally in basketry manufacture. An experiment was carried out cutting up dry rushes. After 30 minutes work no polish developed on the tool edge (figure 33).

After soaking the rushes (as is usually done in basketry), the polish which appeared in only 15 minutes was very well-developed (figure 34) but unlike anything observed on the archaeological microdenticulates.

Reaping, cutting and whittling dry reeds (*Phragmites*) also produced the very well-developed macroscopic polish (figures 29 and 35) commonly attributed to use on plants.

Again this was in no way similar to the traces on the archaeological implements. Such a well developed macroscopic polish, often referred to as "sickle gloss", generally survives even when the flint surfaces have been affected by post-depositional processes. The reason for this phenomenon is not yet very clear but has been empirically observed elsewhere (Unger-Hamilton 1984:266; Levi Sala 1986b). If the Hengistbury microdenticulates had been used to reap *Phragmites*, it is very likely that the polish would have survived. Dry reeds wore down the teeth too heavily while reaping fresh reeds wore them only lightly. As mentioned above, in both cases the polish was very well developed and distinctive, which is not the case with the archaeological specimens.

It is possible that the Hengistbury microdenticulates were used for incising or cutting green, but woody plant material for a relatively limited length of time.

No one task is envisaged but it has been suggested that a particular function might have involved the "roughening" of arrow shafts near the nock-end prior to gluing on the feather flights (C. Bergman pers. comm.).

Unfortunately, the limited experiments so far carried out (using a scraping motion for 30 minutes on soft wood) did not produce the same wear seen on the teeth of the Hengistbury tools. A scraping motion as in de-hairing hide was also used in Experiment 2 (see Appendix 1). Some edge rounding was observed microscopically while no such edge wear was observed on the archaeological implements. Despite the failure to find a precise parallel, it seems that in the task(s) the concave shape of the working-edge clearly mattered and that the observed wear patterns are consistent with work on a material softer than stone, shell, bone or hardwood but harder than meat.

4.1.5 Microscopic Analysis

As mentioned above the evidence from the microscope was rather confusing. Many implements seemed, at x100, to have a polish which could have been mistaken for use-wear. At higher magnifications (x200 and over) which are more useful for interpreting the function of the tool, the microsurface appeared too reflective and did not resemble any experimentally produced polish. Although there seemed to be more polish along the edges, the whole surface of the tool was also polished.

In many instances there was no clear cut evidence that the denticulated edges were the ones used since the "polish" was more concentrated on the unserrated edge. Striations (i.e. linear indicators for direction of use) were rare and when present did not show a consistent pattern from which to infer the motion or direction of tool in use.

4.1.5.1 Post-Depositional Surface Modification (PDSM)

It soon became apparent that the conflicting evidence from the microscopic observation was probably due to post-depositional processes. Their effect on flint tools has been observed from a variety of sites of all ages and geographic settings (Levi Sala 1986b). The problem is much more widespread than can be inferred from the microwear literature. Flint surfaces and/or use-wear polish are altered in such a way as to render difficult, if not impossible, their interpretation according to the Keeley method. Even Keeley (1980:29) observed that patination can alter "the light reflection qualities of the flint so that the microsurface and the microwear cannot be observed with light microscopes". The Hengistbury microdenticulates, when not white patinated, had a macroscopic sheen which has been attributed by many researchers to the effect of burial in the soil. This is sometimes called "soil sheen" or "glossy patina". Most of the artefacts examined also had randomly distributed glossy bright spots. Microscopically they appear like a smooth highly reflective polish. They vary in size from the microscopic to the macroscopic, occurring singly and in clusters on cortical and non-cortical surfaces alike.

Although bright spots often appear on artefacts affected by soil sheen, the two are in no way comparable in their appearance (figures 36 and 38). A more detailed description of these and other features can be found in Chapter 6.

Soil sheen is often a uniform brightness over the whole surface of the flint implement. In certain cases it can be concentrated on the prominent areas of artefacts. It is certainly clear to all that this sheen is caused by natural rather than human action.

Bright spots are a smooth, highly reflective polish localised on flint surfaces. They have been observed on artefacts from many sites of varying ages and locations. There is agreement that they are post-depositional in nature although so far their most common description is that they are enigmatic. On many microdenticulates they occur on the serrated edge, inside the scars, and on the teeth.

4.1.5.2 Condition of the Hengistbury Artefacts

As mentioned above, the Hengistbury artefacts were often patinated and affected by soil sheen and bright spots. The Powell site is situated on an exposed hillside covered by windblown sands with a well-developed podsol formation. The Mesolithic artefacts are found in fine sands with pebbles occurring within the lower levels. There is evidence for vertical movement of artefacts on the site as shown by the conjoining of artefacts separated vertically by up to 39cm (Barton 1981:19). Similar observations have been made of Mesolithic material from other southern British sites in sands, which show a marked vertical dispersal of artefacts and surface alteration of the flint (Maher 1983; Dimbleby 1965).

From a microscopic point of view, the effect of wind blown sands on flint artefacts has also been described extensively by Stapert (1976:14–19), who concludes "it is not impossible that in places where windgloss is highly developed, microscopic traces of the use formerly present are completely or partly obliterated".

To test whether such effects could be observed under modern conditions, Barton and Bergman (1982) undertook a series of experiments involving freshly knapped flint scatters in sands. The artefacts were flaked onto a sand surface and lay exposed over several months. Even after a relatively short period of lying on the surface, being buried and then sometimes subsequently re-exposed, many of the flints were noticeably discoloured and patinated. No soil sheen or bright spots were, however, observed. It should be noted in passing that the experimental site was located on the sea shore in a highly alkaline environment and with conditions quite unlike those prevailing during the Mesolithic, when Hengistbury was essentially an inland location.

Another possibility considered was that the soil sheen resulted from mechanical polishing, for example by the abrasive action of sand on flint. This could have been produced by trampling, bioturbation, wind blasting, or simply the settling of the sediment under pressure (Villa 1977 and 1982, Cahen and Moeyersons 1977, Barton 1981, Rowlett and Robbins 1982, Villa and Courtin 1983 and others).

In fact it is likely that one or several of these processes influenced the Hengistbury site.

Finally, it should be noted that the chemistry of the soil itself can affect flint surfaces. No detailed work on the effects of such processes was carried out in this context. Subsequent work was however carried out and is described in Chapter 6. It confirmed that the development of soil sheen and bright spots can occur through soil chemical processes alone, that it can be concentrated on the edge as in the case of the Hengistbury microdenticulates and that such post-depositional processes can obliterate use-wear traces.

4.1.5.3 Experiments for Mechanical Processes

To investigate the hypothesis of mechanical polishing, an experiment was set up to reproduce the movement of artefacts in sands. Since it was obviously impossible to simulate the length of time of burial (9/10.000 years), we were able to artificially speed up the processes by using a mechanical tumbling device (see also Levi Sala 1986a, 1986b). The sands used were collected from Hengistbury and were exactly the same as those which occurred at the Mesolithic site. The results can be summarised as follows:

Experiments with Fine Sand

A Shaking and tumbling individual experimental flint artefacts with known use-wear patterns at 60 revolutions per minute, in fine sand from the site, with and without water. The length of time and the speed were increased gradually, from 10 minutes to 50 hours (see details of the experiments in sections 6.5.1 and 6.5.2) after which:

a) soil sheen appeared on the artefacts (figure 38),
b) striations present on artefacts used to whittle dry reeds and incise wood had completely disappeared (figures 14 and 16),
c) the surface of the flint away from the edge was scoured clean (figure 16).

Experiments with Gravelly Sand

B Experimental artefacts used on green birch, bracken, rushes, dry oak, fresh willow were tumbled in **gravelly** wet sand, for up to 50 hours at 60 rpm which resulted in:

a) the appearance of varying degrees of sheen intensity,
b) the disappearance of use-wear striations,
c) the addition of polish on the edges without any clues to indicate that there had been two consecutive polishing events (figures 39,40, 41).
d) the appearance of a concentration of polish on either side of the serration (figure 42).

DISCUSSION

The presence of gravel in the sediment in the second set of experiments is undoubtedly the variable that accounts for the different results. It is significant that gravel was also present in the sands at the site. The concentration of polish on the edges after tumbling may mean that the original use-wear polishes were "increased" by movement in the sediment. However, similar concentrations of polish were also noted on the unserrated (and unused) edges of the experimental pieces. This suggests that the edge is more vulnerable to polishing than the rest of the flint surface, a fact empirically observed by other experimenters (Moss pers. comm.). The fact that movement in the sediment can cause polish to concentrate on unused edges indicates that extreme caution should be exercised before uncritically equating a concentration of polish on an edge with some form of use. As mentioned earlier, other experiments (see section 6.5.12) have shown that sheen produced by chemical processes also has a tendency to concentrate on the edges.

These observations are directly relevant to the artefacts from the Hengistbury site. During the original microscopic examination soil sheen was noted on many of the artefacts but these also showed a significant concentration of polish along the edges. It would appear that the above mentioned explanation might account for the occurrence of polish concentration along the edges.

Likewise, the traces which appeared on either side of the serration and were first thought to be hafting traces are now believed to have a different origin. These bladelets are curved and twisted and on many of them the affected areas are quite prominent, and consequently more likely to be polished when moving in the sediment.

Friction Experiments

C This involved repeatedly rubbing flints together to produce macroscopically visible "bright sports" (see section 6.5.3). The bright spots most similar to archaeological ones appeared only when using water as a medium and rubbing for two minutes (figures 43, 44 and 45).

DISCUSSION

Bright spots do not usually have any systematic relation to the edges and should not generally be confused with genuine signs of use-wear, although they can mask parts of the edge (figure 43). In such cases it is important to understand the sequence of development; this can help ascertain whether other traces on the edge are natural or not. When bright spots occur as large, very flat and isolated polish they are easily identified, but if they occur in clusters which are not circumscribed in area and taper off, they can look remarkably similar to use-wear polish.

A very appropriate example is the case of 6 scrapers from the Powell site which were examined to see whether the problems with the microdenticulates were related to function or due to post-depositional factors. In the case of a particular tool J9/4a/30, the traces on the retouched scraping edge could have been mistaken for use polish (figure 46). However, examination of the other parts of the implement showed that bright spots on the proximal end of the ventral aspect were definitely not attributable to use. Interestingly, they appear very similar to the polish commonly described as "sickle gloss" (figure 47). Where these bright spots were less developed they were very similar to the "use" polish on the scraping edge. It is therefore extremely likely that all the surviving polish on this implement was post-depositional in nature.

4.1.6 Concluding Remarks

The disappearance of traces on artefacts showing soil sheen renders the results of microwear analysis in these cases doubtful to say the least. Similar effects have been noted on assemblages other than Hengistbury. Microdenticulates from another Mesolithic site were also examined. The site of West Heath, London, excavated by the Hendon and District Archaeological Society under the direction of Desmond Collins, has a soil environment very similar to that of the Powell site. The artefacts were also either patinated or covered with soil sheen and bright spots. Microscopically too, the flint surfaces were in a comparable condition to the Hengistbury Head material. Refitting of artefacts points to their "considerable vertical displacement" in the sediment (Maher, 1983).

A further example are the artefacts from the Mesolithic site of Meer II, in Belgium (Van Noten 1978) which also underwent extensive vertical and horizontal movement in the sediment and are macroscopically so lustrous that they appear varnished. My examination of two of the Meer artefacts (Keeley 1978) under the microscope showed that the previous analysis may have mistaken post-depositional traces for use-wear.

In the cases where post-depositional surface modifications (PDSM) occurs on artefacts, the number of tools which can be confidently analysed is often reduced to an almost unrepresentative sample and the detailed interpretation of microwear traces on such artefacts is clearly very doubtful. This does not mean, however, that functional analysis on such tools should be abandoned altogether. Tool and edge morphology can suggest possible uses; experimentation with observation of edge wear patterns can narrow down the number of tasks a tool is suitable for or is likely to have been used for. It is no doubt a more modest result than the first microwear studies may have led archaeologists to expect. On the other hand, while the results may be limited in scope they are at least realistic.

In conclusion, therefore, it is only possible to suggest that the Hengistbury microdenticulates were used in a general cutting/sawing motion on a relatively soft material probably a bracken type plant or green wood.

This interpretation is based on the edge morphology and other characteristics coupled with the results of experimental work.

In this instance microwear polish was not able to substantiate the above observations. It was not possible in the sample of 34 artefacts to isolate with certainty, the polish produced by use, if any, from that produced by post-depositional processes.

4.2 Mousterian artefacts from Kebara

4.2.1 The Site and the Artefacts

The Kebara cave site lies in the Carmel hills and has known deposits from the Mousterian to the Epi-Palaeolithic. The history of the various excavations is described in detail in Schick and Stekelis (1977) and its most recent re-excavation is discussed by Bar-Yosef *et al.* (1986).

At the start of my work as a member of the excavation team, it was suggested that I carry out a microwear analysis of certain Mousterian artefacts following the Keeley method (1980). Since retouched and formal tools are rare in the assemblage, as is often noted on such Mousterian sites, interest centred on the determination of which implements were used, how they were used, and what materials were worked.

From hundreds of excavated artefacts from the 1982 season, I sampled 56 artefacts which macroscopically did not appear patinated, burnt or otherwise unsuitable for microwear according to Keeley's guidelines.

My previous attempts at microwear analysis of flint implements from other Israeli sites showed that flint surfaces were affected by post-depositional processes. This seems to result in the disappearance of use-wear polish and/or the addition of other "polishes", not related to use, which confuse the appearance of use wear traces (Levi Sala 1986a).

When the artefacts from Kebara 1982 were examined microscopically, edge-wear on potentially useful edges of 15 tools suggested that these at least seemed likely to have been used. On these 15, however, there was no discrete patterning of polish on the edges which might suggest that the edges were indeed used. There were patches of "polish" which could be attributed either to use on soft materials or to post-depositional processes.

Before abandoning the project completely, an attempt was made to answer the following questions:

a would a more careful approach in excavation and storage techniques eliminate some, if not all, of the post-depositional "background noise";

b could any conclusions be drawn from looking at a particular tool type?

To investigate question a, all the 59 pieces from Square I 18 Level XI 1985 were excavated, washed, bagged individually and examined. Many of the pieces were either patinated, broken and/or patently unusable debitage. To assist with addressing question b, the excavators kindly sent 12 Mousterian scrapers from various squares and levels (see 4.2.3 below).

4.2.2 Post-Excavation Handling

It was decided to investigate whether handling of artefacts can cause traces which might constitute "background noise" when analysing edges. In a paper warning about the effects of careless post-excavation handling, Plisson (1985:99–101) recommends avoiding both the brushing of artefacts on site and the use of paper bags. He claims that paper caused polish to appear on experimental flint implements after only five seconds of contact (*ibid.*:100 Plate 6).

A group of 32 of the Kebara pieces were washed and brushed with nylon brushes on site before being bagged in individual **paper** bags. Another group of 27 pieces were washed with water only and bagged in individual **plastic** bags. No difference in surface preservation and/or modification was observed between the two groups. No special polish was observed on the artefacts which were in contact with the paper bags for several weeks in a knapsack before being brought to the laboratory.

Experiments were carried out in the laboratory by brushing used pieces of Brandon flint with water and sediment for up to five minutes (on-site brushing rarely lasts a whole minute) using a nylon brush. The surface was observed and photographed before and after brushing (figures 48, 50, 52). Neither the polish nor the surrounding unpolished flint appeared to be affected by this cleaning when observed optically at x100 and x200 (figures 49, 51, 53).

As Brandon flint is more fine-grained than the flint found in the cave at Kebara one would expect it to be even more susceptible to the formation of polish and other traces related to attrition and abrasion (Vaughan 1985:28; Holmes 1987:92; Masson 1982:33).

All pieces examined microscopically were cleaned in the laboratory with ammonia-free detergent and by immersion in distilled water in an ultrasonic tank for about 20 minutes. A few pieces (on which encrustations covered the probable used edge) were dipped in HCl 10% for periods of up to 30 minutes, as advocated by Keeley (1980:10–12).

All carbonate encrustations were removed by this cleaning but some more obstinate encrustations were not affected, presumably due to their being of non-carbonate composition. There is evidence of bone dissolution on site and the cave is extensively used by bats.

4.2.3 Scrapers

A number of retouched "scrapers" (12) from various Mousterian levels and squares were also examined in order to compare post-depositional processes on:

1 presumably used, polished surfaces on the scrapers

2 possibly unused unpolished surfaces on the remaining artefacts

The scrapers had been washed and brushed on site and transported for storage in boxes in contact with other flint tools.

As mentioned earlier the scrapers were presumed to have been used. On six of them heavy patina prevented observation of use-wear traces and on some it was not possible to distinguish even the flint micro-structure. It was only possible on these patinated scrapers to record surface condition and edge morphology. On the other six scrapers some traces of weakly developed polish with some striations were present (figure 54). The distribution of the polish on the edge and on the flint structure could have been produced by use on a soft material like meat in the presence of grit (figure 55), and/or by use on separating cartilage from bone (figure 56).

The surface of these six pieces is nonetheless also affected by post-depositional processes which very often produce "polishes" not too dissimilar from the weak polish produced by butchering. This occurs either through contact with other stones and/or movement in the sediment or more importantly following the chemical processes in the soil which affect the flint surface as discussed in section 6.5.10–12 (Levi Sala 1986 a and b).

It is therefore not possible on the basis of polish development and patterning to interpret the use of these tools as butchering knives. In drawing conclusions regarding function it is furthermore advisable to observe a much larger sample where eventual patterning of traces could be used as evidence.

Other features were also examined which are probably more reliable for interpretation of function than polish which has been altered by post-depositional processes. The retouched edges on these artefacts were not automatically assumed to be the active edges. Examination of the morphology and profile of all these edges suggested that they were in fact the most useful. Edge damage was taken into account but of course is not very informative on retouched edges or on tools which have been rather roughly handled.

4.2.3.1 Edge Angle and Edge Morphology

On all 12 scrapers the useful retouched edge was more acute than 50 degrees, made on thin blanks, which were below 6mm thick at the midpoint of the edge. All scrapers had straight, slightly concave to slightly convex edge profiles, which would be useful in a cutting activity like butchering.

4.2.4. Use-wear Experiments

4.2.4.1 Butchery

Butchering experiments carried out with edges quite similar to these "scraper" edges also produced weakly developed polishes as described above (figure 57 see also figures 4 and 21).

Nonetheless, other similar edges were used to scrape hide, wood and bone, to test their suitability for such tasks.

4.2.4.2 Scraping Hide

The edges were too thin and sharp to have been used to scrape hide; the scraping motion cut the hide, a most undesirable result.

The Levallois technique was extensively used in the lithic industry at the site, probably to produce thin blanks. The scalar retouch present on these pieces thins, sharpens and strengthens the edge but does not blunt it; hide scraping is best carried out with a thick high angled and blunter edge such as that found on end scrapers.

The experimental tools showed that these thin edges were useful to de-hair hide. This activity however often results in edge rounding which was not observed on the Kebara scrapers. Only a few traces parallel to the edge were present on the six archaeological pieces while de-hairing hide usually leaves a pattern of faint wear traces perpendicular to the edge. It is not certain that the observed traces parallel to the edge are use-induced in these altered artefacts. If they were, they would be more indicative of cutting than scraping (figure 58).

Furthermore, had these tools been used to de-hair hide as a specialised activity, one would expect other, thicker scrapers, more suitable for processing hide to be found on site; no such tools are represented in the assemblage (Meignen pers. comm.).

4.2.4.3 Scraping Wood

The experiments revealed that the scrapers were inefficient wood scrapers removing only very minute shavings; a thicker, high angled flake could whittle or scrape better and quicker.

4.2.4.4 Scraping Bone

The scrapers proved as inefficient for scraping bone as for working wood. However, as bone is a harder material than wood, the tool was even less efficient, yielding no shavings at all.

4.2.5 Conclusions

4.2.5.1 Function of Retouched pieces

It seems therefore that these 12 "scrapers" are not useful tools to scrape either hide, wood or bone. It appears much more likely, by process of elimination, that these tools represent butchering knives rather than scrapers. There are only 12 scrapers involved in this study and although they were all from one season's excavation their temporal relationship to each other has not yet been studied. It is therefore not possible to make generalisations from a small sample where there is no polish which is clearly use-induced. It is however possible to make a theoretical hypothesis: on a site where artefacts with these characteristics are the most significant, where no other tools which are suitable for hide scraping are found, it might be worth postulating, provided other evidence is also compatible, that hide processing was taking place elsewhere, if at all.

The results of experiments with these tools classified as scrapers typologically would suggest that functionally they are probably cutting tools.

4.2.5.2 Function of Unretouched pieces

Regarding the 59 unretouched pieces, the condition of the flint surface allows only very limited information to be gathered regarding function. They have been divided in three categories as follows:

1) 11 pieces which could have been used. These each have a useful edge and/or edge damage of probable use origin. The condition of the flint surface does not allow more definite statements on the exact function to be made.
2) 28 pieces which were probably not used either because the edges were unsuitable or because there is no evidence of use present. However, it is usually not possible to state with certainty that these pieces were "not used"; especially in view of the condition of the flint surface.
3) 20 pieces which are burnt, too patinated or too broken to study.

These results are not surprising in view of the possibility that square I 18 Level XI is an area where refuse from a nearby hearth was being dumped (Bar Yosef pers. comm.).

It is also quite compatible with the disappointing outcome of efforts to find recognisable use-wear polish on various flint collections from Israel (see section 4.3 and Levi Sala 1986a).

4.2.5.3 "Background Noise" and Post-depositional Processes

As described earlier (section 4.2.2), on-site cleaning with a nylon brush and bagging singly in paper bags does not seem to affect flint condition as observed from these 59 pieces and from the experimental verification.

It was not possible to identify the elements which might have caused such widespread post-depositional surface modification in this assemblage. Percolating water through limestone over a long period of time is, however, likely to affect flint surfaces; laboratory experiments set up to observe this phenomenon over time are described with a discussion of the results in Chapter 6 particularly in section 6.5.5 and 6.5.6. However in view of the difficulty in speeding up such natural processes, these experiments cannot be relied upon to give us conclusive evidence for the understanding of post-depositional processes at Kebara within the time span of this work.

No.	SITE	AGE	LOCATION	PIECES
I	ARAD	CHALCOLITHIC	NEGEV DESERT	5
II	AZARIQ	MUSHABIAN	NEGEV DESERT	10
III	BIR SAFADI	CHALCOLITHIC	NEGEV DESERT	10
IV	FAR'AH II	MIDDLE-PAL.	NEGEV DESERT	20
V	FAR'AH	NEOLITHIC	NEGEV DESERT	5
VI	GHASSUL	CHALCOLITHIC	JUDEAN HILLS	3
VII	GOUGH'S CAVE	UPPER PAL.	NEAR CHEDDAR	3
VIII	GRAR	CHALCOLITHIC	NEGEV DESERT	13
IX	HAYONIM	KEBARAN	CARMEL HILLS	9
X	HAYONIM	NATUFIAN	CARMEL HILLS	3
XI	HAYONIM	MOUSTERIAN	CARMEL HILLS	10
XII	HAYONIM	AURIGNACIAN	CARMEL HILLS	6
XIII	HENGISTBURY	MESOLITHIC	DORSET HEADLAND	45
XIV	JERICHO	PPNA/PPNB	LOWER.JORDAN VALLEY	25
XV	KEBARA '82	MOUSTERIAN	CARMEL HILLS CAVE	55
XVI	KEBARA '85	MOUSTERIAN	CARMEL HILLS CAVE	71
XVII	NAHAL HEIMAR	PPNB	JUDEAN HILLS CAVE	27
XVIII	NAHAL OREN	NATUFIAN	CARMEL HILLS	34
XIX	NETIV HAGDUD	PPNA	LOWER JORDAN VALLEY	15
XX	NEVE DAVID	KEBARAN	N.COASTAL PLAIN	10
XXI	QADESH BARNEA	UPPER-PAL.	N.SINAI DESERT	16
XXII	WEST HEATH	MESOLITHIC	HAMPSTEAD HEATH	57
TOTAL ARTEFACTS EXAMINED				**452**

4.3 Sites sampled

The above is a list of sites from which samples were taken in order to decide whether they would be suitable for microwear analysis at magnifications of around x 200. The sites, which are listed alphabetically, are of widely differing ages spanning a period from the Mousterian to the Chalcolithic. The geographical setting varies from the southeast of the British Isles to the Negev desert. The macroscopic examination of the artefacts from the site of Meer in Belgium and microscopic examination of two of those artefacts has not been included in this list. The two sites from which artefacts were eventually subjected to a functional analysis are, however, included and the analysis is reported under sections 4.1 and 4.2. In Israel itself besides the Negev sites, samples were taken from sites in the coastal plain, the Carmel hills, the Judean hills and the lower Jordan valley. Some of the sites were caves, others were rockshelters, and still others were open air sites. In spite of this diversity all the flint surfaces of the samples showed surface modification to the extent that use-wear polish had either not survived on their edges or could not be isolated from the surrounding alterations.

The numbers of artefacts examined reflect the order in which these assemblages were sampled. At the start of my research the artefacts which are considered "fresh" archaeologically were chosen and consequently large numbers are involved. As my experience in recognising surface modification increased fewer and fewer artefacts appeared suitable even after only a macroscopic examination as for example in the case of artefacts covered with intense sheen.

Chapter 5
Examples of over-interpretations: Cautionary remarks

In contrast to the low level interpretations which have been offered in the preceding chapters, the microwear literature has abounded in analyses which have led to what is, in the writer's view, over enthusiastic conclusions. Space does not allow a complete catalogue of references but the following cases should give an idea of the problem.

5.1 Hafting traces

As far as I am aware there is no known archaeological evidence for stone tools having been hafted during the Mousterian in southwest France. However, Anderson-Gerfaud (1981:41–44) has interpreted as possible evidence for hafting some traces observed on Mousterian tools (Pech de l'Aze IV and Corbiac) away from the used edge. No experiments seem to have been carried out to ascertain whether hafting similar tools would leave similar traces. Moss points out that Anderson's "sample excavated some years earlier was not really geared toward answering questions about distinct Mousterian tool kits" (1983b:20). Moss and Newcomer (1982:292), suggest that no definite traces are produced by sinew hafting (cf. reconstruction by Anderson-Gerfaud 1981: figures 46 and 47). Some faint and "unidentifiable polish" was observed, however, on the "ridges" and "high points" (Moss and Newcomer 1982:292). The experimental work described in section 4.1 has not left any traces which are diagnostic of hafts. See also section 3.4.3. Furthermore Unger-Hamilton *et al.* (1987:280) have found no hafting traces on drill bits wedge-hafted with wood and sinew specifically to test the presence of such traces. Some polish was found on a few pieces, on dorsal ridges *before hafting* and a probable explanation is offered by the authors as follows:

> ...as the tools were retouched on a hide pad, small chips of flint were removed that congregated beneath the tool, so that when pressure was applied during retouch the drill bits were pressed down on to the flint chips causing flint on flint contact.

As there is no evidence for pressure flaking in the Mousterian, similar traces are not likely to occur on the dorsal ridges of Mousterian tools. It is surprising to see "hafting traces" on the ventral aspect as shown by Anderson (*ibid.* figure 38). I have observed the polish described by Anderson on many tools both Mousterian and Mesolithic and am convinced that, on the pieces I have observed, it is caused by PDSM.

Beyries' hafting experiments (1987:80) on replicas of Mousterian artefacts prove that two double scrapers can be hafted and so hafted can chop wood. As there is no description of what traces were left on the experimental pieces by use or hafting, these experiments cannot lead to a conclusion regarding use or regarding the presence of hafts.

5.2 Handedness

On the basis of three "boring implements" used on bone/antler which were interpreted as having turned in a counter-clockwise fashion, Cahen and Keeley (1980:178–179) concluded on experimental evidence that "these patterns are related to handedness" and "we believe we have been able to isolate specifically the work of two craftsmen". Unfortunately, Keeley does not mention whether his reconstruction is based on the use-wear polish or the edge scarring. The flint from Meer is very heavily affected by soil sheen (Levi Sala 1986a; 1986b) and it is doubtful that polish survived to allow such a fine reconstruction of traces to be made with confidence. Commendable though the experimental verification was, there seems to be a great deal of over-interpretation in their conclusion due to the condition of the Meer flint artefacts.

5.3 Duration of work

Many analysts have estimated the length of time that archaeological pieces have been used such as "2000 strokes?" for a Clactonian flake interpreted as a meat knife (Keeley 1980:91) and Vaughan "from a few minutes to around twenty" (1985:60). Analysts have generally explained that their interpretations are based on the degree of development of the observed polish. Length of use is not however the only variable causing polish to be more or less developed. There is a large number of other variables that affect the development of polish namely: material hardness, humidity, pressure, angle of work, edge wear and resharpening (Frison 1968:154). It seems to me therefore very difficult to estimate length of use, taking so many unknown variables into account. It is difficult in blind tests to estimate duration of use when reconstructing an activity on an experimental tool and even more unlikely on archaeological implements.

5.4 Meat knives

Together with many colleagues in the Institute, experiments have been carried out butchering deer, cow, and lamb meat on several occasions and for long periods of time. With one exception none of us seems to have obtained a well developed polish except when flint was used to disarticulate bone and was therefore used on the hard bone or in cases when meat was cut on a hard surface such as a wooden chopping board or a flat stone.

Juel Jensen (1988a:55) confirms that no polish developed after cutting meat for 90 minutes. Meat cutting alone never produces any more than a very faint general "shine" which Keeley describes as a "pronounced greasy lustre" and admits that "it showed little contrast with the unaltered areas of the flint" (1980:53). This alteration occurs as the first stage of polish development on *all* used materials. Even if this stage of polish development were to survive thousands and millions of years of burial and even if one could distinguish this brightness from the subsequent soil sheen, it would be impossible to identify it as "meat polish" since, as mentioned earlier, all used materials produce this generalised brightness in the early stages of use (figures 3 and 4).

It is therefore most disturbing that Keeley should have identified as "cutting meat" (apart from the Clactonian flake mentioned above) four implements from two sites in Koobi Fora, Kenya dated to 1.5 million years ago and made of an undefined "chert" (Keeley and Toth 1981:464-465). The published pictures are at a magnification of only x50 when he himself rightfully stressed that polish should be studied at x100 and better still at x200. The paper makes far reaching conclusions "pertinent to the controversy concerning the diet of early hominids and the role of stone tools in their adaptations". The study deals with traces on only nine "implements" among 54 flakes examined and collected from five different sites. Keeley has himself recommended a much greater rigour in microwear methodology (Keeley 1974 a and b) which is sadly lacking in the Koobi Fora analysis.

In view of the ambiguity and lack of development of the traces left by cutting meat, it might be that those who find too many implements used on "meat" in their archaeological collections are probably failing to recognise the early stages of soil sheen.

5.5 Doubtful Methods

Bueller (1984:202) describes the process by which he learnt to recognise used materials from microwear polish:

1 he examined pieces and photographed "polishes";
2 he learnt microwear;
3 he deciphered photographs.

One must assume, therefore, that when he photographed the traces he could not distinguish those related to use-wear from traces not caused by use. In fact he himself explains this strange procedure in analysis by stating:

> the precise nature of the different kinds of microwear occurrences were not fully understood (Bueller 1984:202).

Yet he concludes (*ibid*:213) that endscrapers were used to saw bone and that burins blows were intended for "hafting, that the edges other than the burin edge were used on "meat, hide, wood and bone" without mentioning any experiments to show the traces which the suggested hafting would leave on burins. These conclusions were based on 11 endscrapers and 12 burins.

Bueller (1983:107–125) has also analysed 10 artefacts per tool type from Ain Mallaha and El Wad, and concluded that El Wad was a big game specialised butchering camp while Ain Mallaha was a home base where small game was hunted in spite of the fact that he rightly noted that "even after prolonged use real polish is never formed [by meat]" (*ibid.*:108).

It is obvious from the above comments how misleading it is for archaeological reconstructions to be based on functional analysis carried out with doubtful methods.

Other examples will be referred to in the following chapters of conclusions whose effects on archaeology are very important and yet are either based on insufficient numbers of artefacts and/or lack a published documentation of the traces on whose basis such conclusions were drawn.

CHAPTER 6

REVISED AIMS

6.1 RESEARCH DESIGN

In view of the problems discussed in Part I, my research took a new direction. Instead of attempting yet again to study the function of archaeological artefacts, it was proposed to subject flint surfaces to controlled and repeatable friction under laboratory conditions, observing the wear traces after various time intervals. The factors considered important in the appearance of wear traces were varied one by one to assess their relative importance. The investigation included experiments to determine the effect of the presence/absence of a fluid medium and its nature including whether it interacts chemically and/or physically with the flint and also the effect of the presence, absence, nature and size of particulate matter on the development of polish and on the formation and type of striations.

The development of polish on flint tools is thought to be affected by the following:

- hardness of the material worked;
- water or other liquids added or contained in the worked material;
- pressure exerted;
- type of flint especially grain size, pattern of the crystalline structure and size and frequency of inclusions;
- presence and grain size of abrasives;
- duration of tool use;

I am not aware of any laboratory work done as yet to investigate these variables.

Mansur-Franchomme (1983) investigated the importance of abrasives and humidity in hide scraping. Her work, though systematic, was based on n uncritical acceptance of the idea of silica gel formation as proposed by Anderson (1980). Consequently her interpretations of the results are arrived at only from that perspective. Furthermore her work was documented mainly by SEM where she observed a certain classification of striations which is not recognisable using optical microscopy.

Other experiments described in the following pages were designed, furthermore, to study the factors, whether chemical or mechanical, affecting flint surfaces during burial and causing alterations to the flint surface (PDSM) as well as alterations to the use-wear polish.

They are grouped together as follows:

section 6.4 experiments to study use-wear and polish formation;

section 6.5 experiments to study post-depositional processes.

In some of the work experiments have had to be repeated with only one variable altered. These are described as Runs and are indicated with Roman numerals. When the same artefact or a group of artefacts are being subjected to successive treatments these treatments are also numbered with Roman numerals and are referred to as Stages.

6.2 POLISH FORMATION

The mechanisms involved in polishing glass have been studied over the years by many scientists interested in the improvement of the method for various industrial purposes in particular for the production of optical lenses. A history and bibliography of this research was compiled and commented on by D.C. Cornish in 1961. Since then two symposia have taken place under the auspices of the U.S. National Bureau of Standards with several papers on the subject of surface finishing of glasses and ceramics (Schneider and Rice (eds.) 1972; Hockey and Rice (eds.) 1979).

Due to the similarity of flint to many glasses and ceramics [they are largely composed of silica and exhibit the same conchoidal fractures] it was thought that the knowledge accumulated on the subject of glass and ceramic polishing could be useful in the study of polish on flint. Curiously enough some of the earliest glass in antiquity, and up to the 17th century A.D., was made using crushed flint as the source of silica (Shepherd 1972:208–210).

For a long time it was hypothesised (Cornish 1961) that glass was being polished in a process comprising two stages:

1st stage: the microscopic asperities on the surface are being flattened.

2nd stage: the material displaced from the asperities fills up cracks, pits and scratches. The "new" surface appears polished.

The two stages described above constitute what is termed "plastic flow", "surface flow" or "plastic deformation".

In 1921 Beilby proposed the following model for the mechanism of the formation of polish on glass:

> The rouge particles, it may be supposed, hardly penetrate below the surface, but coming into almost molecular contact with the sheet of molecules on the surface, drag it off like a skin. The fresh molecular layer left by the removal of the skin retains its mobility for an instant, and before solidification, is smoothed over by the action of surface tension, thus producing the liquid-like surface which is the necessary condition of a perfect finish (Cornish 1961:3).

Already in 1927 Adam stated however that "it is not necessary to assume that a liquefied layer precedes the formation of the amorphous layer...Any method which rearranges the surface molecules will give an amorphous surface" (Cornish 1961:17).

Recent research has led to processes being explained by formulae rather than by figurative descriptions as in Beilby's time. It will no longer be necessary to use words like "a liquid-like surface" which was the cause of much misunderstanding both by scientists and by microwear analysts.

Several mechanisms are involved in polish formation. Friction, wear, lubrication, and adhesion are a group of mechanical surface interactions which are studied together. They are well known but the relative importance of each is not well understood.

A molten layer would indicate that very high temperatures were reached over the whole surface during the polishing process whereas generally "this consists of a moderate temperature rise over the contacting bodies, and superimposed on this there are higher "flash temperatures "at the junctions" (Rabinowicz 1966).

A hydrolysed layer would furthermore presuppose that water interacted chemically with the flint. It is not known whether or not these two processes (high temperature and chemical interaction with water) do take place. That is not to say that they do not occur but they do not appear to be essential for the formation of the polished layer. As the silica gel layer was hypothesised for the presence of polish on flint, it was possible to go further and suggest that residues such as phytoliths from plants can be trapped in that layer during its "amorphous gel" state (Anderson 1980:184). There is no need to assume that the polished layer **on flint** is a "molten" or "hydrolysed" layer even were such a layer shown to occur in glass polishing. An argument against this hydrolysis is the quantity of silica that would have to go in solution and be redeposited in order to obtain a blade as glossy as some Near-Eastern Neolithic "sickles" are. Furthermore, in order to trap phytoliths, the thickness of such a layer would have to be measurable in microns (see 1980:Plate 8b ""sickle gloss"' limit, built up approximately 10μ above flint surface" and Ollendorf 1987:457 "phytoliths of 11–44μ length"). A considerable mass of silica would have to be dissolved and precipitated to give such a thick layer on a blade whose dorsal and ventral surfaces are glossy along the whole edge up to 20mm inland. It may be noted that no evidence of such a thick layer was found under the SEM.

In glass studies when testing for surface defects it is customary to etch the polished surface with hydrofluoric acid (Landingham *et al.* 1979:231–246). However glass is very homogeneous while flint has a complex crystal structure. On glass the acid removes the stressed material preferentially but also appears to remove the flowed material (perhaps because it is under stress or because it is more disordered).

Etching flint with hydrofluoric acid preferentially removes the more structurally disorganised chalcedonic phase (Griffiths *et al.* 1987:47–51). Consequently, it seemed superfluous to carry out this test as it could not answer the question about the nature of the layer which optically appears polished.

In this research the aim was to observe whether, **when polish occurs on flint**, one can attribute it to the phenomenon of plastic deformation alone (which would suggest a mechanical action) or whether silica gel (through hydrolysis) was formed as well. Plastic deformation almost always means the formation of a layer of amorphous silica because the particles on the flint surface are no longer ordered in the same way as in unpolished flint. Amorphous silica in this context means silica lacking the long range periodic structure found in crystalline silica and is to be distinguished from silica gel (see Glossary).

6.3 Post-depositional surface modification

The following experiments were designed to study the processes involved in post-depositional alteration of flint artefacts and their effect on use-wear traces. As mentioned earlier assemblages from many sites in various countries have been found to be affected by post-depositional surface modification (PDSM), a term first used by Holmes (1987:91), and whose abbreviated form PDSM is now widely used in our Institute. These assemblages were consequently found to be unsuitable for analysis by the Keeley method. Most of these implements, when not patinated, had a macroscopic sheen attributed by many researchers to the effect of burial in the sediment. Bright spots, described below, were also very frequent.

The ubiquity of PDSM, especially with sheen and bright spots, prompted the investigation of the processes of its formation. Chemical processes were investigated to test which of the phenomena they were responsible for. The hypothesis that some of these alterations on the flint surface are due to mechanical processes was also tested. The conjoining of artefacts (whether flint, pottery or bone) from several stratigraphic levels on a variety of sites has, of late, drawn the attention of archaeologists to the fact that artefacts can move in the sediment without there being clear stratigraphic evidence of such movement. The vertical and horizontal movement is much more widespread and frequent than previously imagined (Cahen and Moeyersons, 1977; Barton 1981; Villa, 1982; Rowlett and Robbins, 1982; Villa and Courtin, 1983; and others).

6.3.1 Sheen

Sheen has been referred to as "glossy patina" by Rottlaender (1975: 10l), "gloss patina" by Stapert (1976:11–12), "surface sheen" by Plisson and Mauger (1988) and "soil sheen" by Moss (1983:81). Moss 1983) reported that "all archaeological collections have some sort of soil sheen even when the flints seem to be in pristine condition". There was no specific mention of the problem this sheen creates for microwear analysis. Plisson and Mauger (*ibid.*) investigated chemical and mechanical alteration of microwear polishes, pointing out that Keeley (1980) and Anderson-Gerfaud (1981) had to "eliminate from their studies of palaeolithic flint tools not only tools having macroscopic natural alteration traces but also 25–40% more of the tools in the collections which had pronounced microscopic soil sheen".

Keeley himself (1980:28–35) warns against the danger of ignoring naturally caused traces. He excluded the "majority" of Swanscombe Lower Loam artefacts due to "weathering" including a "glaze on their surfaces". He describes the "slight edge damage and localised abrasions" caused by "minor soil movements" which constitute "a problem for the microwear analyst". He does not discuss the effect of soil sheen on microwear polish.

Rottlaender (1975) attempted to produce soil sheen, which he calls "glossy patina", through the action of "weakly acting solvents". He describes their effect as being "like a mechanical polishing". He encountered difficulties in trying to produce this sort of surface modification in a laboratory; when increasing the strength of the solvents to accelerate the process one risked patinating the flint.

Sheen is usually uniform over the whole surface of the flint though it tends to be more concentrated on edges and prominent parts of artefacts. It varies in degree of intensity from very faint to very glossy, as in "wind-gloss" and "desert varnish". The latter are names given to gloss observed on artefacts from very windy environments and/or deserts which appear very glossy all over their surface. The gloss is attributed to the effect of abrasion from wind and sand. In his important review of naturally produced traces on flint artefacts from Dutch sites, Stapert (1976:12-18) distinguishes between the "wind-gloss" and "desert varnish" and "gloss patina" as follows:

"wind-gloss" being colourless and appearing "greasy" and "variable on the same piece" he feels it is "formed mainly mechanically as a result of polishing, though a chemical component may contribute to its formation";

"desert varnish" being dark in colour (presumably from mineral deposition);

"gloss patina" "although it is sometimes suggested that a mechanical effect may be responsible (e.g. polishing in a sand matrix due to soil movements), most authors assume this smoothness is a result of chemical processes in the soil".

It is my experience from surface artefacts collected in the Negev desert that many artefacts have a colourless varnish on them suggesting that these three types may overlap. In any case, he concludes that "it is not impossible that, in places where wind-gloss is highly developed, microscopic traces of use formerly present are completely or partly obliterated" (*ibid.*:19).

In this Chapter some experiments are described producing sheen on experimental artefacts both by mechanical and chemical processes.

6.3.2 Bright spots

Bright spots occur very frequently on archaeological artefacts of a variety of ages and provenances. They are a smooth, highly reflective polish (gloss) on flint surfaces, varying in size from a pin point to the whole artefact being glossy, occurring singly and in clusters.

Bright spots vary: they can be flat, raised, crossed by striations or not striated. Flat bright spots appear as an extreme and localised smoothing of the surface within the area of occurrence, but not so with certain more rounded varieties which have an appearance and a distribution on the flint topography very similar to that of use-wear polish.

Bordes (1950) named them "miroitement" and "plages brillantes". He reviews the various possible explanations put forward for their formation and concludes that mechanical wear is the most logical. The grain size of the polishing agent (sand, loess, and/or soil in springs) could explain their variety though not their pinpoint localisation. Semenov (1964:11) describes them as "single or groups of scintillations, sharply defined stars or luminous veins. Their origin still remains unexplained". Shepherd (1972:120) called the flat ones "friction gloss" and suggested they were easy to reproduce by rubbing two flints together. Stapert (1976:29–39) noted that these bright spots, which he describes in detail, were far commoner on younger material than on older artefacts. He suggested that "friction gloss" may be a misleading name for them. Vaughan (1981: 365) and Moss (*ibid.*) have both tried unsuccessfully to produce the gloss by following Shepherd's suggestion. Vaughan (1981:361:66) has noted a high incidence of bright spots at the Magdalenian site of Cassegros level 10. He calls them "special micropolishes" and has noted a "flat" variety and a "ripply" variety. After SEM investigations at x2000 he suggests that the flat kind were formed by "both friction and water" while he attributes the ripply variety to the "precipitation of ground water silicates" on the surface of the artefacts. In spite of all the hypotheses for their formation they remained unexplained to many: "mysterious" to Moss (1983:81), and "enigmatic" to Stapert (1976:38) and Vaughan (1981:366). Moss (1983:81) called them "bright spots" as Stapert reached no firm conclusion about their origin and this is the name adopted in this thesis.

Holmes (1987) carried out an experiment on modern artefacts of Egyptian flint. After two weeks of trampling, a number of microscopic bright spots appeared, comparable to those observed on her collection of pre-dynastic flint tools.

Bright spots occur on edges, ridges and flat areas of archaeological flint tools. However, they do not have a systematic relation to the probable used edges. Although they can generally be excluded as use wear, especially when isolated, flat and large, they do often mask large stretches of edge (figure 43). Sometimes bright spots cluster at the edge and taper off, mimicking use-wear polishes (figure 47). Furthermore, understanding the way they are formed can suggest to the functional analyst whether the other traces present on the edges are due to use or to post-depositional causes. The work carried out in this thesis has shed light on the processes of their formation. They are discussed in section 6.5 3 and 6.5.12 where experiments have shown that bright spots can be formed both by friction and by chemical processes. The correlation between bright spots and sheen on the same pieces has been noted both on archaeological and experimental pieces (section 6.5.12) suggesting a similarity in the processes responsible for the two phenomena.

6.4 Experiments for use-wear and polish formation

6.4.1 Flint Type Characterisation and Abbreviations

The following characterisations are based on freshly struck flakes and are a summary of observations on several flakes from different nodules. Figures 7, 62, 98, 99 and 225 show photographs produced as a visual aid to these descriptions. Most of the Israeli flint types used are derived from Eocene formations.

BRND Brandon or other East Anglian fine black flint.

P/BAR Potter's Bar grey, medium to coarse, flint.

DRST Dorset pebble flint, buff, medium grained.

ISR/1 Honey coloured flint. Source unknown
macro: very fine grained, easy to knap, smooth texture appears finer than Brandon
micro: does not refract light to the same extent as fine grained Brandon flint. Few inclusions many "ridges" on fracture surface (figure 23). Some Kebara tools appear to be made from this flint.

ISR/2 Negev flint in various shades from light brown to nearly black, often on the same nodule
macro: inclusions, medium to coarse, much grain size variation in same flake
micro: shiny flint, large and small inclusions, large shallow hollows on fresh fracture surface, medium to coarse grained.

ISR/3 Har Qeren flint buff coloured weathered nodule
macro: coarse due to patchy colour NOT banded
micro: medium grained with large inclusions. Glare at x200.

ISR/4a Nahal Tzin flint buff coloured
macro: matte not shiny like ISR/1 and Brandon but seems fine grained
micro: coarse with deep "ridges" or "waves" rare opaque inclusions. Even in areas of finer grain not tightly packed. Very hard to polish and photograph. Glare at x 200.

ISR/4b Nahal Tzin very dark flint almost black
macro: very shiny and fine grained
micro: very shiny, with deep "ridges" and shallow hollows with grey crystal inclusions.

ISR/5a Galilee "flint", grey coloured mottled with large spots
macro: texture sugary, not smooth like flint
micro: some medium grained areas separated by coarse ones. Glare at x 200.

ISR/5b Galilee chert, banded dark buff to light grey
macro: rougher than ISR/5a in texture
micro: extremely coarse—lighter areas more so than dark ones - crystals at greatly differing heights separated by large hollow spaces— very difficult to polish and photograph —appears very porous.

6.4.2 Abbreviations Used in the Lists of Experiments

B/FACET Burin facet
BLDLT Bladelet
BRND Brandon flint.
CVX Convex
CCV Concave
DRSL Dorsal
DST Distal
E.A. Edge Angle
LL Left lateral edge
MACRO Macroscopic—which can be seen with the naked eye
M/D Microdenticulate
MICRO Microscopic
MIN Minutes
NR Near
P Polish
POLISHED BY Flint mechanically polished by worked material
POLISHED ON Used by hand on worked material
PERP Perpendicular
PRX Proximal
RET Retouch
RIDGES "Fissures" or "feathers" on thin edges of flakes arising from the mechanical properties of the flint microstructure. They appear microscopically as ridges on the flint surface (see figure 23 for a graphic illustration).
RL Right lateral edge
SERR Serration
SIN Sinuous (combination of CCV + CVX/STRT on same edge)
SM Surface modification
STRT Straight
USC Ultra-sonic cleaning
VNTR Ventral
WAVES Undulated topography caused by percussion ripples. Ripples are a feature of conchoidal fracturing and form concentrically from the point of percussion (see figure 23 for a graphic illustration).

In the following pages experimental polish has been described according to the stages of its development suggested by Grace (1989:92). It is graphically illustrated in figure 63 and summarised below:

A polish occurring as individual elements clearly separated from each other within a matrix of an unpolished surface;

A+ larger individual elements but not yet linked together;

B polish linked but the majority of the observed area is unpolished;

B+ a linked distribution where the polished area is approximately equal to the unpolished area.

I have subdivided Grace's C polish into three stages; see figures 116, 117 and 118 for these stages which I define hereunder:

C when most of the observed area is polished;

C+ polish linked to cover most of the microtopography with rare empty spaces;

C++ (all over) almost blanket polish.

6.4.3 List of Artefacts in Use-Wear Experiments

No.	Type	Flint	Use	Material	E.A.	Edge	Min
EXP/1	M/D	DRST	cutting	dry rushes	55	CCV SERR	30
EXP/1	M/D	DRST	cutting	wet rushes	55	CCV SERR	15
EXP/3	M/D	DRST	reaping	Phragmites	38	CCV SERR	60
EXP/4	M/D	BRND	scraping	soft cedar	50	CCV SERR	30
EXP/5	M/D	BRND	whittling	dry beech	45	CVX SERR	05
EXP/5E	FLAKE	BRND	whittling	green willow	65	NO RET	20
EXP/6	M/D	BRND	whittling	dry beech	50	CCV SERR	40
EXP/7	M/D	P/BAR	sawing and whittling	dry reeds	35	CCV SERR	10
EXP/8	M/D	P/BAR	cutting and whittling	dry reeds	38	SIN SERR	20
EXP/9	M/D	BRND	whittling	dry reeds	35	STR SERR	10
EXP/10	M/D	P/BAR	cutting	cow hide	40	CCV SERR	15
EXP/11	M/D	P/BAR	sawing	shell	42	CCV SERR	15
EXP/12	M/D	BRND	cutting	bracken	40	CVX SERR	10
EXP/13	BLADE	BRND	debarking	green alder	60	NO RET	15
EXP/14	M/D	P/BAR	cutting	cow hide	45	CVX SERR	5
EXP/15							
(1)	BLADE	BRND	de-hairing	deer hide	50	NO RET	45
(2)	M/D	BRND	de-hairing	deer hide	45	STR SERR	5
(3)	BLADE	BRND	de-hairing	deer hide	40	STR SERR	20
(4)	M/D	BRND	cutting	deer hide	50	CCV SERR	15
EXP/17	M/D	BRND	cutting	meat	40	CCV SERR	30
EXP/18	M/D	BRND	butchering	cow leg	40	CCV SERR	60
EXP/19	M/D	BRND	reaping	bracken	45	CCV SERR	12
EXP/23	M/D	BRND	reaping	bracken	37	CCV SERR	20
EXP/26	M/D	BRND	cutting	deer hide	45	CCV SERR	25
EXP/27	M/D	BRND	sawing	green birch	45	CCV SERR	10
EXP/28	M/D	BRND	sawing	green oak	55	CCV SERR	5
EXP/29	M/D	BRND	de-hairing	soaked, scraped hide	65	CCV SERR	35
EXP/31							
(1)	BLADE	BRND	de-hairing	deer hide	38	NO RET	20
(2)	BLADE	BRND	cutting	deer hide	40	NO RET	15
EXP/52	BLADE	BRND	sawing	cow bone	30	NO RET	10
EXP/53	BLADE	BRND	butchering	deer	50	NO RET	15
EXP/54	BLADE	ISR/2	chopping	deer heart	55	NO RET	15
EXP/55	SCRAPER	BRND	scraping	deer hide	70	PRX RET	15
EXP/56	BLADE	ISR/2	butchering	deer	55	NO RET	15
EXP/57	BLADE	BRND	butchering	deer	55	NO RET	20
EXP/58	BLADE	BRND	sawing	wet antler	60	NO RET	10
EXP/59	BLADE	BRND	incising	dry oak	55	NO RET	30
EXP/60	BLADE	BRND	scraping	wet antler	55	NO RET	20
EXP/61	E/SCRAPER ON BLADE	BRND	butchering	meat/bone	85	RET +	
					65	NO RET	60
EXP/62	SCRAPER	P/BAR	scraping	cow hide	95	RET DST	20
EXP/63	BURIN/SCRAPER	BRND	scraping	DRY PINE	90	BURIN BIT	15
EXP/64	BLADE	BRND	cutting	fresh hide	55	NO RET	10
EXP/66	M/D	BRND	de-hairing	deer hide	60	CCV SERR	40
EXP/67	M/D	BRND	whittling	wet antler	65	CVX SERR	45
EXP/68	SCRAPER	ISR/2	scraping	deer hide	90	RET DST	25
EXP/69	SCRAPER	ISR/2	scraping	deer hide	90	RET DST	10
EXP/70	SCRAPER	ISR/2	scraping	deer hide	70	RET RL	20
EXP/73	SCRAPER	BRND	scraping	deer hide	75	RET RL	30
EXP/74	FLAKE	BRND	cutting	deer hide	55	NO RET	10
EXP/75	BLADE	ISR/1	whittling	dry oak	70	NO RET	65
EXP/78	SCRAPER	ISR/3	whittling	green alder	45	RET LL	60
EXP/80	BURIN	ISR/2	shaving	wet antler	80	B/FACET	50
EXP/81	DENTIC	ISR/3	sawing	wet antler	75	RET DST	65
EXP/82	BLADE	ISR/3	reaping	rushes	75	DENTIC RL	60
EXP/83	SCRAPER	ISR/3	scraping	green alder	75	DST RL	20
EXP/87	BLADE	ISR/1	tar coating	rush basket	75	NO RET	15
EXP/88	BLADE	ISR/2	butchering	meat/bone	65	NO RET	15
EXP/107	M/D	BRND	scraping	goat hide	55	CCV SERR	35

6.4.4 List of Artefacts in Polish Formation Experiments

Tool No.	Flint type	Experiment description	Section
EXP/71	BRND	Rubbing hide, One hour	3.6.4
EXP/72	BRND	Rubbing hide, One hour	3.6.4
EXP/79	ISR/2	Buffing, wet hard buff	6.4.8
EXP/84	ISR/2	¼µ grit, on wheel	6.4.8
EXP/85	BRND	¼µ grit, on wheel	6.4.8
EXP/89	BRND	¼µ grit, scraping	6.4.8
EXP/90	BRND	Buffing, dry wool buff	6.4.8
EXP/91	ISR/2	Grooving, dry birch	6.4.6
EXP/92 (1)	ISR/2	Grooving, birch + water	6.4.6
EXP/92 (2)	ISR/2	Grooving, birch + oil	6.4.6
EXP/93	ISR/2	Grooving, dry birch + talc	6.4.6
EXP/94	ISR/2	¼µ grit, on wheel	6.4.8
EXP/95	ISR/2	Scraping, ¼µ grit	6.4.8
EXP/96	ISR/2	Grooving, birch + talc	6.4.6
EXP/97	ISR/2	¼µ grit, on wheel	6.4.8
EXP/98	ISR/4b	Wet chamois, on wheel	6.5.5
EXP/101	ISR/4a	¼µ grit, on buff	6.4.8
EXP/103	ISR/4a	¼µ grit, on buff	6.4.8
EXP/104	ISR/2	Whittling, wood	6.5.5
EXP/108	BRND	Buffing, wet hard wool buff	6.4.9
EXP/109	BRND	Buffing, hard wool with wet chamois	6.4.9
EXP/117	ISR/5b	Scraping, hide	6.4.7
EXP/118	ISR/2	Scraping, hide with ochre	6.4.7
EXP/119	BRND	Scraping, hide with ochre	6.4.7
EXP/120	ISR/4b	Scraping, hide with ochre	6.4.7
EXP/121	ISR/4b	Scraping, hide with ochre	6.4.7
EXP/123	BRND	Scraping, hide with ochre	6.4.7
EXP/124	ISR/3	Scraping, hide with ochre	6.4.7
EXP/125	ISR/2	Scraping, hide with ochre	6.4.7
EXP/126	ISR/4b	Scraping, hide with ochre	6.4.7
EXP/127	BRND	Scraping, hide with ochre	6.4.7
EXP/128	BRND	Scraping, dry/wet chamois	6.4.5
EXP/129	BRND	Scraping, hide with ochre	6.4.7
EXP/136	BRND	Brushing, ¼µ grit	6.4.8
EXP/137	ISR/2	Weight variation on flint	6.4.11
EXP/138	ISR/2a	Weight variation on flint	6.4.11
EXP/139	BRND	Weight variation on flint	6.4.11
EXP/140	BRND	Buffing, wet dust-free goat hide	6.4.10
EXP/141	BRND	Buffing, dry dust-free goat hide	6.4.10
EXP/142	BRND	Polished on and by cow hide	6.4.12
EXP/143	BRND	Polished on goat hide	6.4.12

6.4.5 Experiments on the Effect of Water on Polish Formation

Aims

It has been reported that green plants produce polish faster than dry ones (Unger-Hamilton 1983:244; Anderson-Gerfaud 1981:46). Although the difference was attributed to the presence of their juices it was not clear what role they played in the development of polish. Furthermore it has been hypothesised that water was interacting with the silica in the flint causing a silica gel layer to form in a process of hydrolysis. It was suggested that this silica gel layer is what we observe optically as polish. To observe and try to understand the importance of fluids such as water in the process of polish formation, several experiments were carried out, starting with the ones described below.

Experiments

1 A basket made of rushes was cut in half. A flint blade **EXP/1** was used to shred one half of the basket into very small bits for 30 minutes. The other half of the basket was soaked in water overnight and then shredded in the same way and for the same length of time.

2 A turntable was covered with a piece of chamois leather. While the turntable was revolving it was scraped manually with a flint artefact **EXP/128** for five minutes while it was dry and five minutes after wetting it. The turntable was used in order to speed up the development of polish.

Observations

1 Cutting the dry rush basket produced hardly any polish on **EXP/1** (figure 33).
Cutting the wet rushes produced a very well developed polish even after 15 minutes (figure 34/8) and after 30 minutes it was a totally linked C polish even after immersion in 20% NaOH to remove possible residues (figure 37).

2 After five minutes on the dry leather no polish whatsoever appeared on the flint edge of **EXP/128**, while on the wet leather, B+ well linked polish was produced after scraping for five minutes (figure 59).

Conclusions

It is obvious that the presence of water had greatly enhanced polish formation. The experiments described in section 6.4.6 show that this is not due to a process of hydrolysis as the same results were obtained when using liquids (other than water) not known to cause the dissolution of silica. In other words the enhancement of polish development was due to the presence of a liquid medium rather than to the chemical properties of water as opposed to any other liquid.

6.4.6 Wood-Grooving Experiments

Aims

The purpose of these experiments was to monitor the role certain variables play in the development of polish on flint edges. The factors considered important in the appearance of polish such as abrasives, water, other fluids and edge angle, were varied and the results observed microscopically.

Experiments

A plank of dry birch was grooved by hand with flakes removed from the same core of ISR/2 medium grained flint. The edge angle of the tools was recorded and the additives were water, hydrocarbon oil and talcum powder used successively. Each back and forth movement was counted as one stroke. Every artefact was used for 3000 strokes. Observations were made at various stages of work to document visually and photographically, the edge wear and the development of the polish.

The experiments and the optical observations have been summarised in Table No.1.

Discussion

Fluids

Polish from grooving dry wood with **EXP/91** formed slowest (figure 74). The polish produced by grooving birch with **EXP/92(1)** constantly wetted with water was slightly more invasive than that on **EXP/92(2)** dipped in oil (figures 72 and 73). As the edge angles were similar (35°) and the flint was the same, penetration into the wood might be expected to be comparable. Repeating the experiment showed that the water seems to be softening the birch (oil does not), allowing the edge to penetrate a little deeper in the plank.

Abrasives

The most interesting polish was the one produced by grooving birch with a flake, **EXP/93**, whose edge angle was 45° and using talcum powder as a soft abrasive (figure 60). Some well developed polish similar to that on the other artefacts appeared on the very edge. Further away from the edge, however, rather than the polish appearing first on the high points of the microtopography and then linking up (as had occurred when grooving dry birch), the introduction of the soft abrasive produced a surface polished both on the high and low points of the microtopography. The overall appearance of the surface after polishing was smoothed and smeared as is sometimes observed with PDSM. The distribution of the polish on the microtopography was consistent with that which might be expected from working a soft material which, by moulding itself better to the flint microtopography, polishes the low points at the same time as the high points. There were, however, striations, and edge scarring indicative of a hard material.

Striations/Edge Scarring

To ascertain whether the striations on **EXP/93** were caused by the microchips from the edge scars, the experiment was repeated with a burin **EXP/96** using the ventral edge of the facet to groove. Tools with greater edge angles are considered less likely to produce microchips. The edge angle was 80 degrees and no macroscopic edge scarring was produced, nor were there striations though a linear distribution of polish did result from the back and forth motion used in grooving the wood (figure 61). This observation suggests that microchips from edge scars may be coming between the flint and the worked material and causing the striations.

Cleaning

An attempt was made to remove any talcum powder that may have invaded the interstices of the flint on **EXP/93** and **EXP/96**. The artefacts were immersed in a 1% HCl solution for 10 minutes in addition to brushing with a nylon brush in warm water and detergent followed by an ultrasonic cleaning in distilled water. The polish remained unchanged. Although superficial residues from the powder had been removed, one cannot, however, exclude the possibility that some talcum remained in the interstices of the flint.

Conclusions

This series of experiments suggests that the role water plays in polish formation is probably not solely a chemical one, since very similar polish is obtained when using a hydrocarbon oil which is very different chemically from water in terms of polarity. The basic difference in the polish produced by 3000 strokes grooving **dry** wood alone or with water or oil is in the speed of its formation and its distribution on the edge.

The development of the unusual polish on **EXP/93** and **EXP/96** suggests that soft abrasives can cause a change in the polish distribution on the microtopography and emphasises the importance of abrasives in polish formation above and beyond the hardness of the material being worked.

The polish produced by this experiment is similar to certain traces found on some implements involved in a trampling experiment at Klithi (see section 6.5.4 cf. figure 154 and figure 60). Contact of trampled pieces both with other stones and with sediment seems to have produced, in localised spots, the same polish distribution as contact with wood and talcum powder.

These experiments seem to confirm an observation already made on other experimental tools, namely that edge scarring is responsible for striations. It need not necessarily be the only cause as the experiments described in the following pages will show.

Table 1 Hide Scraping Experiments (listed according to flint type)

Tool no.	Flint Type	Edge angle	Ochre	Medium	Hide type	Polish	Striations	Edge wear	Cleaning	Comments
119	BRND /fine	90	yellow unsifted	water	fallow deer	Micro: well developed C / narrow band Macro: thin bright band	fine, numerous	severe edge rounding	bio-detergent /nylon brush USC	Macroscopic well developed polish. No glare as in EXP /118 which (except for flint type) has been used under the exact same conditions. Polish developemnt comparable. Tool efficient
123	BRND /fine	90	red 212/213 µ	water	juvenile red deer	Micro: very well developed C++. Macro: very bright wide band	very fine, numerous	distinct edge rounding	detergent/nylon brush/USC	Fine ochre + fine grained flint + yound deer hide caused extremely well developed polish. Polish from hard hammer retouch completely polished over. Very efficient edge.
127	BRND /fine	75	red <212 µ	water	goat	Micro: very well developed C++/comet tail pits. Macro: bright narrow band	very fine, numerous	very severe edge rounding	detergent /USC	Very fine flint + very fine ochre + thin goat hide caused maximum development of polish with comet tail pits.
129	BRND /fine	85-90	yellow <212 µ	Hexane	goat	Micro: very well developed C++. Macro: bright bank on dorsal	very fine	severe edge rounding	detergent /USC	Edge rounding and polish development suggest tool was in contact with the hide but hexane (a volatile oil) was evaporating too fast and not blending the ochre into a paste so colouring the hide was not adequate in spite of a very efficient edge.
118	ISR/2	90	yellow unsifted	water	fallow deer	Micro: well developed C. Macro:thin band	fine, numerous	severe edge rounding	bio/detergent /nylon brush/10% HCl/20%NaOH/USC	Glare on polished surface not withstanding nylon brush, biological and chemical cleaning to remove possible ochre from hollows of the coarse flint. Flint inclusions, pre-use caused some glare now increased. Wider edge preferable
125	ISR/2 medium to coarse	90	yellow <212 µ	water	juvenile red deer	Micro: well developed C+. Macro: none	very fine, numerous	slight edge rounding	detergent/biotex overnight/USC	Due to coarse flint, polish well developed on few points of fine flint, hence not macroscopic. Edge angle and morphology very efficient. EXP/125 coarser than EXP/118 same flint type.
124	ISR/2 medium to coarse	70	yellow <212 µ	water	cow	Micro: little developed A+. Macro: none	none	slight edge rounding	detergent/biotex overnight/USC	Very coarse weathered flint nodule. Cow hide very coarse fibre. Not sufficiently broken down in 20 minutes scraping? Ochre not penetrating hide cf. EXP/126. Higher edge angle preferable.
126	ISR/4b medium to fine	75-80	yellow 212/300 µ	water	cow	Micro: little developed A+. Macro: none	none	very slight edge rounding	detergent/biotex/USC	Flint not coarse but naturally shiny. Lack of polish development possibly due to cowhide. In EXP/1442 cowhide caused little polish development even in mechanical buffing. Tool efficient.
121	ISR/4b medium	60-70	yellow 425 µ	water	cow	Micro: little developed A+. Macro: none	rare	slight edge rounding	biotex 40 min/NaOH 20%/USC	Coarse inclusions in flint, rough textured cowhide probably caused lack of polish development and rare striations in spite of coarse ochre. Higher edge angle preferable.
120	ISR/4b medium	45-50	red 212/300 µ	water	fallow deer	Micro: well developed C+. Macro: pronounced gloss, narrow band	very fine, numerous	severe edge rounding	detergent/nylon brush/USC	Fine ochre on smooth hide caused well developed polish, severe edge rounding and fine striations. Edge too thin and low angled cutting hide.
117	ISR/5b very coarse	90	none	none	fallow deer	Micro: little developed A+. Macro: none	none	very slight edge rounding	detergent/nylon brush/USC	Coarse flint and absence of ochre possible cause of polish not well developed.

6.4.7 Experiments Scraping Hide with Ochre

Aims

The aim of the following experiments was to observe how striations and polish development are affected by ochre grain sizes keeping certain variables constant duration of use in particular (see below). There was no attempt to make a statistical analysis as this would require a far larger number of experiments. Mansur-Franchomme (1983:143) carried out an experiment scraping hide with fine and coarse sand and found that certain striation types occur according to the type of sand used. The use of ochre in connection with hide is well documented among Australian aborigines and is often postulated for prehistoric times. It was consequently decided that ochre with a measurable grain size should be used in these experiments.

Experiments

Hide was scraped with ochre from a naturally occurring lump that had been pounded into powder. Unsifted powder was used as well as ochre sieved through a specific mesh size. The ochre was made into a paste by adding distilled water except in one case when hexane (a hydrocarbon oil) was used instead. The scraping was carried out with the pieces of hide resting on a board covered with a thick plastic cover to avoid any contact between the tool and the wooden surface. Brandon flint and three types of Israeli flints were used on deer, goat and cow hide. High angled and low angled scraping occurred according to the requirement of the task often dictated by the edge morphology of the tool. The work aimed at replicating the way hides would be coloured rather than the way they would be processed after butchery, which would require much longer periods of scraping and would probably not include ochre at that stage.

The following variables have been kept constant:

- all hide was raw, that is not chemically tanned,
- all hide was soaked and scraped while damp,
- all hide was scraped for twenty minutes only on the non-hairy side.

Other variables which could influence polish development were also recorded:

1. flint type,
2. edge angle,
3. hide type,
4. liquid medium.

It was impossible to cover all the permutations in the time available. A range of experiments was conducted that should at least throw some light on the factors affecting polish development, even if not elucidating every aspect.

Three tables were compiled to show the various experiments and their results grouped according to:

Flint Types Table No. 2

4 tools of Brandon flint;
7 tools of various Israeli flints.

Ochre Grain Sizes Table No. 3

2 tools with unsifted ochre hence with a mixture of grain sizes;
1 tool with ochre of a grain size held within a 425μ sieve;
3 tools with ochre of grain size between 212 and 300μ;
4 tools with ochre of a maximum grain size of 212μ.

Edge Angles Table No. 4

5 tools with a 90° edge angle;
6 tools with edge angles below 90 and decreasing to 45/50°.

Edge rounding was viewed by optical microscopy and very roughly classified in five categories: "very slight", "slight", "distinct", "severe", and "very severe" in ascending order of observed rounding of the edge.

Some comments resulting from the microscopic observation of the edges are given on the tables. An attempt was made to observe the polished flint surface and the striations by SEM. The conclusions drawn from these observations are discussed below. Some abbreviations were necessary in compiling the tables and are given in full in sections 6.4.1 and 6.4.2. See also figure 63 for a diagram illustrating the stages of polish development referred to in the tables and explained in section 6.4.2. On all edges only the maximum stage of polish development has been indicated. Generally polish is best developed at the point of maximum contact and less developed polish appears elsewhere usually away from and along the edge.

Observations

The following are intended simply as statements of what was observed microscopically and macroscopically after the experiments. The discussion and interpretation of these observations appear under a separate heading.

1 Polish Development

a. The same degree of polish developed whether water or oil (hexane) was used as a medium (see **EXP/129** and **EXP/127** Figures 91 and 93). The same was true when scoring wood see experiment in section 6.3.6. Other experiments with hexane were not carried out as it tends to evaporate very quickly and does not allow the ochre to develop into a paste.
b. The coarseness of the hide fibres affected the development of polish. The finer the grain of the hide the more developed the polish and vice versa (cf. **EXP/120** and **EXP/126** figures 85 and 90 where the same flint type [ISR/4b] and the same ochre grain size [212/300μ] produced vastly differing stages of polish). **EXP/126** scraping cow hide developed A+ polish while **EXP/120** scraping fallow deer produced macroscopic C+ polish and severe edge rounding. All three artefacts used on cow hide developed only A+ polish in spite of the presence of ochre in three different grain sizes from coarse to fine. There were also three different flint types from medium/coarse to medium/fine in grain size. See section 6.4.12 for further

Table 2 Hide Scraping Experiments (listed according to decrreasing grain size of ochre)

Tool no.	Ochre	Edge angle	Flint	Medium	Hide type	Polish	Striation	Edge wear	Cleaning	Comments
118	yellow unsifted	90	ISR/2 medium	water	fallow deer	Micro: well developed C. Macro: thin band	fine, numerous	severe edge rounding	bio-detergent /nylon brush /10% HCI /20% NaOH /USC	Glare on polished surface notwithstanding thorough cleaning to remove possible ochre from the hollows of the coarse flint. Pre-use glare from natural flint inclusions now increased. Wider edge would be preferable.
119	yellow unsifted	90	BRND /fine	water	fallow deer	Micro: well developed C, narrow band. Macro: thin bright band	fine, numerous	severe edge rounding	bio-detergent /nylon brush /USC	Macroscopic well developed polish. No glare as in EXP/118 which (except for flint type) was used under the same conditions. Polish development comparable. Tool efficient.
121	yellow 425 μ	60-70	ISR/4b medium	water	cow	Micro: little developed A+. Macro: none	rare	slight edge rounding	biotex 40 min /NaOH 20% /USC	Coarse inclusions in flint, rough textured cowhide probably caused lack of polish development and rare striations in spite of coarse ochre. Higher edge angle preferable.
120	red 212/300 μ	45-50	ISR/4b medium	water	fallow deer	Micro: well developed C+. Macro: pronounced gloss narrow band	very fine, numerous	severe edge rounding	detergent/nylon brush/USC	Fine ochre on smooth hide caused well developed polish, severe edge rounding and fine striations. Edge too thin and low angled cutting hide.
126	yellow 212/300 μ	75-80	ISR/44b medium to fine	water	cow	Micro: little developed A+. Macro: none	none	very slight edge rounding	detergent/biotex /USC	Flint not coarse but naturally shiny. Lack of polish development probably due to cowhide. In EXP/142 cowhide caused little polish development even in mechanical buffing. Tool efficient but hide not coloured.
123	red 212/300 μ	90	BRND /fine	water	juvenile red deer	Micro: very well developed C++. Macro: very bright wide band	very fine, numerous	distinct edge rounding	detergent/nylon brush/USC	Fine ochre + fine grained flint + young deer hide caused extremely well developed polish. Polish from hard hammer retouch completely polished over. Very efficient edge.
125	yellow <212 μ	90	ISR/2 medium to coarse	water	juvenile red deer	Micro: very well developed C+. Macro: none	very fine, numerous	slight edge rounding	detergent/biotex overnight/USC	Due to coarse flint polish well developed only on a few points of fine flint, hence not macroscopic. Edge angle and morphology very efficient. EXP/125 coarser, EXP/118 same flint type.
124	yellow <212 μ	70	ISR/3 medium to coarse	water	cow	Macro: little developed A+. Micro: none	none	slight edge rounding	detergent/biotex overnight/USC	Very coarse weathered flint nodule. Cowhide very coarse fibre, not sufficiently broken down in 20 minutes scraping? Ochre not penetrating hide cf. EXP/126. Higher edge angle preferable.
127	red <212 μ	75	BRND /fine	water	goat	Micro: very well developed C++, comet tail pits. Macro: bright narrow band	very fine numerous	very severe edge rounding	detergent/USC	Very fine flint + very fine ochre + thin goat hide caused maximum development of polish with comet tail pits.
129	yellow <212 μ	85-90	BRND /fine	hexane	goat	Micro: very well developed C++. Macro: bright band on dorsal	very fine	severe edge rounding	detergent/USC	Edge rounding and polish development suggest tool was in good contact with hide but hexane (a volatile oil) was evaporating too fast and not blending the ochre into a paste. In spite of an efficient edge the hide was not adequately pigmented.

Table 3 Hide Scraping Experiments (listed according to decreasing edge angle)

Tool no.	Edge angle	Flint type	Ochre	Medium	Hide type	Polish	Striation	Edge wear	Cleaning	Comments
119	90	BRND fine	yellow unsifted	water	fallow deer	Micro: well developed C, narrow band. Macro: thin bright band	fine, numerous	severe edge rounding	bio-detergent /nylon brush /USC	Macroscopic well developed polish. No glare as in EXP/118 which (except for flint type) has been used under the same conditons. Polish development comparable. Tool efficient.
123	90	BRND fine	red 212/300µ	water	juvenile red deer	Micro: very well developed C++. Macro: bright wide band	very fine, numerous	distinct edge rounding	bio-detergent /nylon brush /USC	Fine ochre + fine grained flint + young deer hide caused extremely well developed polish. Polish from hard hammer retouch completely polished over. Very efficient edge.
117	90	ISR/5b very coarse	none	none	fallow deer	Micro: little developed A+. Macro: none	none	very slight edge rounding	bio-detergent /nylon brush /USC	Coarse flint and absence of ochre possible cause of polish not well developed.
118	90	ISR/2 medium	yellow unsifted	water	fallow deer	Micro: well developed C. Macro: thin band	fine, numerous	severe edge rounding	bio-detergent /nylon brush /10% HCI /20% NaOH/USC	Glare on polished surface notwithstanding thorough cleaning to remove possible ochre from the hollows of the coarse flint. Pre-use glare from natural flint inclusions now increased. Wider edge would be preferable.
125	90	ISR/2 med-coarse	yellow <212µ	water	juvenile red deer	Micro: well developed C+. Macro: none	very fine, numerous	slight edge rounding	detergent/biotex overnight/USC	Due to coarse flint, polish well developed on a few points of contact hence not macroscopiic. Edge angle and morphology very efficient.
129	85-90	BRND fine	yellow <212µ	hexane	goat	Micro: very well developed C++. Macro: bright band on dorsal	very fine	severe rounding	detergent/USC	Edge rounding and polish development suggest tool was in good contact with hide but hexane (a volatile oil) was evaporating too fast and not blending the ochre into a paste. In spite of an efficient edge the hide was not adequately pigmented.
126	75-80	ISR/4b med-fine	yellow 212/300µ	water	cow	Micro: little developed A+. Macro: none	none	very slight edge rounding	detergent/biotex/ USC	Flint not coarse but naturally shiny. Lack of polish development due probably to cowhide. In EXP/142 cowhide caused little polish development even in mechanical buffing. Tool efficient, hide not coloured.
127	75	BRND fine	red <212µ	water	goat	Micro: very well developed C++ comet tail pits. Macro: bright narrow band	very fine, numerous	very severe edge rounding	detergent/USC	Very fine flint + very fine ochre + thin goat hide caused maximum development of polish with comet tail pits.
124	70	ISR/3 med-coarse	yellow <212µ	water	cow	Micro: little developed A+. Macro: none	none	slight edge rounding	detergent/biotex overnight/USC	Very coarse weathered flint nodule. Cowhide very coarse fibre, not sufficiently broken down in 20 minutes scraping? Ochre not penetrating hide cf. EXP/126. Higher edge angle preferable.
121	60-70	ISR/4b medium	yellow 425µ	water	cow	Micro: little developed A+. Macro: none	rare	slight edge rounding	biotex 40 min /NaOH 20% /USC	Coarse inclusions in flint, rough textured cowhide probably caused lack of polish development and rare striations in spite of coarse ochre. Higher edge angle preferable.
120	45-50	ISR/4b medium	red <212/300µ	water	fallow deer	Micro: well developed C+. Macro: pronounced gloss narrow band	very fine, numerous	severe edge rounding	detergent/nylon brush/USC	Fine ochre on smooth hide caused well developed polish, severe edge rounding and wide fine striations. Edge too thin and low angled cutting hide.

Table 4 Wood Grooving Experiment

Tool no.	Material	Edge angle	Macro edge wear	Micro edge wear	Maximum polish	Striations	Comments
EXP/91	dry birch	35	large snap fractures, edge looks denticulated	flake scars <5x10mm (+5 large snaps)	linked all and all over (B+ to C) at 3000s on micro elevations	rare	Polish only after 1200 strokes. Edgewear subsequently removed polished edge not allowing polish to link and develop. Thin edge.
EXP/92 (1)	dry birch + water	45	3 flake scars on ventral	flake scars >5x10mm	at 3000 strokes linked and all over (B+ to C)	rare	Polish formed after 1000 strokes mainly on convex part of edge but more invasive than from dry birch.
EXP/92 (2)	dry birch + oil	40	none	flake scars <5x10mm	at 3000 strokes linked and all over (B+ to C)	none	No real polish formed before 1500 strokes. Penetration slight. Polish not invasive but essentially similar to use on birch (with and without water).
EXP/93	dry birch + talc powder	45	1 large snap fracture, 3 tiny snap fractures, 3 tiny flake scars	flake scars >5x10mm	all over (C) only on edge, smeared surface away from edge	many	More edge wear than either birch with water or birch with oil. Different from dry birch alone. Polish all over on very edge. Elsewhere smoothed and smeared surface even after thorough cleaning.
EXP/96	dry birch + talc powder	80	none	one flake scar on tip, very few tiny flake scars, <55x10mm	All over (C) only on edge, smeared surface away from edge	none	Polish development similar to EXP/93. Higher edge angle less edge wear. No striations.

Birch plank - 1 stroke = back and forth movement
Perpendicular working angle
All artefacts from ISR/2 medium grained flint

experiments investigating the importance of hide fibres.

c The two tools with the best polish development (**EXP/127** and **EXP/129**) also had the most severe edge rounding. Edge rounding was severe on all tools with macroscopic polish. All the tools having no macroscopic polish had slight or very slight edge rounding as shown on **EXP/ 117, EXP/121, EXP/124, EXP/125**, and **EXP/126.**

d The best developed polish (C++ on **EXP/127** and **EXP/129** figures 91 and 93) was produced by the finest grain size ochre (š microns). Slightly less polish developed on artefacts used with unsifted ochre (C on **EXP/118** and **EXP/119**). By comparing two pairs of artefacts on which all the variables were the same except ochre grain size, the following variation is observable:
EXP/118 with unsifted ochre
EXP/125 with fine ochre produced C and C+ polish respectively;
EXP/119 with unsifted ochre
EXP/123 with fine ochre produced C and C++ polish respectively (figures 84 and 87).
On **EXP/**117, used as a control scraping fallow hide **without ochre**, very little polish developed, although this artefact is admittedly of a very coarse flint type (figure 82).

e In the most linked stages such as C+ or more, all polished surfaces had "comet tail pits".

2 Striations

Most of the tools had relatively fine striations when observed under an optical microscope. **EXP/121** used with 425μ ochre had no striations: there was little developed polish on rather coarse flint. In spite of using the SEM, no distinct correlation of striation type with ochre grain size was observed. More than one type of striation was present on each tool. There were also too many striations for them to be measured individually. Therefore only a very general classification such as "fine", and "very fine" striations has been made from optical observations and these also did not correlate clearly with ochre grain size. See section 7.1.6 for a discussion of striations in the understanding of polish formation.

3 Flint Type

On **all** the Brandon artefacts polish was very well developed. On the Israeli flint types the development of polish did not appear to be solely a function of flint grain size.

4 Edge Angle

Table 3 is arranged by grouping the artefacts according to edge angles to observe whether edge rounding was affected by them. High angled and low angled edges were rounded.

5 Hide Type

In the case of **EXP/126** and **EXP/120** both of the same flint type and used with the same ochre grain size, polish developed very well on **EXP/120** used on fallow deer hide and very little on **EXP/126** used on cow hide (Figures 85 and 90). As working time was kept constant it was possible to observe that the **speed** of polish development is faster with the finer grained hides of deer, goats and of young animals than with the coarser grained hides of adult cattle.

6 Edge Rounding

All the tools in these experiments developed some edge rounding. The Brandon flint tools developed more edge rounding than the Israeli flint types with the exception of **EXP/120** which developed severe edge rounding. All the artefacts with severe edge rounding had C or more polish on the edge. Artefacts with slight edge rounding developed only A+ polish except for **EXP/125** with C+ polish development and slight edge rounding.

Discussion and conclusions

The following is a discussion of the results of these experiments. While some tentative inferences might be drawn, one should be cautious not to over interpret the results until a far greater number of experiments has been conducted. Comparisons are made here between examples where only one parameter was varied.

1 Macroscopic Polish

Some of the artefacts in these experiments developed a polish visible with the naked eye while others did not, even those which microscopically had a very well linked polish (see **EXP/118** and **EXP/125** figures 83 and 89). A possible explanation of this phenomenon is that when flint is of varying grain size the finer grained areas of well developed polish are separated by areas of coarse flint where the polish is very scarcely developed. **EXP/118** and **EXP/125,** though flakes of the same flint type, differ widely in their homogeneity: **EXP/125** has five different flint textures on the edge. On fine flint the same polish is visible macroscopically due to its being a continuous band.

2 Striations

The absence of a pattern correlating ochre grain size to striation type need not be interpreted in the sense that grain size of abrasives is unimportant in the formation of striations. It is suggested rather, that during work, the ochre particles themselves are being crushed causing a variety of striation types to form.

In the Keeley method of describing polish there is much use of the words "smooth" and "rough". Not all analysts agree on which polish is smooth or rough. In these experiments I have noted, optically, that when the striations were very fine the polish appeared smooth and conversely when the striations were not as fine the polish appeared rough. An interpretation of this, admittedly a very subjective observation, is that when the finer striations are very close together the polish appears to be smooth. When, on the other hand, the striations are larger and more widely spaced polish appears rough. Unfortunately one cannot see the roughness and smoothness of polish in the SEM and it was not possible to measure the depths and widths of all the striations.

3 Flint Type

In these experiments the observations did not yield a clear cut relationship between polish development and grain size. The fact that all Brandon flint implements developed C polish or more, while Israeli flint types had a variety of development stages, not related purely to their grain sizes, need not be interpreted as meaning that there is no relation between flint grain size and polish. All Brandon tools were used on finer hides. Unfortunately the limited amount of the finer raw hide available restricted the effectiveness of this experiment, forcing me to introduce the cow hide as an added variable. Cow hide could account for the lack of polish development on **EXP/126**, while polish developed on **EXP/125** and **EXP/120** used on deer hide. A glance at figures 82-93 however will show visually the influence of the flint structure on the distribution of polish.

4 Edge Angle

It would seem that there was no correlation between edge angle and edge rounding. The highest edge angles were found, however, to be the most efficient tools to spread the ochre paste (formed by ochre and water) and colour the hide. This operation normally follows the processing of hide (which takes place after butchering) and there is no longer a need for the removal of flesh or other matter from the hide. The lowest edge angles on relatively thin edges cut into the hide. In the case of thick cow hide, where it is necessary to break down the fibres before the colour is absorbed into the hide, a flint tool is necessary; otherwise a smoother bone tool with a naturally high edge angle would probably be more appropriate for this task. I did not carry out an experiment to verify this suggestion.

5 Hide Type

The structure of the hide fibres seems to be an important factor in polish development. Microwear literature has generally stated that different types of hide produce the same "type" of polish. What is referred to here is not the "type" of polish but the speed of development. In twenty minutes of work deer hide produced C+ polish and cow hide only A+ polish all other variables being comparable (**EXP/120** and **EXP/126** figures 85 and 90).

6 Edge Rounding

The presence of some rounding on all the pieces is probably due to the use of a hard wooden counter to support the hide during working. Other experiments carried out scraping hide which is stretched on a frame did not cause rounding after such a relatively brief scraping time. It is probable that the presence of the ochre contributed to the attrition even though there was no clear relation between the ochre grain size and the extent of rounding. All Brandon flint tools developed rounding while not all the Israeli flints did. The explanation, however, is probably not due to flint grain size.

EXP/120, which is the only Israeli flint tool to show rounding, was used on fallow deer. It is of the same flint type and of comparable homogeneity to **EXP/121** and **EXP/126** used on cow hide. The edge rounding on **EXP/120** is probably due to scraping a finer type of hide. There is little doubt that the conditions which cause well developed polish to form are also **likely** to be those that cause edge rounding. But the evidence emerging from these experiments does not allow for a one to one relationship between the two types of wear to be made.

7 Cleaning/Residues

It will be noticed that certain artefacts were subjected to a more powerful cleaning process than others. This was prompted by the fact that during optical microscopy these artefacts seemed to have a very strong glare not observed before use. It was originally thought that some residue either from the hide or more likely from the ochre was still adhering to the flint surface. These artefacts were cleaned with a nylon brush and subsequently soaked overnight in a biological detergent. They were also soaked in a 20% w/v sodium hydroxide solution and subsequently in 10% hydrochloric acid. This cleaning procedure did not remove the glare. None of the Brandon artefacts presented this problem and were consequently not subjected to this cleaning method. It is argued therefore that the cause of the glare is more likely related to the coarseness of the flint. It may be that ochre particles had entered the many hollows present on most Israeli flint and could not be removed by the chemical cleaning or by brushing with a nylon brush.

8 Ochre Grain Sizes

In these experiments ochre grain size seemed to influence polish development but it did so less than hide. The best polish was developed when the finest ochre grain sizes were used and this may be due to the fact that the fine ochre with a liquid medium creates a slurry like that used in glass polishing (Cornish 1961:3). This slurry may be removing the asperities from the flint surface and smoothing it. The smoothness of the resulting surface can be seen from the SEM pictures figures 80 and 81. On the other hand the coarse grained ochre may either not produce a slurry and/or may cause contact between its grains and the flint to occur only on widely spaced points. Further experiments using 425μ ochre on deer hide and Brandon flint could clarify the effect of ochre grain size. It should be borne in mind that ochre is a considerably softer material than flint. The coarsest ochre was also a very ineffective colouring agent.

9 Colouring Hide with Ochre

For whatever purpose it was done, ochre is supposed to have been used in prehistoric times during the processing of hides. It is suggested here that prehistoric people may have pounded the ochre until a finely grained powder was obtained as it would yield a more homogeneous and effective pigmentation of the hide. Though ochre cannot be said to "dye" hide but rather to colour it, an attempt was made to test the durability of such a pigmentation. The juvenile red deer hide scraped with 212–300μ red ochre was very evenly red. It was subsequently soaked in water for several hours without losing its colour. One may reasonably suppose that in a prehistoric situation where a garment was soaked by rain or while wading in a river, the pigmentation would presumably not wash away.

10 "Comet Tail Pits"

These have been observed both optically and by SEM in these and other experiments on polish produced by diamond paste, hide and wool (figures 66; 70 and 64). They are not, therefore, diagnostic of plant polish as suggested by Witthoft (1967) and Keeley (1980:61). They are discussed in more detail in sections 7.1.6, 7.1.9 and 7.2.7.

11 Duration of Use

Vastly different stages of polish development were obtained by scraping hide for the same length of time. It is obvious that the amount of polish observed on an implement is the result of the interaction of many variables the most important of which, according to these experiments are: flint type, hide type, and the presence of an abrasive. Fluids are also crucial but were generally kept constant in these experiments which were all carried out on damp hide. In the one instance where hexane was used rather than water to blend the ochre into a paste, little difference between the resulting polishes was found (cf. **EXP/127** and **EXP/129** figures 91 and 93). It would seem, therefore, incorrect for functional analysts to reconstruct the length of time a prehistoric artefact was used, on the basis of the amount of polish observed on the edge when so many of the variables are unknown.

12 Tools with Multiple Uses

Many functional reconstructions have been made suggesting that a given archaeological tool has been used on different materials on the same edge where one polish is supposed to overlie the other (Vaughan 1985:40 and others.). I had sometimes noticed that polish from hard hammer retouch would still be visible if the edge was subsequently used briefly and/or on a soft material. In this series of experiments, all the scraper edges were retouched with a hard hammer and carried polish from contact with the retoucher. This polish was, however, in all cases polished over by scraping the hide and it was no longer possible to distinguish the two polishing episodes on **EXP/123** (figures 94 and 95).

6.4.8 Mechanically Produced Polish

Aims

In order to test whether polish is characteristic of the worked material as claimed by Keeley (1980) several tests were carried out polishing flint mechanically in conditions which might isolate the important factors governing polish development. These experiments do not aim to replicate activities likely to have been carried out in prehistoric times but to mechanically polish flint in such a way as to control the variables. Besides making the replicability of the experiments easier, the mechanical polishing cuts down the element of boredom which might induce error. Scoring wood while counting 3000 strokes may take less time but is more monotonous than polishing flint for 45 or even 60 minutes with a buff.

Experiments

A ¼μ diamond paste was used to polish flint artefacts by various mechanical means.

1. A turntable rotating at 150 rpm and covered with a Buehler Microcloth (a synthetic short pile velvet-like cloth) was coated with ¼μ diamond paste. Experimental flint artefacts: **EXP/85** (Brandon flint), **EXP/84**, **EXP/94**, **EXP/97** (all ISR/2 flint) were polished by holding them against the turntable in a scraping motion at an angle of about 60° at approximately 17cms from the centre. After 30 seconds the polish was macroscopic, very well developed, and totally linked up like "sickle gloss" (figures 111, 112, 113, 114). Continuing the experiment up to 2 minutes did not seem to cause any more polish to develop.
2. Other artefacts, **EXP/89** and **EXP/95**, were used by hand to scrape the surface of the stationary turntable. The resulting polish was also very well developed like "sickle gloss" after only one minute (figure 115).

   The turntable was unfortunately too fast to observe the earlier stages of polish formation. Furthermore, due to the rigidity of the surface, the flint area in contact with it was very limited and consequently, in all the above cases, very well developed polish formed very fast only on the very prominent parts of the edge. It was decided, therefore, to find a different way of polishing enabling a more invasive polish to form more slowly in order to permit observation of the early stages of polish development.
3. A small hand-held electric motor was fitted with a soft woollen felt buffing wheel turning at high speed. The buff was coated with 1/4μ diamond paste. Two flint blades of ISR/4a flint, **EXP/103** and **EXP/101**, were polished. Each blade was held in one hand while the buffing was done by the other hand. When the buff was quite worn out and was hard rather than soft, the linked-up appearance of the polish developed much more slowly (figure 116).

When a new soft buff was mounted the polish started again to link up rapidly (figure 117). The soft felt buff produced polish after a few minutes. After 30 minutes buffing (figures 118 and 119), the experimental blades were as macroscopically glossy as archaeological sickle blades (figure 6).

4 The electric motor in the previous experiment was fitted with a nylon brush coated with ¼μ paste as above. The brushing did not produce any polish on **EXP/136** from the same flint after 30 minutes in spite of the diamond paste which was even applied on the flint itself.

5 A new soft woollen buff was used to polish a flint blade **EXP/90** without adding water or diamond paste. No polish developed after 60 minutes dry buffing.

Observations

The following comments are itemised according to the experiment numbers appearing above:

1 once the polish had completely linked up on the microtopography and had become macroscopically glossy, there was no readily visible increase in the amount of polish;

3 a) in the experiments with diamond paste on the sickle-like blades, the tools were hand held with the flint resting on the index finger. Such high speed buffing, even with a soft wool buff, warmed up the flint, but the increase was probably no more than a few degrees. Had a high temperature been reached, even in very localised areas on the surface of the flint, the wool buff would presumably have to have reached the same temperatures. There was, however, no evidence of singeing;

b) the contact with soft, yielding wool and hard, unyielding wool produced very different rates of polish development;

c) the very well developed polish produced by the diamond paste appeared optically smoother than archaeological "sickle gloss";

5 In the case of the wool buff used dry on the flint, the same heating phenomenon occurred although no polish developed.

Conclusions

The following interpretations and conclusions can be drawn from the above observations:

The experiments described above under point 1 have shown that there is a maximum degree to which a flint surface can be polished. Increasing the duration of work beyond this stage has little if any effect upon the further development of the polish. The interpretation of polish as a smoothing of the surface is consistent with this observation. Once the surface is smooth there are no more asperities to remove and more polish linkage may occur only after an extremely long time.

In the experiment described under 4, when brushing the flint surface with diamond paste, it would appear that the nylon bristles may have slid across the flint surface without sufficient contact pressure to polish it. If this interpretation is correct, it highlights the importance of pressure as a variable in polish development.

The difference in the development of polish between the soft wool buff and the worn down hard buff (in the experiment at 3) suggests that even when an abrasive polishing agent is used the pliability of the contact material is very crucial. This points to the important difference in polish development which one might expect when a soft material such as hide is scraped on a hard backing as opposed to when the hide is scraped when stretched on a frame.

Regarding the theory of the silica in flint reaching a semi-molten state or being transformed into gel by the presence of localised high temperatures and water (Anderson-Gerfaud 1982:152–153, Witthoft 1967:385), the experiments described in points 3 and 5 above suggest that even if high temperatures develop in localised areas the total amount of heat involved is far short of that required to produce melting over any significant proportion of the surface. It seems highly unlikely that increases in temperature during use could be of such a magnitude as to melt or even soften the flint since neither the cloth on the rotating turntable nor the woollen buff were singed and, being in intimate contact with the flint, similar heat would be dissipated in them as that suggested to be responsible for melting the flint. On the other hand, for a silica gel to form, temperatures have to stay moderate at all times, for it would decompose at higher temperatures. More water would be required in the composition of a gel than would be available from the flint structure and yet in the experiments where polish clearly formed, no additional water was used. A process of dissolution and redeposition of the silica in flint might occur. In this case, however, substantial quantities of water would presumably be necessary (the solubility of silica being very low), yet none whatsoever was used in any of the above experiments.

The high silica content of grasses has been thought by many to be the main cause for the development of the macroscopic gloss on Neolithic sickles (Spurrell 1892; Curwen 1930–1941; Witthoft 1967; Kamminga 1979; Anderson 1980; Anderson-Gerfaud 1981–1986; Vaughan 1981, etc.). However, neither of the experimental blades discussed in point 3 above had come in contact with any external source of silica, the diamond grit being pure carbon. The greater smoothness observed optically on the polish produced by the diamond paste as opposed to that seen on archaeological sickle blades may be due either to the presence of the oil in the paste, or to the particle size of the diamond being much smaller than the sizes of phytoliths. The latter are supposed to be the agent in grasses which polishes Neolithic sickles to such a high gloss. A combination of both factors (oil and small particle size) could be the cause of the polish appearing smooth and sleek under the optical microscope. It is not intended here to argue that the gloss on ancient blades was not caused by silica in the materials cut (nor to suggest that prehistoric people were using ¼μ diamond paste!).

As the same kind of macroscopically glossy polish can also be produced by ¼μ diamond paste on a soft wool buff, and hide scraping, polish would seem then not to be specific to the worked material (figure 85 **EXP/120**).

See section 6.5.13 in which **EXP/103** was subjected to processes likely to occur in post-depositional circumstances with a discussion of the results.

6.4.9 Wool and Hide Mechanical Experiments

Aims

During the work involved in this thesis it was noticed that polish distribution and speed of linkage seems to be related to material "hardness" in the sense of pliability. To test the hypothesis that the rigidity of materials rather than their structure was a significant variable in the development of polish on flint, the following experiments were carried out.

Experiments

Two flakes knapped from the same core of fine Brandon flint were polished as follows:

EXP/108 was buffed for 60 minutes with a hard wool buff frequently dipped in distilled water.

EXP/109 was buffed for 60 minutes with a hard buff covered with several layers of soft chamois leather also frequently dipped in distilled water.

Observations

The same stage of linked polish development occurred on both edges after 60 minutes of work (figures 96 and 97). The observations made halfway through the work (namely at 30 minutes on each flake) showed that almost no polish had yet formed. The contact of the buff with the flint caused edge scarring such as occurs in contact with hard unyielding materials.

Conclusions

Although wool and hide are quite similar in their chemical composition they are very different in their structure and might have caused different "polishes", had polish been material specific. Furthermore the wool was very hard pressed making it relatively rigid and unyielding even when dipped in water.

This experiment suggests that in spite of the flake being in contact with soft damp chamois, the hard unyielding backing was the more important variable. The slow formation and link up of the polish are typical of unyielding materials. The effect of a hard backing on edge wear when working damp hide has also been discussed in section 6.4.7 in the experiments scraping hide with ochre on a wooden counter.

6.4.10 Polishing Flint with Dust-Free Hides

Aims

The following experiments attempted to elucidate the abrasive factor which seems bound to be present when flint surfaces are polished by soft materials such as damp hide.

Experiments

Goat hide was washed with ammonia-free detergent and brushed with a nylon brush as free of dust and dirt as possible. Separate squares were then used to cover the buffing wheel described in section 6.4.8.

Two flakes were buffed as detailed below. To avoid microchips from the edge of the artefacts coming between the hide and the flint, no contact was allowed between the buff and the edges of the flakes. The hide squares used in this experiment were photographed before and after use with both optical and scanning electron microscopy.

Two flakes of fine Brandon flint were used in this experiment as follows:

EXP/140 was buffed for 45 minutes with a soft wool wheel covered with a clean piece of hide dipped frequently in distilled water. The flake was also repeatedly dipped in water to remove hide particles from its surface during work. The flake was hand held with fingers protecting the edge and the buff touching only a previously circled area on the dorsal surface. There was no apparent increase in temperature probably due to the frequent immersions in water.

EXP/141 was buffed for 45 minutes with a similar wheel covered with a dry piece of hide previously cleaned as above. Although this flake was also rinsed frequently to remove any hide particles, it was each time allowed to dry thoroughly before resuming work to avoid introducing any dampness in the experiment. While the flake was drying the buff was carefully covered with plastic so that it should remain dust-free. The flake was also hand held to preclude contact with the edges and to polish a previously circled area on the ventral aspect. During work the flake became barely warm to the touch.

Observations

Optical Microscopy

EXP/140 Polish started to link up (stage B) on the flint surface after only 15 minutes of work. At the end of the 45 minutes a very well developed C+ macroscopic polish appeared with "comet tail pits" and very fine striations on the area of maximum contact. Figures 75 and 76 show the goat hide surface before and after the experiment suggesting that flint particles have become embedded in the hide.

EXP/141 Polish started to form (stage A+) only after 25 minutes of buffing. At the end of 45 minutes polish had only linked up to a reticular pattern (stage B), except for one prominent point on the microtopography where a spot of C+ polish appeared with one "comet tail pit". Figure 77 shows that in this experiment as well, the goat hide seems spangled by flint crystals.

Scanning Electron Microscopy

EXP/140 Flint particles appear embedded in the hide square used to buff the flint (figure 100 and 101). Figure 102 shows a dot mapping of the distribution of silicon on the hide surface in figure 100 by energy dispersive X-ray analysis. An X-ray spectrum of elements present on the goat hide surface was taken before use (figure 103). After use a new spectrum shows the increase of silicon on the surface of the hide (figure 104).

Discrete areas of the flint surface have been smoothed by the high speed contact with hide (figure 64). The smoothness however is restricted to more localised areas on the surface than in scraping hide by hand where the edge is involved (figure 79). Figure 66 shows a "comet tail pit" on the surface of the artefact. Figure 64 on the other hand is of a straight striation appearing like a groove on the flint surface with unsmoothed crystals at its bottom and smoothed edges.

EXP/141 Similar results have been obtained from SEM and X-ray examination of the hide square used on this artefact. Flint crystals appear embedded in the hide (figure 106, 107 and 105) and the flint surface appears smoothed compared with SEM pictures of unused flint (figure 225). The elevated area of the microtopography on which C+ polish had been observed optically appears in figure 65) as a large raised asperity which has been extremely smoothed. The one "comet tail pit" observed optically appears in figure 70.

Discussion and conclusions

EXP/140 As the artefact was frequently dipped in water it is doubtful that any significant heating could have occurred. Very well developed macroscopic polish formed nonetheless. As no edge flaking was allowed to occur, the abrasive element in this experiment had to come exclusively from the asperities removed from the flint surface itself. "Comet tail pits" developed yet again, in association with C+ well developed polish. The presence of striations documented optically and with the SEM in the absence of dust, dirt and added abrasives suggests that the asperities of the flint are themselves being dragged across the surface. The area appearing in figure 108 is similar in its flatness and polish to certain bright spots produced by the rubbing together of two wet flint flakes. This similarity might be due to contact having occurred between a larger asperity and the area being polished. The flint particles embedded in the hide have seemingly been acting as a fine abrasive causing macroscopic polish to develop. In the experiment producing bright spots described in section 6.5.3 it was noticed that while contact between the edges of the two flakes produced a very well linked and invasive polish, contact between the edge of one and the bulb of the other, produced a much less linked up and more diffuse polish. In this experiment, as the edge was not involved, the polish was at a C stage on the flat areas of the microtopography only. On the more rounded parts of the surface it was diffuse A+ and not linked up at all (figure 109).

EXP/141 Buffing with dry hide caused polish to develop much more slowly. Although there was one spot of C+ polish most of the area did not go beyond the reticular pattern in a B+ stage (figure 110). This confirms the results from other experiments in this work that the presence of water promotes a speedier and better linked polish on the flint surface. These results do not support, however, the hypothesis of a silica gel layer since polish forms even in the absence of water. There is some water, no doubt, in the flint interstices and the hide fibres. It is unlikely, however, to be present in sufficient quantities to dissolve significant amounts of silica or cause significant hydrolysis. In contrast to **EXP/140**, a moderate increase in temperature occurred during work without water. It is unlikely, however, to have been a significant variable in the development of polish since even better developed polish formed on **EXP/140** which was not allowed to heat up. As expected the edge and the flint microtopography affect the distribution and linking up of polish. The amount of pressure exerted when the edge is used is likely to be greater than in rubbing even at the high speeds involved with a mechanical buff. The edge is generally more vulnerable as its particles are free standing on at least one aspect, hence more easily removable (D.R. Griffiths pers. comm.).

In conclusion, these experiments suggest that polish is a smoothing of the flint surface produced by the removal of surficial asperities which are sometimes dragged across the surface flattening it. Some removed asperities can become embedded in a soft worked material to act as an abrasive. Water seems to be a useful medium rather than an essential variable, while heat did not appear to be crucial to the development of polish. Pressure seems, on the other hand, to be significant since a more developed polish was produced when edges are involved. "Comet tail pits", as seen on the SEM, need not be indicative of contact with plant silica since they appear whenever a very well developed polish forms. Their shape is probably only apparent when the surrounding surface is very well polished. On the SEM they appear to be a type of striation in the sense of a groove on the flint surface. Striations seem to be caused by crystals being dragged across the surface grooving the flint. "Comet tail pits", on the other hand, seem to occur when larger crystals or groups of crystals have been displaced.

6.4.11 Weight Variations on Polished and Unpolished Flint Artefacts

Aims

In view of the disagreement among microwear practitioners regarding the mechanism which produces polish on used flint surfaces (see Chapter 1), an attempt was made to monitor the weight of flint artefacts over a period of time in the hope of eventually measuring any increase or decrease in weight which might occur on a very highly polished artefact. The aim was to measure any addition from the used material onto the flint surface or any removal of material from the flint itself through abrasion.

Experiment

Three freshly knapped artefacts of different flint types were used:

EXP/137 ISR/2 Blade weighing 10.7028 gr. before cleaning;

EXP/138 ISR/4a Blade weighing 6.4171 gr. before cleaning;

EXP/139 BRND Blade weighing 5.9706 gr. before cleaning.

The three artefacts were weighed with electronic scales (which are accurate to a thousandth of a gramme) repeatedly over a period of 5 months. They were weighed

- after washing,
- after drying in a warm room overnight,
- after being stored on an open shelf for two weeks, and
- after being kept in a sealed plastic bag for two weeks.

In each case several weighings were made on the same day.

Macroscopic observations

Although the first decimal digit stayed constant, the second digit fluctuated on the heavier **EXP/137** and in all three pieces the last two digits were highly variable before the flint was used in any way. The fluctuation is a result of very localised atmospheric changes even over a span of seconds. It was not expected that eventual additive polish from the used material or reductive polish from abrasion would be measurable in centigrams (first and second digits). An attempt was nonetheless made to polish **EXP/139** with wetted goat hide on a buff until the polished area was macroscopically glossy. The weight fluctuated between 5.9702 and 5.9690 gr. during each weighing as it had done before being polished. Several other artefacts polished mechanically later, on different materials, were systematically weighed before and after polishing and on none (including **EXP/139**) was any weight increase or decrease observed which could be measured in centigrams.

Conclusions

It was observed that weight fluctuations (depending on the contamination from handling grease, and the gain or loss of humidity from the atmosphere even when kept in a sealed plastic bag) are greater than a possible variation due to deposition of material or loss of a layer from the surface. Polishing the artefacts to a macroscopically glossy finish does not seem to alter the weight of an artefact by an amount which is significant compared to weight changes due to adsorption and desorption of water and surficial contamination. It is suggested that this kind of approach to the understanding of the nature of polish formation is not possible in view of the weight fluctuations observed in flint both in this experiment and in a similar one, carried out for other purposes by Dr. D.R. Griffiths (pers. comm.).

6.4.12 Experiments on the Role of Hide Fibres

Aims

In the experiments described in section 6.4.7, all three artefacts used to scrape cow hide developed only A+ polish in spite of the presence of ochre in three different grain sizes from coarse to fine. There were also three different flint types from medium/coarse to medium/fine in grain size (see Tables 1 to 3). The following experiments were carried out to better observe the role the structure of the hide fibres plays in polish formation.

Experiments

Two fine Brandon artefacts were used to scrape cow hide and goat hide respectively without ochre for 90 minutes each.

EXP/142 scraping cow hide

EXP/143 scraping goat hide

Observations

No macroscopic polish developed on either artefact in spite of using fine Brandon flint for such prolonged scraping even if in the absence of ochre. Microscopically there was A polish on **EXP/142** and A+ on **EXP/143** (figure 120 **EXP/142** and **121 EXP/143**). Under the SEM, in fact, the surface of **EXP/143** used on goat appeared smoothed particularly on the edge (figure 126 shows the edge as in 122).

Another attempt was made to buff an area of **EXP/142** not involved with the scraping experiment, using a square removed from the same cow hide before use. The hide-covered buff was dipped in water repeatedly as in the experiments described in section 6.4.10. Buffing was continued until the hide square was torn (45 minutes figure 129). Very little A+ polish developed inland while on the very edge some better developed polish B+/C was visible (figure 123). When the area was examined by SEM (figure 127) at x 6000 it appeared extremely smoothed similar to the surface of a glossy Neolithic sickle blade (compare figure 127 with figure 216). The cow hide appears very coarse under the SEM (compare figure 100 goat hide with figure 129 cow hide). It is important to note that this is the appearance of the two hides after being smoothed by 45 minutes buffing flint.

Discussion and conclusions

In the hand scraping experiments described above the goat hide produced a marginally better developed polish than the cow hide.

EXP/142 is a thick flake with an unretouched straight edge having an edge angle of 65°.

EXP/ 143 is also a thick flake but the edge used to scrape hide was straightened and strengthened by retouch which produced an edge angle of 85°.

As particles from the edge do act as an abrasive during work enhancing the formation of polish, one would have expected **EXP/142** to develop more polish having an unretouched edge which is more liable to cause edge scarring. It may be argued therefore that the hide fibres being as coarse as they are impeded good contact between the flint and the hide resulting in a very little-developed polish.

On the other hand the very smooth surface which resulted from buffing mechanically may be due to both increased pressure and increase in the duration of work. It is presumed that pressure is much higher when a buff is used than when one is scraping by hand. While it is not intended to attempt an equation of the duration of use when polishing mechanically with the duration of scraping by hand, one can safely estimate that 45 minutes at 60 rpm amount to a much longer duration than 90 minutes hand scraping.

In conclusion, this experiment seems to confirm the observations made in section 6.4.7 that cow hide is very slow to produce polish even on such fine grained artefacts as the two carefully chosen for this experiment. The very smooth surface on the buffed area suggests that even with materials which do not produce polish easily a well developed polish can result from prolonged use.

6.5. Experiments for Post-Depositional processes

List of artefacts in Post-depositional experiments

The artefacts bearing Δ were manufactured by Dr. R.N.E. Barton and examined for macroscopic edge damage before and after the shaking experiments. They are therefore listed separately and differentiated by a Δ from the artefacts numbered by me with similar numbers.

Tool No.	Flint type	Experiment description	Section
EXP/1Δ	BRND	shaking	6.5.1
EXP/2Δ	BRND	shaking	6.5.1
EXP/3Δ	BRND	shaking	6.5.1
EXP/4Δ	BRND	shaking	6.5.1
EXP/5Δ	BRND	shaking	6.5.1
EXP/6Δ	BRND	shaking	6.5.1
EXP/7Δ	BRND	shaking	6.5.1
EXP/8Δ	BRND	shaking	6.5.1
EXP/9Δ	BRND	shaking	6.5.1
EXP/10Δ	BRND	shaking	6.5.1
EXP/11Δ	BRND	shaking	6.5.1
EXP/12Δ	BRND	shaking	6.5.1
EXP/13Δ	BRND	shaking	6.5.1
EXP/14Δ	BRND	shaking	6.5.1
EXP/15Δ	BRND	shaking	6.5.1
EXP/16Δ	BRND	shaking	6.5.1
EXP/17Δ	BRND	shaking	6.5.1
EXP/18Δ	BRND	shaking	6.5.1
EXP/19Δ	BRND	shaking	6.5.1
EXP/20Δ	BRND	shaking	6.5.1
EXP/21Δ	BRND	shaking	6.5.1
EXP/22Δ	BRND	shaking	6.5.1
EXP/23Δ	BRND	shaking	6.5.1
EXP/24Δ	BRND	shaking	6.5.1
EXP/25Δ	BRND	shaking	6.5.1
EXP/26Δ	BRND	shaking	6.5.1
EXP/27Δ	BRND	shaking	6.5.1
EXP/28Δ	BRND	shaking	6.5.1
EXP/29Δ	BRND	shaking	6.5.1
EXP/30Δ	BRND	shaking	6.5.1
EXP/31Δ	BRND	shaking	6.5.1
EXP/32Δ	BRND	shaking	6.5.1
EXP/1	DRRS	tumbling	6.5.2(b)
EXP/2	BRND	tumbling	6.5.2(a)
EXP/E5	P/BAR	tumbling	6.5.2(b)
EXP/9	BRND	tumbling	6.5.1 & 6.5.2(a)
EXP/11	BRND	shaking	6.5.1
EXP/12	BRND	tumbling	6.5.2(a)
EXP/16	BRND	shaking	6.5.1
EXP/19	BRND	patination	6.5.10

TOOL NO.	FLINT TYPE	EXPERIMENT DESCRIPTION	SECTION
EXP/20	BRND	shaking	6.5.1
EXP/21	BRND	shaking	6.5.1
EXP/22	BRND	shaking	6.5.1
EXP/24	BRND	shaking	6.5.1
EXP/25	BRND	shaking	6.5.1
EXP/26	BRND	shaking	6.5.1
EXP/27	BRND	tumbling	6.5.2(a)
EXP/29	BRND	sheen	6.5.12
EXP/52	BRND	patination	6.5.10
EXP/59	BRND	tumbling	6.5.2(a)
EXP/65 A&B	BRND	friction	6.5.3
EXP/75	ISR/1	tumbling	6.5.2(b)
EXP/76 A&B	BRND	friction	6.5.3
EXP/77 A&B	BRND	friction	6.5.3
EXP/86	ISR/4	buried in Kebara	6.5.5
EXP/98	ISR/4a	buried in Kebara	6.5.5
EXP/99	ISR/4a	sheen	6.5.7
EXP/100	ISR/4a	patination	6.5.10
EXP/102	ISR/2	patination	6.5.8
EXP/103	ISR/4a	PDSM on a highly polished blade	6.5.13
EXP/104	ISR/2	buried in Kebara	6.5.5
EXP/105	ISR/2	patination	6.5.8
EXP/110	BRND	soil process	6.5.6
EXP/111	ISR/2	soil process	6.5.6
EXP/112	ISR/2	soil process	6.5.6
EXP/113	BRND	soil process	6.5.6
EXP/114	ISR/2	soil process	6.5.6
EXP/115	ISR/2	soil process	6.5.6
EXP/116	BRND	soil process	6.5.6
EXP/128	BRND	buried in Kebara	6.5.5
EXP/135	ISR/4a	patination	6.5.9
EXP/138	ISR/4a	sheen	6.5.7
EXP/139	BRND	patination	6.5.8
EXP/144	BRND	sheen	6.5.7
EXP/145	P/BAR	sheen	6.5.7
EXP/146	ISR/4b	sheen	6.5.7
EXP/147	BRND	patination	6.5.11a
EXP/148	BRND	patination	6.5.8
EXP/149	BRND	patination	6.5.11b
EXP/150	ISR/2	patination	6.5.11a
EXP/151	ISR/2	patination	6.5.11b
EXP/152	ISR/2	patination	6.5.11a
EXP/153	ISR/2	patination	6.5.11b
EXP/154	ISR/1	sheen	6.5.12
EXP/155	BRND	sheen	6.5.12

MECHANICAL PROCESSES

6.5.1 Shaking experiments

Aims

These experiments were planned to observe the effect on flint surfaces of the movement of implements during burial such as occurred in the sediment at Hengistbury Head.

Experiments

An Endecotts Test Sieve Shaker was used. One implement at a time was buried at a depth of approximately 6–8 cm in a plastic bowl measuring 30 cm in diameter filled with either sand sieved through 0.30 mm mesh or gravelly sand (which is the naturally occurring matrix on the Hengistbury site) and the sand and implement were shaken for varying lengths of time. Artefacts of East Anglia flint were made into microdenticulates (see section 4.1 for description) by R.N.E. Barton. The 40 pieces involved (see list in section 6.5) were observed microscopically and documented photographically before and after shaking. Initially artefacts were shaken for 2, 5, 10, 20, 30 and 60 minutes. The edge damage on unretouched unused bladelets was minimal even after 60 minutes. Except for a few random scratches (figures 124 and 125)Aims, no visible surface modification occurred on the flint surfaces especially on the pieces shaken for short periods. Subsequently artefacts were shaken for 20, 30, 60, 90, and 120, minutes.

The sieved dry sand and the gravelly sand were combined with water and salt as follows:

RUN	MATRIX
I	sieved dry sand
II	gravelly dry sand
III	sieved dry sand and salt
IV	sieved sand and water
V	gravelley sand and water
VI	gravelly dry sand and salt
VII	gravelly sand, salt and water

Macroscopic and microscopic observations

On artefact EXP/9 in run II (shaken for 10 minutes in dry gravelly sand) linear features, which were present on the tool edge after whittling dry reeds, were partially removed (figures 14 and 15).

One of the artefacts in run VII (EXP 12_), developed a bright spot (figure 130) (see section 6.3.2 for definition) after 90 minutes in gravelly wet sand and salt. Subsequently other runs were carried out decreasing and increasing the proportion of salt to the sand in order to monitor whether the salinity of the sediment was a crucial variable. Apart from these two instances no other surface modifications were observed in any of the runs and no other bright spot occurred. Bright spots were produced experimentally (as described in section 6.5.3) by rubbing two wet flints together. See also section 6.5.12 for other experiments producing bright spots.

Discussion and conclusions

The experiments suggest that movement in the sediment of up to 2 hours is not likely to cause much edge damage or surface modification. Increased salinity of the soil did not seem to the cause of bright spots. The importance of wet sand and the presence of gravel was to be further investigated (see section 6.5.2). Longer periods of movement in the sediment were also to be investigated in view of the difficulty in estimating the length and severity of movement in nature over thousands of years. Residues from worked materials "adhering" to the surface of the flint after cleaning with a detergent are removed by even brief movement in the sand (ten minutes) and are therefore not likely to survive archaeologically where artefacts have moved in the deposit.

6.5.2 (a) Tumbling Experiments

Aims

The results of the experiments in 6.5.1 suggested that it was necessary to subject artefacts to more lengthy movement in the sediment in order to obtain results. In fact the brief shaking time did not alter the flint surface enough to produce sheen or any other generalised PDSM comparable to what is observed on archaeological artefacts. A tumble-polishing kit for jewellery was used to simulate movement in the sediment. It comprises two plastic barrels measuring 8 cm in diameter and 8.5 cm in depth which are rotated at 60 rpm. The time needed to smooth stones for jewellery is given in weeks rather than hours with very rough grits. The length of 50 hours eventually needed in the experiments described below need not, therefore, seem excessive. Furthermore, 60 rpm, though not as slow as the movement of artefacts in a natural situation, is not a violent rotation. Shipman and Rose (1983:77), investigating "the effect of sedimentary abrasion" on bones, used a geologic tumbling barrel at a rate of 1800 rpm which ran for up to 35 hours.

Experiments

Two pairs of experimental artefacts of Brandon flint were buried in the sediment, one in each barrel, and tumbled for 50 hours.

(a) Wet and dry fine sand
EXP/9 a microdenticulate used to whittle and cut dry reeds, in Barrel 1 filled with **wet** sand;
EXP/2 an unused microdenticulate, in Barrel 2 filled with **dry** sand.

(b) Wet gravelly sand
EXP/12 a microdenticulate used to cut bracken for 10 minutes in Barrel 1;
EXP/27 used to incise green birch for 10 minutes in Barrel2.

Macroscopic and microscopic observations

(a) Wet and dry fine sand

Macroscopic soil sheen appeared marginally earlier and was more developed on the artefact shaken in wet sand (EXP/9) than on the one in dry sand (EXP/2). All striations on EXP/9 used to whittle dry reed for 10 minutes (figure 14) were removed after 50 hours of tumbling (figure 16). Only the well developed reed polish could still be observed on some parts of the edge, although its texture was now dulled. Abrasion tracks left by manufacture on the unused microdenticulate (EXP/2) (figure 131) disappeared after tumbling (figure 132) and the surface of the flint was scoured by the fine sand.

(b) Wet gravelly sand

On EXP/12 only faint sheen developed. Striations parallel to the edge and present before tumbling disappeared (figure 30 and 133). The weakly developed polish produced by cutting bracken was increased in development and invasiveness by contact with wet sand and gravel, and its distribution on the edge was, in places, now transversal (as from scraping) rather than longitudinal (as from cutting). Edge damage scars were added on both edges. On EXP/27 the sheen was macroscopic and best developed on the prominent areas of the bladelet. The faint polish produced by incising green wood (figure 39) now appeared in various areas of the implement to be at different stages of development having the characteristics of two different "polishes" as described in the Keeley method (figures 41 and 40).

DISCUSSION AND CONCLUSIONS

Wet and dry fine sand EXP/2 and EXP/9

The removal of all striations might suggest either as indicating that the surface was now more polished hence these linear features are no longer observable or more likely as indicating that the "striations" were actually residue "adhering" to the surface which was removed by the movement in the sediment. In any case the absence of these linear indicators makes a confident reconstruction of the direction of motion in use more difficult. Furthermore, in the Keeley method, the presence and absence of striations and their types are often used as criteria for distinguishing between polish "types".

Wet gravelly sand EXP/12 and EXP/27

There was no evidence on these two artefacts functionally, to indicate that there had been two polishing events. Polishes produced by cutting bracken for ten minutes or by incising birch for the same duration are not well developed and it is likely that subsequent polishing by the gravelly sand **polished over** or removed the earlier traces. The reconstruction of motion would now be misleading as the striations caused by the tumbling were sometimes in the opposite direction to those produced by use. It is interesting that two implements of the same flint quality and subjected to identical experimental conditions developed sheen of a totally different intensity. In the following experiments this phenomenon was observed several times suggesting that the effect of post-depositional processes might develop differently on each flake even if from the same core.

6.5.2(b) Tumbling experiments

Aims

The experiments described in 6.5.2a were repeated with some alterations in order to investigate the following two points:

a The possibility that the striations on **EXP/9** removed by movement in the sediment could be residue from bracken. The implement had been cleaned with detergent and warm water followed by immersion in distilled water in an ultrasonic tank. As such a residue would probably not have survived on an archaeological artefact (see section 3.4.6), it was decided that the artefacts involved in the following set of experiments would be immersed for 20 minutes after use in a solution of 20% (w/v) NaOH in water intended to remove organic residues, as suggested by Keeley (1980:11). The appearance of the use-induced traces was recorded photographically before and after this cleaning process.

b The tumbling time of 50 hours was not arbitrary but was dictated by the appearance of well developed macroscopic sheen on the artefacts. However it might be that 50 hours tumbling in the sediment could be mimicking an extreme situation rather than a commonplace occurrence. It was therefore decided to repeat the experiments for tumbling for artefacts of different flint types in wet gravelly sand two for 21 hours and two for 15 hours as described below:

1 EXP/E5 a rather coarse Potter's Bar flint flake used for 20 minutes to whittle fresh willow ;

2 EXP/59 a fine-grained Brandon flint bladelet used for 30 minutes on dry oak ;

3 EXP/1 a microdenticulate of medium-grained Dorset pebble flint used for 30 minutes to cut wet rushes (soaked dry rushes);

4 EXP/75 an unretouched bladelet of ISR/1 fine-grained flint used for 65 minutes to whittle dry oak .

Cleaning

After washing with warm water and detergent, EXP/E5 had a macroscopic gloss on the used edge several months after use. This gloss is often observed on experimental working of green wood or other woody plants but is usually removed by washing (see Chapter 3 for discussion of residues). Microscopically, the appearance of the gloss (figure 8) is not in any way similar to use-wear traces and the direction of the "striations" does not seem to reflect the direction of use. EXP/59 and EXP/75 did not develop gloss, probably due to their use on dry wood. EXP/1 used on wet rushes had well developed polish but no macroscopic gloss.

EXP/E5 After cleaning for 20 minutes in the NaOH solution, the area of macroscopic gloss disappeared and only a few "striations" survived (figure 9), while use-wear polish on other areas of the edge remained unchanged (figure 134). Microscopically, these remaining striations are quite similar to those on **EXP/9** (cf. figures 14 and 9) and their direction is quite compatible with the whittling movement of the tool.

EXP/59 The appearance of the polish was not affected by 20 minutes in the NaOH solution (figure 135).

EXP/1 The same is true of this artefact as can be seen by comparing figures 34 and 37.

EXP/75 The polish produced by whittling oak for 65 minutes was also examined after cleaning with detergent and again after 20 minutes in the NaOH solution. There was no change in the appearance of the polish (figure 136).

Macroscopic and microscopic observations

Tumbling in wet gravelly sand for 15 hours

EXP/1 Very little sheen developed on this artefact after tumbling except on the prominent parts. Edge damage occurred on both edges and the scars changed the profile of the edge (cf. figures 34 and 137). New micropits appeared that were not present before use.

EXP/75 Again very little sheen developed except on the edges. Much macroscopic edge damage, which could mistakenly have been seen as use-induced, occurred on the unretouched cutting edge. New polish was added (figure 138), making the flint surface very similar to archaeological artefacts affected by microscopic **PDSM** while appearing very fresh to the naked eye.

Tumbling in wet gravelly sand for 24 hours

EXP/E5 After tumbling no sheen developed on the coarser parts of the artefact, while it did on the finer-grained edges. The "striations" which had survived the NaOH bath have now been removed (figures 9 and 10). Scars caused by edge damage changed the profile of the used edge.

EXP/59 Macroscopic sheen developed but was, predictably, not as intense as on the artefacts tumbled for 50 hours. The appearance of the polish caused by work on wood (figures 135 and 139) was now totally altered (figure 140). Striations parallel to the working edge, produced by incising wood, and surviving the NaOH bath (figure 139) were no longer present, presumably polished over or away by the contact with the gravelly sand. These striations were replaced by polish distributed perpendicularly to the working edge. The flint surface was now polished all over as in archaeological examples of **PDSM**. The edge damage scars had removed areas of well developed use polish and striations. Micropits appeared which were not present before tumbling (figures 140 and 141).

Discussion and conclusions

These experiments suggest that fine-grained flint develops sheen faster than coarse-grained flint. This is the same phenomenon empirically observed by many analysts regarding the development of use-wear polishes. It would seem to suggest a similarity between the processes of formation of sheen and polish, in which the abrasion factor is very important if not predominant.

The Israeli flint used in this experiment is fine grained but the surfaces have a large number of fissures. The effect of tumbling on this flint was comparable to other fine-grained flints such as the black chalk flints from East Anglia.

"Striations" which survived the 20 minute NaOH bath were removed by movement of the artefacts in the sediment and are presumed to be "adhering residues" (see section 3.4.6). The new polish formed is similar to that observed on archaeological artefacts with soil sheen. The reconstruction of tool use is hindered by the removal of the well developed polish through edge damage, and the appearance of striations unrelated to the direction of motion in use. These new striations presumably reflect the relative motion of artefact and sediment in the tumble polisher.

Micropitting (observed under the optical microscope) which occurs after tumbling with gravelly sediment could be interpreted as the result of the removal of large crystals (or a group of crystals) from a particular spot causing the pit to form. These micropits also occur when flint tools are used on hard materials. I had noticed that archaeological artefacts with PDSM often had a large number of micropits. These experiments have also produced frequent micropitting (figures 137, 141 and 138) while no such micropitting was observed before the shaking experiments (figures 139, 136 and 34).

It is interesting to note that the polish on EXP/1, which was very well developed like sickle gloss, was not affected in the same way as was polish from other materials. Nowhere was it "increased" as had occurred with the polish produced by cutting bracken and incising wood. It disappeared only where a new scar removed that part of the edge, and its brightness was slightly dulled by the abrasion. This is an encouraging confirmation of what had been observed empirically, that is that "sickle gloss" is the only polish that survives on the archaeological artefacts which I have examined so far, particularly from the Near East.

Unger-Hamilton (1984:156) also noted that while polish on other artefacts "may have been removed...polishes on most of these [sickle blades] were relatively intact".

A probable explanation for the survival of 'sickle gloss' is that it represents one of the latest stages of polish formation where the polished area is much greater than the unpolished. This makes it very difficult to "increase" the polish by mechanical means. Furthermore there are consequently fewer areas of open fissures through which chemical processes can penetrate and attack the flint surface (see section 6.5.12). When the polished area is totally smooth and cannot be polished further it may, however, be scratched by a coarse abrasive or by a material as hard or harder than flint (see section 6.5.13).

6.5.3 Friction Experiments

Aims

The following experiments were carried out to ascertain whether certain bright spots were due to friction, rubbing two pieces of flint together with pressure in the "same small area", as suggested by Shepherd (1972:120), particularly with water as a medium, as suggested by Vaughan (1981:366).

Experiments

1 Two flakes of Brandon flint (EXP/65 A and B) were carried for two months in a plastic bag, in a small pocket, tightly rubbing against each other.
2 Two other flakes of the same flint (EXP/76 A and B) were rubbed together wet with pressure as mentioned above for five minutes.
3 The experiment was repeated using two flakes **(EXP/77A and B)**, rubbing them together with water for 2 minutes each, first rubbing together the areas of fine-grained flint on the bulbs and then rubbing one edge against another.

Microscopic and macroscopic observations

1 EXP/65 No bright spots, macroscopic or microscopic, were observed, although macroscopic polish formed with linear features similar to the polish observed on EXP/11 sawing shell (figure 142 see also figure 125).
2 EXP/76 After 5 min, intense macroscopic gloss appeared on the edge, as observed on archaeological artefacts. Microscopically the polish was very bright and flat, crossed by striations as described by Stapert (1976:32) but with an untypical crushing of crystals (figure 143).
3 EXP/77 On the bulb the polish was diffuse, uncharacteristic of bright spots but more similar to soil sheen (figure 144). On the edge the bright spots were almost identical to the archaeological ones (figures 44 and 45), with macroscopic gloss and no crushing of crystals.

Conclusions

The above observations suggest that

a friction between two flint surfaces produces the same polish as is produced by use on very hard materials such as shell;
b water is an important element in localising the polish so that "bright spots" appear. It may serve to retain displaced fragments of flint so that they promote wear whereas in the dry situation they may be free to leave the area of flint to flint contact (see sections 3.4.5.2 and 6.4.5 for a discussion and experiments regarding the importance of fluids in the development of polish on flint);
c the pressure produced by rubbing a flat area with an edge is greater than that produced by rubbing two flat areas on the flint surface in that the force is concentrated on a far smaller area;
d the microscopic distribution of the flat bright spots produced in these experiments seems, like all polish, to depend not only on duration of contact but on the microtopography of the flint.

In conclusion it seems that the hypothesis, proposed by many, that bright spots are post-depositional in nature has been confirmed by these experiments. They have been produced here, as suggested by Shepherd (1972:120), through friction of flint with other flint, although their typical localisation occurred only when water was used. The polish formed by rubbing flints on the ventral bulb had a diffuse distribution on the microtopography very similar to soil sheen. This suggests that contact between flints in the presence of water in the sediment could produce either bright spots or a diffuse polish such as soil sheen depending on the distribution of the areas in contact and the duration of wear. Experiments described in section 6.5.12 show that chemical processes can also cause bright spots and soil sheen to develop on the same artefact.

6.5.4 Trampling Experiment

Aims

The following experiment was carried out to observe the effect on artefacts discarded on the surface or buried just below the surface when they are trampled by a group of people as might occur in a prehistoric camp site. It was aimed at monitoring movement of the artefacts on the surface and in the sediment and to observe the correlation, if any, of such movement to scratches, edge scarring, soil sheen and/or bright spots.

The experiment was conducted at the site of Klithi in Greece by Dr. M.H. Newcomer who also knapped the flint artefacts. My thanks go to Dr. Newcomer and the site excavators for allowing me to carry out the microscopic work on these artefacts.

Stages

I From 24–27 Aug. 1985; 280 flints (with bones) were laid out in two layers with 140 flints per layer in the same metre square. The location of each artefact was recorded three dimensionally to monitor vertical/horizontal movement. The lower layer was covered by a 5 cm thickness of sterile soil which consisted of sieved sediment from the Klithi shelter, while the top layer was left exposed at ground surface.
II Every day for 7 days 20 to 30 people walked over the square, some with shoes others with sandals.

III On 4th Sept. 1985, after 7 days of trampling, one quadrant of the square was excavated. Both upper and lower levels were excavated with brush and scoop. The new position and orientation of the pieces were plotted. The sterile level was still preserved but was compacted to 2–3 cm. A plastic cover was put on over the remaining three quadrants of the square weighted down by 5 cm. sediment.

IV In August of 1987, two years after deposition, the rest of the metre square was excavated. The sterile layer was more compacted but the upper and lower layers of flint were still clearly divided. Roots had grown between the artefacts and around the plastic sheet. There was little difference in the extent of movement between large and small pieces.

Microscopic observations

Two freshly struck flint pieces which were kept in separate plastic bags at the start of the experiment to use as controls, were examined before observing the trampled pieces. The flint used in this experiment was obtained from pebbles gathered from the river near which the site is located. The quality of this flint was very shiny and the freshly struck surface has microscopically an appearance similar to Brandon flint when it is affected by the early stages of use. There were many inclusions, fissures, and ridges and some linear features, presumably from movement against the core during manufacture.

Four pieces (2 each from the upper and lower levels) were randomly chosen for microscopic examination. The U and L in both the lists below stand for upper layer and lower layer:

U	172	1987	rare ridge crushing
U	200	1987	some ridge crushing
L	69	1985	some ridge crushing
L	70	1987	some ridge crushing

No surface modification (SM) was observed, some random post-depositional (PD) "striations" (STR) were present but none were significant.

A further 15 pieces were chosen either because they had moved more than 2 cm in one direction or they were inverted and had undergone movement which was possibly significant. Very small pieces were excluded even if they had moved because they would not be likely candidates for functional analysis.

The dates appearing after the tool number in the following list indicate the year of excavation. The pieces excavated in 1985 have been trampled and buried for only seven days while those excavated in 1987 have been buried for two years. The following are some comments on each piece:

TOOL NO	FRESHNESS	POLISH STR ED SM AND OTHER
L 78 1987	much ridge crushing	Random STR. Initial P VNTR.
L 83 1985	some ridge crushing	Random STR. Initial P VNTR. Strange macro P on DRSL.
L 89 1987	faint ridge crushing	PERP STR + VNTR. crushing
L 91 1987	no ridge crushing	White salty deposit. No SM.
U 142 1987	no ridge crushing	Much ED thin bladelet. Curved cracks.
U 152 1987	no ridge crushing	No ED. Rare STR. No SM.
U 153 1987	ridge/VNTR crushing	No SM.No ED. RL random STR.
U 165 1987	heavy ridge crushing	VNTR ED and random STR. No SM.
U 166 1985	no ridge crushing	ED and random STR. No SM.
U 168 1985	no crushing	Some ED and random fresh STR. No SM.
U 171 1987	heavy drsl crushing	Many STR. Some P no SM.
U 176 1987	crushing LL + DST	P NR crushed edges. No SM. Random+PERP STR. No P VNTR/RL.
U 177 1987	some crushing	No SM. DRSL Abrasion DRSL.
U 188 1985	heavy Curved cracks.	No SM. DRSL/VNTR Random +PERP STR. crushing
U 189 1987	faint ridge	ED rare. No SM. crushing Rare STR traces of salty deposit as L91.

Discussion and conclusions

A) It was noted that the pieces excavated in 1985 had very few signs of abrasion (trampling traces figures 145, 146, 152) on them even when they came from the upper level. In fact, out of 15 pieces chosen on the basis of the extent of movement or inversion, only three belonged to those excavated in 1985. This strongly suggests that movement occurs over the course of time and is not particularly confined to the earliest stages of burial or to the duration of trampling. It may be that the longer pieces are buried the more they are likely to be subjected to movement and to consequent post-depositional traces occurring even after trampling itself had ceased.

B) The flint used in this experiment was knapped from heavily rolled river pebbles. These pebbles were also the source of the archaeological flint implements present on the site. The crushed ridges are preserved on some of the artefact edges even after knapping (figure 147). The drawback for the use of this raw material in a trampling experiment is that one is not then able to observe edge damage caused by trampling if it occurs on a previously crushed edge.

C) It was intriguing to note that some pieces appeared to have badly "trampled" edges while others were extremely fresh like the control pieces, and the difference could neither be attributed to the level in which they were buried nor to the length of burial time. Once the connection between the heavy edge and ridge crushing and the river pebbles used as cores (point B above) was understood, it was apparent that the trampling had not caused much edge damage on the freshly knapped edges. None of the crushing could however be confused with use-wear but had earlier been erroneously attributed to trampling. It seems therefore that one week of trampling by 20 to 30 persons a day is not likely to cause damage to flint tools that cannot be isolated during microwear analysis. In a prehistoric situation animals might also have been trampling the site and, in the case of a permanent base camp, a much longer period of trampling might have occurred.

D) It was not surprising that soil sheen or other surface modification did not occur after two years of burial. The other experiments (section 6.5.5) carried out also suggested that a period of one to two years is not long enough for chemical and mechanical processes to create the effects which are normally observed on archaeological implements. It was surprising however that no bright spots had occurred as they can be produced by friction between two pieces of wet flint for a very short time (section 6.5.3). Although there was dew overnight, the trampling experiment took place during a dry period so this may be a reason why bright spots were not formed. On the other hand on a few artefacts which have been in the lower level for two years but not directly affected by trampling, a diffuse, not well developed polish was observed (figure 148) similar to the polish produced in section 6.5.3 by rubbing two artefacts with their bulbar area in contact (figure 144). When little pressure can be exerted well developed polish cannot form very fast. This lack of pressure and water may explain the absence of bright spots.

   Unexplained salty deposits were observed on two artefacts one from the upper level and one from the lower level. They were not removed by thorough cleaning consisting of brushing with a nylon brush with biological detergent followed by ultrasonic cleaning. They had earlier been wiped with acetone and methylated spirit (figure 149).

   Some very strange curved cracks were observed on the ventral aspects of some pieces (figure 148). They occur both on the pieces with heavily crushed edges and on those whose edges are not crushed. One can only speculate that they occurred inside the pebble when it was being crushed in the river.

E) The importance of water or another liquid medium in the development of polish has been noted empirically by many microwear analysts and has been investigated in this project. The only traces of polish on the examined artefacts (figure 150) were noted on or near the crushed edges. One can only speculate that the polish might have formed, not during burial, but rather as a result of movement of the pebbles in the river.

F) Although none of the traces found on the Klithi experimental pieces could be mistaken for use-wear, it was interesting to note that on one of the artefacts a polish appeared very similar to that produced by grooving wood with talc (compare figure 151 with figure 61). Its distribution on the edge was localised and was easily identified as non use-wear. A disturbance more prolonged than a two year experiment might have produced a more confusing distribution. Although edge damage on edges not affected by crushing was relatively light, had certain pieces been used any edge damage could have caused the partial or total removal of the used edge, especially on thin blades.

CHEMICAL PROCESSES

6.5.5 Effect of Natural Soil Processes

Aims

In order to study the effect of natural soil processes on flint surfaces, it was planned to bury artefacts in Kebara cave. An area was chosen near the limestone cave wall with water dripping. The artefacts were cleaned with warm water and detergent and buried at a depth of 50 cm. The aim was to observe whether soil sheen would occur after a relatively brief period of burial in a limestone cave with percolating water. Furthermore it was intended to observe the difference (if any) of the effect of burial on various flint qualities and on various stages of polish development. The buried artefacts were:

EXP/128 Black Brandon flint blade polished by damp chamois;

EXP/86 Buff ISR/4a flint blade polished by damp chamois;

EXP/98 Buff ISR/4a flint blade polished by damp chamois;

EXP/104 Brown ISR/2 flint flake polished by whittling wood.

Microscopic observations

After 400 days there were scratches on the flint surface (figure 154) and slight damage on the edge presumably from the settling down of the sediment or from the effect of the trowel. No macroscopic or microscopic soil sheen had developed. Polish seemed unaffected except when it had been removed by edge damage. In some cases polish was slightly diminished microscopically after burial as occurs when superficial residues are removed by biological cleaning or by a dilute NaOH solution (figures 59 and 153).

Conclusions

The artefacts were reburied again in October 1988 to be retrieved at the end of the excavation (in several years) as it is obvious that a period of one or two years is too short to produce significant changes. It was interesting to note that residues still "adhering" to the surface had been removed after such a brief period of burial.

6.5.6 Effect of Naturally Occurring Soil Processes

Aims

The following experiments were set up to observe

1 whether it is possible to produce post-depositional surface alterations on flint without greatly accelerating the process by using concentrated solutions and/or heat.

2 to check the development of eventual alterations on:

a) different flint types (Israeli and Brandon),
b) polished and unpolished flint,
c) water whether stagnant or dripping.

EXP/110 HH soil + stagnant distilled water
Brandon flint polished by chamois leather

EXP/113 HH soil + distilled water dripping[1]
Brandon flint tool polished by chamois leather

EXP/116 Peat + decaying leaves
Brandon flint freshly struck unused

EXP/111 Kebara soil + stagnant distilled water + $CaCO_3$
ISR/2 flint polished by chamois leather

EXP/112 Kebara soil + stagnant distilled water + ACO_3
ISR/2 flint polished by chamois leather

EXP/114 Kebara soil + dripping distilled water + $CACO_3$
ISR/2 flint polished by chamois leather

EXP/115 Kebara soil + stagnant distilled water + $CaCO_3$
ISR/2 freshly struck flint unused

The experiment was conducted at room temperature[2].

An attempt was made to reproduce as far as possible the conditions at Hengistbury and Kebara and therefore the sediment from each site was used in conjunction with flint from that area and calcium carbonate was added to experiments involving Israeli flints to simulate the limestone percolations in Kebara.

Microscopic observations

The artefacts were examined on several occasions over a period of two years. Early checks showed no effects. The intervals between checks were lengthened as it became obvious no effects could be observed in the short term.

After two years the pieces show initial slight PDSM as evidenced by the deepening of ridges and shallow hollows (figures 155 and 156) and an increase in glare especially at x200 making both microscopy and photography very difficult.

Macroscopic observations

Water was passing faster through the box containing Kebara sediment than through the one containing Hengistbury Head sediment. This seemed surprising because Kebara sediment is very fine grained while Hengistbury Head sediment is a coarser grained sand. Both tools were relatively thin and inserted length-wise. The passage of water was rendering the Kebara sediment much more compact. In fact the water was flocculating the fine sediment into small lumps ca.1–2 cm long and half a centimetre in diameter. This was noticed after 450 days of percolation. The lumps are very hard and can be mistaken at first for pebbles, but they crumble with pressure from the fingers.

I have also encountered them when excavating at Kebara though they are larger in the cave sediment. Their small size in this experiment could be a consequence of the considerably shorter time span and the confined area of the vessel containing EXP/114. It is probable that water was percolating in the cave for long periods of time throughout its history. This experiment suggests that the presence of these lumps in a particular layer can be taken as confirmation of this.

Conclusions

The conclusions to be drawn from the microscopical observations are that some dissolution and/or redistribution of silica from the surface has started to occur. It is not possible to make a more definite statement although in the following experiments it can be seen that glare often precedes the development of sheen and patination. It is obvious from this experiment that the replication of PDSM as it appears naturally on the artefacts from Kebara can only be expected after a much longer period of time.

6.5.7 Effect of Naturally Occurring Chemicals

Aims

The following experiments were inspired by Rottlaender (1976:56; 1975:108) who reported producing "gloss patina" (his expression for soil sheen) on flint surfaces accompanied by 0.3% weight loss after leaving freshly struck flint artefacts for three weeks in peat with decaying leaves and water at 50° C.

My experiments were planned to compare:

1 the speed of development and/or intensity of soil sheen on four different flint types;
2 the different effect (if any) on polished and unpolished surfaces of flint from the same nodule.

EXP/99 ISR/4a flint polished by wet hide 20 minutes weight gr.4.08

EXP/138 ISR/4a flint freshly struck unused weight gr. 6.41

EXP/144 Brandon flint freshly struck unused weight gr. 5.17

EXP/145 P/Bar flint freshly struck unused weight gr. 9.80

EXP/146 ISR/4b flint freshly struck unused weight gr. 4.39

The artefacts were immersed in peat with decaying leaves and distilled water in separate covered beakers

[1] slow droplet 5 days a week. Runs dry over weekend and holidays. Not too dissimilar from cave conditions where wet and dry cave conditions where wet and dry periods alternate and percolation does not occur always in the same spot. In HH rain and dry weather also alternate today. Dripping water in this experiment is nonetheless probably running through faster than in nature and this may have accelerated the post-depositional processes to some extent. It is possible that otherwise even the modest results obtained could not be expected within the timespan of such a project.

[2] about 20° C during working hours, which could go 2–3° lower on weekends and 2–3° higher on hot summer days. Conditions not too dissimilar from seasonal variations in nature except for the absence of heat peaks as in Kebara in the summer and possible freezing as in Hengistbury today. Cave conditions are, however more temparate than those in the open air.

heated continuously at temperatures between 50–60° C. at approximately pH 6.

Stage I: Macroscopic and microscopic observations

No sheen developed after eight weeks possibly due to the fact the equipment could not be maintained at a constant temperature without supervision. It may be that when the temperature went above 50° C the micro-organisms in the soil were killed. They might have an important role in the patination process. At this time garden compost was added to introduce new micro-organisms and the temperature was lowered to under 50° C.

The artefacts were removed after 4 months (8 weeks of which were in the new compost) although no macroscopic soil sheen had appeared.

Stage II Macroscopic and microscopic observations

After four months no macroscopic or microscopic sheen developed and no weight loss occurred on any of the artefacts. Some alteration of the flint surfaces had taken place as evidenced by the increased glare which inhibited focusing on all the artefacts even with a green filter (figure 157 and 158). The polish on EXP/99 was not removed or diminished but it too was affected by extreme glare.

Discussion and conclusions

Time did not allow me to continue this experiment for more time with a temperature below 50° C, to ascertain whether glare would be followed by soil sheen as suspected. Glare appeared on experimental implements which developed sheen and is often observed on archaeological artefacts affected by PDSM. Had sheen developed after the addition of garden compost it might have suggested that micro-organisms played an important role in the dissolution of flint surfaces during burial. Future research will be necessary to investigate this interesting possibility. For a graphic illustration of the mechanism which is causing glare to appear at the same time as PDSM and particularly together with soil sheen on all flint artefacts see figure 167 and section 6.5.12.

CHEMICAL PROCESSES

6.5.8. Patination with Naturally Occurring Chemicals

Aims

The following experiment was inspired by Curwen (1940:435–437) who produced patina on artefacts immersed for 22 months at room temperature in "garden mould" with chalk and water and decaying grass. The experiment was repeated to observe:

1. the difference (if any) in speed and intensity of patination on flints of very different type and grain size;
2. the effect of patination on polish;
3. if patination is accompanied by weight loss and if so to measure its extent.

EXP/139 Brandon flint weight 5.97 gr. polished by wet hide

EXP/148 Brandon flint weight 14.91 gr. unused

EXP/102 ISR/2 flint weight 2.66 gr. polished by wet hide

EXP/105 ISR/2 flint weight 5.03 gr. unused

Artefacts were immersed in a Pyrex dish (20 cm diameter) filled with peat, leaves decaying in water, urine and calcium carbonate.

Curwen's experiment was approximated in the ingredients. As the term "garden mould" was rather vague, peat was used for six weeks and garden compost was added for the second half of the experiment. Decaying leaves were used instead of decaying grass and calcium carbonate was used instead of chalk. Curwen's experiments patinated flint in 22 months (approx. 96 weeks) at room temperature (20° C.). It was assumed that the reaction speed doubles approximately with each 10° C. rise in temperature.

By conducting the experiment at 50° C it was hoped that the onset of patination might occur much sooner, for example in approximately 12 weeks. The heated solution was at pH 7.30/8.0. No patination occurred after 14 weeks. The artefacts were removed, cleaned in the normal way (biological detergent, methylated spirit, acetone, ultrasonic tank), weighed and observed microscopically.

Macroscopic and microscopic observations

Macroscopically there was neither patina or sheen on any of the artefacts nor was there any weight loss. The experiments, however, yielded some results which will be described separately for each artefact.

EXP/139 Certain areas of the ventral aspect were covered with brown encrustations not removed by two hours in a biological detergent. No attempt was made to remove them chemically. Microscopically the polish on the edge seemed very slightly diminished in the areas where it was less developed. There was much glare on all the flint surface which seemed more irregular and etched (figure 159 and 160).

EXP/148 Macroscopically there were stains and encrustations as on EXP/139 (figure 166). These were not removed by prolonged soaking in a biological detergent but eventually disappeared after 30 minutes in lukewarm 10% w/v HCl. Tracks of little developed polish left on the edge by hard hammer retouch were diminished (figures 161 and 162) and there was much glare on the flint surface during microscopy.

EXP/102 Glare, which started to appear after five weeks and continued to increase, was interfering with microscopy even at x100. The surface seemed etched after 12 weeks. Some white crystals formed in the shallow hollows of the flint surface. Lower lying points of the microtopography appeared “polished” after 14 weeks in areas not polished by hide (figures 163 and 164). The very well developed C++ polish on the prominent parts of the edge seemed intact but showing glare at x200.

EXP/105 Microscopically here too there was much glare and the hollows seemed much deeper, making the flint surface appear more pockmarked than before (figure 165). New white crystals appeared as on EXP/102.

Discussion and conclusions

Glare formed on all artefacts whether of the fine Brandon flint or the coarser ISR/2 variety. It seems to indicate that asperities from the flint surface have been removed by dissolution. On both the ISR/2 artefacts the shallow hollows seemed etched deeper making the surface appear more pockmarked and causing much glare. It is in these deepened hollows that the first white crystals appeared. For a more detailed discussion of glare observed on artefacts affected by PDSM (see section 6.5.12 and figure 167).

The brown encrustations were at first thought to be organic residues from the decaying plant matter. As they were not removed by the prolonged soaking in a biological detergent, an attempt was made to soak EXP/148 only in lukewarm 10% w/v HCl which removed the deposits suggesting that they were derived from the calcium carbonate present in the mixture. It was interesting to note that such encrustations were preferentially deposited on the finer Brandon flint. On the ISR/2 flint, however, the first white crystals appeared probably heralding the spread of patina while none formed on the Brandon flint.

The well developed C++ polish on EXP/102 was not altered as seems to be the case with “sickle gloss” and other well developed polish. It is probable that the polish seals off the flint surface, inhibiting attack by any chemicals that might dissolve the flint. Less developed polish inward from the edge on EXP/102 and the little developed polish from the hard hammer retouch on EXP/148 were diminished.

A polish, very similar to PDSM (as it is observed on certain archaeological implements) appeared on EXP/102. The same polish was observed on EXP/93 and EXP/96 (figure 60; section 6.4.6) grooving wood with talc and on the trampling experiment (figure 151, section 6.5.4). This seems to occur when the flint surface is polished both on the protruding asperities and the low lying crevices. This has been noticed both with chemical polishing and with mechanical polishing when soft abrasives are used.

6.5.9 Effect of Laboratory Chemicals

Aims

To observe the effect of sodium hydroxide on unpolished flint artefact EXP/135 ISR/4a flint was treated in three stages:

Stages

I Immersed in 20 wt. % aqueous solution of sodium hydroxide for 54 hours. For 12 hours of this time it was heated at 50° C : No macroscopic or microscopic changes occurred.

II Immersed again for 37 days without heating : again no change occurred to the surface of the flint.

III Exposed to the air for 15 days to see if damage became apparent only on drying: no change was visible either macroscopically or microscopically

Discussion and conclusions

It was surprising that no patination occurred in spite of the long immersion in heated sodium hydroxide though heating was for only 12 hours at 50° C. Some researchers have noted patination to occur on flint immersed in NaOH for shorter periods (Schmalz 1960:47; Roger Grace pers. comm.). The flint involved (ISR/4a) is microscopically very coarse and loosely packed in spite of its fine and smooth appearance to the naked eye. Due to this coarseness this flint was expected to be more susceptible to patination than fine grained and tightly packed varieties of the Brandon flints.

6.5.10 Patination with Sodium Hydroxide and Heat

Aims

The aim of this set of experiments was to patinate flint and observe the correlation of patina with

1 areas of polished and unpolished flint
2 polish from materials of different hardness.

The following experimental flint artefacts were used:

EXP/19 Brandon flint microdenticulate polished on bracken,

EXP/52 Brandon flint flake polished on bone,

EXP/100 ISR/4a flint blade unused.

EXP/19 and EXP/52 immersed in 30 wt % aqueous NaOH for 274 hours at 50° C

EXP/100 immersed in 20 wt % aqueous NaOH for 320 hours at 90° C

Macroscopic observations

EXP/19 and 52 The artefacts were examined daily but no patination occurred during first nine days.

During the following two days (a week end) it was not possible to examine the pieces. After these two days totalling 11 days or 274 hours in solution both pieces were completely patinated .

EXP/100 Several degrees of patination could be seen on the one piece (Figure 172). On some parts of the artefact the flint colour was preserved as if unpatinated.

Microscopic observations

EXP/19 and EXP/52

1 Extreme leaching of silica occurred as was deduced from the lighter weight. As it was not possible to

replace water from evaporation between the 9th and 11th day it may be that the concentration of the solution was increased and caused the abrupt and strong patination. A slower patination would probably be more useful to monitor the effect of patina on polished and unpolished flint surfaces. The serration disappeared on EXP/19. Edge damage removed parts of the edge on EXP/52 which is smaller and thinner than EXP/19 (see figure 242). The surfaces had a pockmarked structure (figure 169) at high magnifications and became very reflective.

2 No polish survived on either tool. Polish both from use on bracken and bone disappeared completely in this powerful process.

EXP/100 The areas of the artefact which macroscopically seemed unpatinated were microscopically covered with blue/grey patina crystals. The less patinated areas appear similar in coarseness to cortex while the more patinated ones are similar to the pockmarked effect on EXP/19 and 52. Heavy patina attacked the edges first and particularly the cortex on the right lateral edge which was highly leached out. The patination on this artefact was not as extreme as on those (EXP/19 and EXP/52) in the more concentrated solutions, in spite of the higher temperature and the longer time. When a small flake was removed from the edge, the underlying surface appeared little affected (compare figures 170 and 171).

Conclusions

It is not possible to attribute the different results obtained in these experiments directly to the concentration of the solution as the flint type was different. It might seem from these experiments that Brandon flint is more susceptible to patination and alteration than some of the coarser Israeli flints (see also section 6.5.9, 6.5.11(b) 6.5.12 and 6.5.13). Further work needs to be done to see whether patination occurs consistently and to understand the processes involved. A further series of experiments (see section 6.5.11 a) was carried out with less concentrated solutions and a more controlled alteration of the variables to observe the process at a slower pace.

6.5.11[a] Patination with Sodium Hydroxide, Heat, and Calcium Carbonate

Aims

The following experiment was carried out in order to replicate the one described by Schmalz (1960:47) in which flint was patinated with 0.5 M aqueous NaOH at 78° C. in the presence and absence of calcium carbonate ($CaCO_3$).

The purpose was to observe:

1 whether sheen develops in the initial stages before patination;
2 the effect of calcium carbonate on the speed of patination;
3 if different flint types patinate at significantly different speeds;
4 the difference (if any) between the effect and/or speed of patination on polished and unpolished flint areas;
5 whether polish is degraded by light patination;
6 the amount of weight loss and the differences, if any, on different flint types.

The following artefacts were immersed in NaOH 0.5 M solution and heated:

- two brown ISR/2 flint artefacts EXP/150 and EXP/152 from the same core each polished in a discrete area by damp hide and ochre.

EXP/150 weight gr. 2.87 in 0.5 M aqueous NaOH (pH 12/13)

EXP/152 weight gr. 2.05 in 0.5 M aqueous NaOH + excess $CaCO_3$ (pH 11);

- one black fine Brandon flint bladelet very different from ISR/2 flint was polished by damp hide with ochre in a discrete area

EXP/147 weight gr. 1.02 in 0.5 M aqueous NaOH (pH 12/13).

The equipment could not maintain a specific temperature above 50° C. without constant supervision.

The solution was heated as follows:

a four days at 80–90° C. for several hours during the day and at room temperature overnight. Artefacts were examined microscopically at intervals of 24, 72 and 96 hours;
b seven days at 50–60° C.;
c three more days at 50–60° C. for EXP/147.

Microscopic and macroscopic observations

EXP/150 After 24 hours rare white patina crystals were visible microscopically (figure 173). By 96 hours glare was very strong and polish was much diminished (compare figure 174 to figure 175). By 11 days macroscopic and microscopic patination had occurred and the surface was covered with pits (figure 176).

EXP/152 Glare started to appear after 72 hours as the first blue/white patina crystals appeared. By 96 hours there was a proliferation of them on some edge and ridges. After 8 days the shallow hollows became deeper (figure 177) and the surface seemed macroscopically patinated by a thin veil. Patina "polish" appeared where no use polish existed. In 11 days the patination was complete both macroscopically and microscopically and the polished edge was completely altered (compare figure 178 to 179).

EXP/147 The process on this artefact was much slower and one could actually see the polish diminish at each observation (figure 180, 181 and 182).

The following comments can be made as a result of the observations and are given in the same order as the proposed aims:

1 No soil sheen occurred in the stages preceding patination, either microscopically or

macroscopically. In section 6.5.11 (b), however, where the same experiment was carried out with Na_2CO_3, the process was much slower and soil sheen did develop but there was no patination contrary to Schmalz (1960:47).

2 Calcium carbonate speeded up the process. In 11 days **EXP/152** was completely white, while **EXP/150** without CaCO3 was mottled white and the colour of the flint is still visible under the patina. Except for the weight of the pieces, all other variables were the same. Presumably the layer of patina formed in 11 days in the presence of $CaCO_3$ is much thicker than the one formed in its absence.

3 Brandon flint patinated at a slower rate than the coarser ISR/2 flint. Mottling was more pronounced in ISR/2 flint since it appeared as a deepening of existing shallow hollows and other pits which are more common in coarse flint. The blue/white crystals seemed to form preferentially inside them. Contrary to Schmalz's observation that "a light blue white patina appears on all surfaces after 24 hours in 0.5 M NaOH at 78° C" (*ibid*:47), in this experiment (where only the temperatures were different) no light patina occurred "on all surfaces". Rather, edges and selected areas of the piece were patinated first, followed by a mottled stage on both flint types and subsequently a total white cover.

4 Patina attacked the unpolished areas first and polished surfaces were the last to be affected. However, edges were also more vulnerable than the rest of the surface.

5 Polish was degraded by patination on all three artefacts. In the stages of light patination (at least the kind of relatively fast artificial patination induced in the laboratory), some polish was still visible but glare suggests that the unpolished flint areas in between the polish were affected, making it difficult to observe the polish and focus on it at x200.

6 Weight loss occurred as follows :
EXP/150 went from gr.2.87 to 2.73, a loss of 5%
EXP/152 went from gr.2.05 to 1.81, a loss of 13.25%
EXP/147 went from gr.1.02 to 0.84, a loss of 21.5%

Discussion and conclusions

This weight loss suggests that silica has been leached from the flint interstices as observed by Schmalz (ibid:46). The weight percentage of loss was, predictably, greater for the smaller pieces as they have a greater ratio of surface area to volume. The heaviest piece lost only 5% of its weight and the lightest piece lost 21.5% of its original weight. However **EXP/147** was left in solution for three more days in order to patinate to the same degree as the other two pieces and **EXP/152** was in a solution of NaOH with $CaCO_3$ which may have accounted for its faster and seemingly deeper patination. Patination seems to be a complex process probably involving removal of interstitial matter, precipitation of new crystals or a combination of both. In the initial stages, it does subjectively appear, that there is an element of crystallisation of new material from the solution. It is clear from these experiments that polish on patinated flint artefacts cannot be used to reconstruct tool function.

6.5.11[b] Patination with Sodium Carbonate, Heat, and Calcium Carbonate

Aims

The same experiment as in 6.5.11(a) was repeated with a solution of Na_2CO_3 as suggested by Schmalz for the same objectives and heated in the same way to observe what, if any, difference in patination occurs between artefacts immersed in NaOH and Na_2CO_3. Artefacts used were:

- two Brown ISR/2 flint artefacts from the same core as 6.5.11(a), each polished in discrete area by damp hide and ochre;
  EXP/151 weight gr. 1.48 1 M aq. Na_2CO_3 (pH10/11),
  EXP/153 weight gr. 2.02 1 M aq. Na_2CO_3 + excess $CaCO_3$ (pH11);
- one black fine Brandon flint bladelet very similar to **EXP/147** but very different from ISR/2 flint was polished by damp hide with ochre in a discrete area;
  EXP/149 weight gr. 0.7987 in 1 M aq. Na_2CO_3 (pH 11).

Macroscopic and microscopic observations

EXP/151 ISR/2 flint

After 5 weeks, macroscopic soil sheen developed on the dorsal central ridge but not on the ventral aspect. Microscopically the surface appeared polished as on archaeological artefacts with slight PDSM (figures 184 and 185). The polished area was not affected. The piece was in the solution for two months. On the ventral aspect microscopic sheen but no macroscopic sheen developed (figure 183).

EXP/153 ISR/2 flint

Macroscopically no sheen developed after 5 weeks but the brown flint became greyer in colour. Microscopically soil sheen developed on the dorsal aspect. There was an increase in the blue/grey crystals which were present on the piece before the experiment (figures 186 and 187) and whose proliferation preceded the onset of patination in the previous experiment (cf. section 6.5.11 (a)). After two months in solution there was a slight smoothing of the flint surface.

EXP/149 Brandon flint

After 9 weeks macroscopic sheen started to form on the ventral aspect, spreading to the dorsal after 12 weeks. Microscopically the surface was beginning to appear "polished", particularly on the edges where there was no experimental polish. The left lateral edge up to the dorsal spine was eventually as glossy macroscopically as the right lateral which was polished by hide with ochre. This new polish was very similar to PDSM on archaeological artefacts (figure 188). The experimental polish, which was very well developed, was affected by glare and it too appeared microscopically altered (figure 189 and 190). This glare was seen to precede the onset of patination in the previous experiment section 6.5.11 (a).

Discussion and conclusions

This experiment seems to suggest that Na2CO3 does alter flint but at a much slower rate than sodium hydroxide. This was noted by Schmalz (1960) who, inexplicably however, obtained patina after 120 hours. In this experiment, as well as in all the experiments where a comparison between different flint types was made, Brandon flint seemed to be more vulnerable to sheen than the coarser ISR/2 flint. This is compatible with the fact that use polish also forms faster on the finer flint than on coarser varieties. It strongly suggests that sheen is a smoothing of the flint surface by whatever means.

6.5.12 Sheen and Patination with Distilled Water, Heat and Calcium Carbonate

Aims

The aims of this experiment were to observe the development of soil sheen as described by Rottlaender (1975:107) who produced "glossy patinas" by embedding flint in ground marble, adding water and heating to 80° C for three weeks. The same procedure was followed here except for the following:

- the substitution of $CaCO_3$ for the marble;
- the fluctuation of the temperature between 80
- and 90° C, as the equipment available to me could not be maintained at 80° C precisely.

The piece was observed periodically. The solution was regularly topped up with distilled water to replace that lost by evaporation. The distilled water available had a pH of 9.

EXP/29

A Brandon flint microdenticulate, macroscopically not homogeneous but with many grey medium coarse areas (weight 4.25 gr.), was embedded in a layer of $CaCO_3$ and covered with a buffer solution of pH 9 in a glass beaker and heated to 80–90° C.

The experiment proceeded in several stages which will be numbered with Roman numerals below. Initially one implement (EXP/29) was involved then two others were added (**EXP/154** and **EXP/155**). Subsequently, **EXP/29** was removed when it was covered with sheen, then **EXP/155** was removed when covered partly with sheen and partly with a blue patina veil. The last to be removed was **EXP/154**. It seemed therefore clearer to describe the observations in the sequence in which they occurred.

Stage I

Macroscopic and microscopic observations

After three weeks no macroscopic soil sheen was visible on **EXP/29** but microscopically a shine started to appear on the edges, the high points of the microtopography and the dorsal ridges. A few days later and particularly after one month, the shine was distinctly like polish: better developed, concentrated on the edges and decreasing inland (figure 192). In view of the necessity of using a glass beaker due to the high temperatures involved, each time the microdenticulate was examined, it was very carefully replaced by being lowered with plastic tongs in the beaker and allowed to lie with the dorsal aspect in contact with the layer of calcium carbonate rather than the glass. It initially seemed possible that the polish might be caused by friction with the calcium carbonate. When the microscopic "polish" developed on the edges of the ventral aspect as well, it was more obviously due to the effect of the solution (figure 193). Nonetheless, to avoid the doubt that the polish was caused by contact of the flint with either the calcium carbonate or the glass beaker, two new pieces were introduced in the solution:

EXP/154 ISR/1 flint unretouched blade weight 1.1O85 gr.

EXP/155 BRANDON flint retouched flake weight 4.26 gr.

Both the new artefacts were mechanically polished in one area by damp deer hide. The artefacts were hung from a string suspended in the solution in a large 5 litre beaker without touching the glass, each other or the calcium carbonate. EXP/29 was also hung in the same way so that from that point onwards there was no contact with calcium carbonate or with the glass.

Stage II

Macroscopic and microscopic observations

After a total of one month in solution and six days since being hung, **EXP/29** showed a further increase in the "polish" development and distribution. The "polish" was now quite similar microscopically to PDSM on artefacts affected by soil sheen (figures 194; 191; 195 and 54). The macroscopic sheen had increased on the dorsal aspect and developed on the ventral. In the early stages of sheen development (from 3 weeks onwards) it was mainly on the edges, gradually decreasing inland like use-wear (figures 198 and 199). Even when the tool became macroscopically shiny all over and microscopically there was "polish" inland, the edges had more developed "polish", as would be the case in use-wear and as it was on the Hengistbury Head artefacts (figures 200 and 201). After 50 days in total the artefact was removed as it was feared patination might follow, having started to occur on **EXP/155** (see below). At this stage the flint surface seemed much more irregular, like etched flint, with fissures seeming deeper and rounder. There was much glare and it was difficult to focus even at x100.

Stage III

Macroscopic and microscopic observations

EXP/155 Macroscopic sheen developed after 25 days but simultaneously an initial veil of blue patina appeared both macroscopically and microscopically particularly on the dorsal aspect. A cluster of bright spots appeared on the ventral aspect immediately adjacent to the edge (figure 202). Further away near the bulb of percussion an area of macroscopic sheen was visible which microscopically has a distribution on the microtopography, near the edge very similar to use-wear (figure 203) and in the other areas appearing as flat striated patches like bright spots (figure 204). The polish produced by hide seemed intact at x 100 (figure 205) but at x200 there was severe glare which interfered with focus and observation. The artefact was removed to avoid more severe patination developing.

EXP/154 After 28 days glare began to develop on the unpolished areas of the flint surface. It made observations at higher magnifications very difficult. Even after 2 months the polish produced by the damp hide seemed still intact. The microscopic glare, however, was increased seriously interfering with microscopy and photography at x200 (figure 196 and 197). The appearance of the flint was of discrete high points with hollows in between. No macroscopic sheen developed but microscopically some "polish" appeared on previously unpolished areas (figure 206). Time did not allow continuation of the experiment any further.

Discussion and conclusions

EXP/155 of fine Brandon flint started to patinate rather quickly (25 days). Rottlaender (1975:109) suggests that an attempt to speed up the reaction for development of soil sheen "results in a white patination." Bright spots also appeared at the same time on this artefact. **EXP/29** made of a less homogeneous Brandon flint had not started to patinate even after 50 days. It did, however, develop diffuse sheen but no bright spots. **EXP/154** on the other hand, after two months, had only started to develop some microscopic sheen and non-use "polish" but no macroscopic gloss.

Although no marked weight loss occurred on the artefacts in these experiments, soil sheen, bright spots and patination seem to be surface modifications which are the result of silica dissolution from the flint surface as suggested by Rottlaender (*ibid.*).

Alkaline conditions, heat and water seem to be necessary for this process to occur in laboratory conditions within a relatively short time. Flint structures seem to influence the speed at which the process occurs and probably the type of alteration which will occur. Of the two Brandon artefacts, the coarser **EXP/29** developed pronounced sheen, the much more homogeneous **EXP/155** developed light patina with some bright spots and sheen at the same time and **EXP/154** was slowest to develop any of the alterations. It is made of ISR/1 flint which is macroscopically very homogeneous but microscopically very different from Brandon flint. It might have also a different diagenesis and resulting microstructure which could be responsible for the difference in results.

It is interesting that in all three cases polish produced mechanically by buffing with hide seemed to survive the development of soil sheen on the rest of the artefact. Although there was only a small band of polish restricted to the edge, it was nonetheless a very well developed C polish with very few areas of unpolished flint which can be readily attacked by the solution. Some small scale dissolution did occur, however, as is evidenced by the glare at x 200. Glare may be due to the fact that in these unpolished spaces interstitial chalcedony is preferentially dissolving leaving relatively isolated lepispheres prominent on the surface. Areas further inland from the intentionally polished edge appeared "polished" as a result of the experiment. It is possible that they were being smoothed by some of the dissolved silica being redeposited in their interstices as suggested by Rottlaender, whose explanation is worthy of being reported in full:

> [Glossy patination] occurs, if the concentration of the chemical agents are too weak to penetrate into the holes and the fissures of the flint. So they just remove the prominences of the flaked surfaces, because of their higher potential energy. Since the fissures and holes have a significantly lower potential energy than the normal surface, they act as pitfalls for silicic acid occurring free in solution. By this mechanism the surface is smoothed and polished. (1976:56).

It may at first appear that there is a contradiction in the observed irregularity of the flint surface in the presence of glare and the hypothesised smoothing of the surface which results in "sheen" or "glossy patina". The same process may be resulting in two separate phenomena:

1. If gloss is the result of a dissolution of silica asperities from the flint surface, some of these particles must be removed from the edges of fissures and hollows giving the surface a greater disparity between the high and low points. This is making it more difficult to bring the whole microscopic area being observed into focus at once (figure 167).
2. When deposition of the dissolved silica occurs it will contribute to a smoothing of the surface on a scale not visible at these magnifications (figure 210 and 211).

Rottlaender, who uses the expression "glossy patina" to describe what microwear analysts refer to as "soil sheen", explains why artefacts with soil sheen rarely patinate. This explanation may also apply to the reason that well developed polish degrades very slowly as noticed in this series of experiments. In fact he says (1975:109),

> ...it is so difficult to produce this sort of patination in a laboratory experiment. The energy differences are

too low for it to be a rapid process, and any increase in the activity of the solvent results in a white patination...So it becomes clear why a glossy surface once formed is able to protect the flint to a certain degree: a vigorously acting solvent finds no open fissures by which to penetrate into the flint.

This is also a probable explanation for the phenomenon of glossy sickle blades surviving on all types of sites where other artefacts are either patinated or have surfaces affected by PDSM and showing no surviving use-wear polish. It may also suggest a similarity between the end result of the polishing processes whether the mechanism be chemical or frictional, as both glossy artefacts called sickle blades and those covered with well developed soil sheen rarely develop patina.

6.5.13 Effects of Post-depositional Processes on Highly Developed Polish

Aims

To understand the effect of possible post-depositional disturbances on very well developed and macroscopically glossy polishes it was decided to submit an artefact with such a polish to scratching, tumbling and chemical attack.

Experiments

EXP/103, a blade of ISR/4a flint, was polished mechanically (see section 6.4.8) with ¼µ diamond grit until it was macroscopically as glossy as a "sickle" blade. It was then subjected to the following processes:

1. Three scratches were made by another flint flake across the polished and unpolished surfaces: two dry scratches and one with water.
2. The blade was tumbled for 36 hours in wet sediment from an excavation. In the 8 cm diameter plastic drum one small pebble was introduced with the blade. The drum revolved at approximately 60 rpm for 36 hours.
3. The scratched and tumbled artefact was then immersed in a 20% wt. aq. solution of sodium hydroxide for five days and heated to between 50 and 60° C.

The tool was examined and photographed at several intervals during these experiments.

Macroscopic and microscopic observations

Observing the resulting traces microscopically before and after tumbling it was noted that :

Experiment 1

- a A real scratch was not easily produced by one flint implement on another **unpolished** flint without a great deal of pressure. The traces which were left were smears (figure 207) which were removed during tumbling.
- b When great pressure was brought to bear in scratching in the presence of water, the **unpolished** part of the blade was grooved (figure 208) and the edges of the grooves, being raised, were polished during tumbling (figure 209).
- c On the highly polished flint, on the other hand, the dry scratch was easily formed as if the surface had become more brittle (figure 212). During tumbling the scratch became wider (figure 213) and more **unpolished** flint was revealed.
- d With water it was slightly easier to scratch both the unpolished and the polished flint. The scratch made on the polish with water had neater edges and on certain areas it seemed to have left a depression (figures 214 and 219) without breaking the polish.

Experiment 2

As a result of tumbling, edge damage appeared on the artefact but no striations scratched the polished surface which was not altered in any other way.

Experiment 3

Immersion in a heated sodium hydroxide solution for a prolonged period did not patinate the flint implement either on the polished or on the unpolished surfaces.

Discussion and conclusions

Figures 212; 213; 215 suggest that the polished area seems to have a very thin, discrete surface layer (see also figure 220) that is brittle and may be detached from the underlying flint by scratching and/or contact with sand and gravel.

Many archaeologists and use-wear analysts have noted empirically that "sickle blades" survive with their gloss almost intact in many sites where all the other pieces are either patinated or have other heavy surface modifications. It was therefore at first surprising that this glossy polish which seems to protect the flint surface from chemical attack during burial (and was not affected by 36 hours of movement in the sediment) could be brittle. A possible explanation seems to be that while very well developed polish can be expected to shield the flint from chemical agents, it is vulnerable to scratching by a material as hard as flint. The hardness and softness of materials is only important in as much as one scratches one material with another.

The dry scratches were observed by SEM (figures 218; 68; 69) which showed that the process involved is the same as that suggested for glass polishing. The raised edges produced by scratching flint with flint are a result of plastic deformation of the surface. The subsequent flattening of their edges in the sediment is a smoothing process. A more detailed explanation of this process is given in section 6.2, where its importance in understanding processes of polish formation is discussed.

These experiments also suggest those striations which are grooves in a very well developed polish (figure 64) are mostly formed during use rather than by post-depositional processes. The streak of polish produced by two dry flint implements briefly streaking each other (figure 207) was removed by movement in the sediment. This suggests that it may have been a silica residue rather than polish which is a smoothing usually increased by further tumbling (see 6.5.2).

It was noticed in these experiments that water made the scratching of flint easier, that the scratches had smoother edges and in certain cases a depression rather than a scratch was formed. Scratches are a plastic deformation in which the surface has been ruptured. It is suggested that water facilitates the adherence of the two sliding surfaces against each other and eases the smoothing of the surface yielding the smooth edges. In the case of the depressions it seems that the water has promoted the formation of plastically deformed areas which did not rupture the surface.

It is unfortunately not possible to draw any conclusions from this experiment regarding the resistance of macroscopically glossy polish to chemical attack, as no patination occurred on any of the unpolished parts either. The flint used in this experiment (ISR/4a) is the same as the one involved in section 6.5.9 where patination was also attempted in vain. In this experiment the desired result was not achieved in spite of a longer period of immersion in a solution heated at a slightly higher temperature. It may be a function of the flint type (see also **EXP/154** in section 6.5.12) but it may be also due to the lack of a catalyst such as calcium carbonate which speeded up the patinations in 6.5.9 and 6.5.11 [a]. Although this might suggest that cleaning flint artefacts with sodium hydroxide for brief periods is safe, I would recommend otherwise until further work has been done to ascertain which flint types are more or less vulnerable.

6.6 Limitations Of Research

6.6.1 Quantification

These experiments are not quantitative in their measure of change. Like all microwear experiments they rely on the human eye and what it perceives through the microscope, together with human memory, and any notes and photographs that may be taken. The perception of change or the lack of it over a series of observations separated in time, is inevitably subjective. The only "documentation" possible that is in any way objective is through the photographs taken at each successive observation. Photographs, however, are rarely strictly comparable because the slightest difference in orientation, focus, illumination, exposure, developing or printing conditions can make a picture show more or less of a certain feature. Furthermore the limited depth of field in the optical microscope is responsible for the fact that small differences in the orientation of the piece or plane of focus can affect the appearance of a feature and small movement of the focusing knob can show "features" which can mislead the human eye. Hence extreme time, care and expertise are required to "document" a spot on a tool under the **same conditions in successive observations** and even then it may not always be possible.

6.6.2 Edge Damage

In spite of the extreme care taken by the experimenter, a fraction of a millimetre can be lost from the edge of a stone tool during handling. This is especially true of post-depositional experiments in which an implement is introduced into a mass of sediment and/or decaying matter and observed microscopically over time, thus involving repeated removals and cleaning. This minute microscopic edge damage is only noticed because of the photographic documentation. It means however that:

a) when investigating polishes, the small area of polish that was being monitored may be lost totally or partially;
b) documentation of the exact spot through time becomes more difficult because of the loss of some "landmarks" on the microscopic edge morphology;
c) that the transfer of information about use-wear from experimental pieces to archaeological implements is problematic, as the latter are bound to have lost some of their original edge during burial.

6.6.3 Duration of Experiments

During the repeated removals of artefacts from the soil (in the case of post-depositional experiments of naturally occurring processes) one may be interrupting a process that is under way and would, if allowed to develop, have an effect on the flint surface. This has been suggested to me by the fact that some organic particles from the decaying matter are stubbornly adhering to the flint requiring prolonged soaking in biological detergent and ultrasonic cleaning and even 10% HCl before they are removed (see section 6.5.8 and figure 166). To avoid this pitfall one could, ideally, have a large number of artefacts involved in each experiment and remove them successively for periodic microscopy while leaving the others undisturbed. This strategy presents two obstacles:

a) an increase in the number of artefacts in each experiment is much more time-consuming;
b) estimating the number of pieces required. The longer the experiment lasts the larger the number of pieces required as it is not possible to know in advance the length of time each experiment will run. Even experiments modelled on other people's work have often taken much longer than anticipated (cf. 6.5.11 and 6.5.12).

6.6.4 Pressure

It is certain that the load factor is important when monitoring the variables involved in polish formation on flint surfaces. We have however not been able to devise experiments to **measure** the load factor objectively. The experimenter's hand pressure is not readily measurable or constant and attempts to use the same hand pressure are inevitably limited in success.

6.6.5 Rigid materials

Experiments described in section 6.4 have shown that polish as developed and as macroscopically glossy as so-called "sickle gloss" can be produced by a variety of materials such as diamond paste, on a soft backing, damp hide with ochre, and just wet wool. These are all yielding materials even though their abrasive qualities are very different. it would have been interesting to find a way to produce this kind of total linkage of polish with rigid materials such as wood, bone and/or antler but this proved elusive. It may be that more prolonged polishing than used here might yield interesting results. Polishing flint artefacts on a rigid turntable is not satisfactory, however, in that two rigid surfaces have only limited areas of real contact. This is rendered more difficult when one of the materials has a topography as irregular as flint. "Putting two solids together is rather like turning Switzerland upside down and standing it on Austria - the same area of intimate contact will be small" (Dowson 1979:424; 495).

6.6.6 Use of the SEM

The use of an SEM in this research was undoubtedly of great help and allowed insight into processes which could not have been achieved by optical microscopy alone. The need to coat flint artefacts with a layer of conductive material in order to obtain a satisfactory image was however a hindrance due to the irreversible nature of the coating (see Chapter 3). It was of particular hindrance in the case of experiments in which repeated observations over a period of time are necessary in order to observe a process at work. In experiments such as those described in sections 6.5.6 to 6.5.8 the only alteration observed optically was glare. Observation by SEM in an attempt to ascertain whether or not silica dissolution might be occurring progressively would have been very valuable. Once coated for SEM observation, however, one would no longer be able to continue the experiment.

6.6.7 Team Work

This kind of work is best done a team of experimenters as in the case of Plisson and Mauger (1988) and others. This approach cuts the length of time required, diminishes the subjectivity and allows for a higher number of artefacts per experiment.

6.6.8 Materials Science

Expertise in the materials sciences on the part of the functional analyst is probably a valuable asset when investigating surface modification, polishing and the like. As this researcher did not possess such expertise this thesis was closely supervised by Dr. D.R. Griffiths to whom I acknowledge my gratitude and scientific indebtedness.

CHAPTER 7
CONCLUSIONS

In Chapter 6 the revised aims of this research programme were described in detail and can be summarised here as:

a) the study of experimentally produced wear traces on flint surfaces in order to assess the relative importance of each of the variables considered significant;
b) the replication of post-depositional processes in order to observe their effect on the surfaces of flint artefacts.

At the end of each set of experiments a discussion and interpretation of the results has been given (sections 6.4 and 6.5). It is therefore proposed here to give only an overview of what has been observed and some concluding remarks on each of the aspects of the dual problem.

7.1 PROCESSES OF POLISH FORMATION

7.1.1 Previous Models

A mechanism has been proposed by Anderson-Gerfaud (1980:184) that polish is caused by flash temperatures occurring at localised points of contact between the flint and the worked material causing a molten or semi-molten layer to occur on the surface of the tool and that then solidifies as silica gel. Following the experiments carried out in this project there are several objections to be made to the mechanism as it is described above:

A For a surface layer of silica gel to form, silica from the flint surface would very probably have to dissolve **in water** and be redeposited as a gel. There is insufficient water in the flint structure to provide that necessary in a gel structure. In many of the experiments carried out (sections 6.4.5; 6.4.6; 6.4.8) no water whatsoever was used nor any other source of hydroxyl groups yet polish developed as macroscopically glossy and microscopically smooth as "sickle gloss". It thus seems unlikely that polished flint surfaces on flint artefacts are comprised of a silica gel layer.

B If high flash temperatures were to develop any silica gel present on the surface of the flint at that time and place would decompose due to the evaporation of the water in its composition. Consequently the idea of high flash temperatures as the mechanism of polish formation and the idea of the polished surface comprising a substantial layer of silica gel are mutually exclusive (Dr. D.R. Griffiths pers. comm.).

C High flash temperatures can only occur at the points of intimate contact between the flint and the worked material. Had such temperatures reached the melting or softening point for flint (1500 degrees C.) then any organic worked material would be charred or singed since a similar amount of heat would be dissipated in the worked material as that supposed to melt the flint surface. This is particularly true in the case of implements whose whole edges are highly polished. This would have meant **repeated** flash temperatures eventually having occurred at all points on the surface to give an even polish.

We have, however, no evidence of burnt materials in any of the hand experiments including the experiment where hide was scraped with ochre and macroscopic polish developed on several tools. Temperature rise was noticed only on implements polished mechanically where a high speed buff was used with or without diamond grit. Yet there were no charred substrates in any of the mechanical polishing experiments, nor any smell of burning. While these observations do not preclude the occasional occurrence of high flash temperatures it seems implausible that the total heat generated by such events would be such as to melt the surface of the flint.

7.1.2 The Proposed Model

The experiments described in this thesis have yielded evidence for a smoothing of the flint surface through a combination of plastic deformation and removal of asperities. These processes appear to be taking place during work on various materials.

In addition to these smoothing processes material may be removed from the flint surface by scratching. Scratching of flint with a diamond point has shown (figures 68 and 69) a lateral chip formation as occurs in glass polishing (Malkin *et al.* 1979: Figure 9 [b]) suggesting the similarity between these two processes. When scratching flint with flint, however, a smoother scratch occurred (figure SEM 218) although cracking of the surface was still present. This has also been observed in glass polishing where Kirchner *et al.* (1979:27) have noticed that at lower loads smoother tracks are formed. Most worked materials are not, however, as hard as flint and are not likely to cause such a large tear on the flint to occur. The reduction of friction in the presence of a liquid medium may be helping to avoid the trails of drag-induced cracks and subsequent fracture and the plucking out that might occur under dry friction. The contrast between wet and dry abrasion is illustrated by scratching a polished flint artefact with another flint (section 6.5.13 figure 218 and 219). Contrary to figure 218 which shows that the polished flint surface was broken in the absence of water, in figure 219 the tracks left by the flint in the presence of water are smooth and no tear in the surface is visible.

The asperities removed from the flint surface being hard quartz crystals, affect the flint surface by inducing a plastic deformation at low loads. The importance of the flint asperities in the formation of polished surfaces on flint was observed in the experiments described in section 6.4.10 in which the flint crystals were found embedded in the hide used to buff the flint in almost dust-free conditions and without any contact between the hide and the edges of the flint tool (figures 100 and 101). The removal of the asperities from the flint surface has also been documented by SEM on polished experimental artefacts (figures 66 and 70) and on a Neolithic sickle blade (figure 221). The presence of a liquid medium may act to retain the abrasive flint particles near the point of contact between tool and workpiece so that they form a slurry which promotes wear by plastic deformation even while reducing friction. In the absence of a fluid, on the other hand, these particles may be free to leave the region of flint to flint contact and smoothing in the form of use-wear polish is noticeably slower to develop. An exception can occur when the dislodged particles are trapped by the worked material which acts as a flint abrader in which case polish will develop faster even in the dry situation. Experimentally the trapping of flint particles in a worked material has only been observed by SEM on damp hide (section 6.4.10).

7.1.3 The Role of Water

The role that water may be playing in polish formation has until now been observed, primarily in natural circumstances, due to its presence in various materials (e.g. the water contents of grasses and other plants and the dampening of hides). In the experiments described in the preceding Chapter, water has been shown not to be essential to the formation of very well developed polish since such polish was produced either in the presence of non-polar liquids or in the total absence of liquids. Water (as opposed to other fluids) does nonetheless promote the formation of polish by softening the worked material and making contact and adherence between the two surfaces more intimate (see section 6.4.6 wood grooving experiments).

The softened materials may also retain removed flint asperities more readily than the hard dry ones. A certain small amount of hydration of the flint surface may be taking place in the presence of water but this does not alter the appearance of the quartz crystals suggesting that no silica gel is formed.

7.1.4. The Role of Abrasives

Polishing on glass has been seen to follow a general model comprising the formation of linear plastic depressions that are fine enough to appear smooth, the greater deformation experienced by the high points. As a consequence of the stresses produced by the penetration of polishing abrasives, pre-existing microcracks will develop into lateral cracks (Evans 1979). Cornish (1961) has noted that in polishing glass, added abrasives act as a slurry enhancing the removal of asperities from glass and contributing to the smoothing of the surface. This has also been observed in the experiments scraping hide with ochre (section 6.4.7) in which the finer ochre grain sizes produced the best developed polish under the optical microscope which corresponds to the smoother surface by SEM (figure 71).

In the wood-grooving experiments (section 6.4.6) the addition of talcum powder as a soft abrasive produced a polish which seemed to have affected both the high and low points of the microtopography as had occurred in the trampling experiment (section 6.5.4 cf. figures 61 and 151). This underlines the importance of the role abrasives play in the development of polish on flint artefacts.

7.1.5 The Role of the Flint Structure

Implements from same flint type have reacted differently to the same experimental conditions. This was observed in the series of experiments on polish formation where invariably the finer flint developed polish faster and in a more linked stage of development for the same length of time than the coarser varieties (see sections 6.4.6 and 6.4.7). As polish seems to be a smoothing of the flint surface it follows that there is less to level off when the particles are smaller and more closely packed than in the opposite circumstance. In the experiments for post-depositional processes the same was observed as seen in section 6.5.2. Each flint flake is sometimes different from the one flaked before it or after it even when they are removed from the same core (see section 6.5.2 (a) EXP/12 and EXP/27). Both in experiments for chemical and for mechanical PDSM, the coarser parts were less affected than the finer grained ones suggesting a similarity in these apparently different processes which affect flint surfaces. It also suggests that the processes involved in polish formation and post-depositional surface modifications such as sheen and bright spots are similar.

7.1.6 Striations

The experiments carried out in this work suggest that striations (in the sense of grooves in the polished surface as opposed to linear features see Glossary and figures 64 and 65) are formed principally by microchips removed from the used edge during work. This is suggested in view of the absence of striations where very high angled and retouched edges are used.

In the hide scraping experiments in section 6.4.7 striations formed even on edges which were high angled naturally or through retouch. This may be due to removed asperities from the flint surface (rather than microchips from the edge) being trapped in the hide which deforms under pressure of the tool and brings the trapped asperities into contact with the flint surface. It is possible that ochre particles will be blended in the slurry formed by the surface asperities thus increasing abrasion and causing a higher quantity of striations to form.

Under the SEM the striations that Mansur-Franchomme (1983) aptly described as "fern-like" formed on EXP/127 used on hide with ochre as described in section 6.4.7. These striations (figure 67) illustrate the process by which they are formed namely the removal of surface particles which are dragged across the surface and the consequent smoothing of the area on which they are dragged. The fern-like appearance may be due to cracks formed on either side of the main track by drag.

Some features which were perceived as straight striations optically were seen under the SEM to have been the result of two different processes. In figure 64 it is clear that a groove in the polished surface was formed by the removal of one or more crystals. In figure 65 the straight "striation" observed optically was seen as a rather big surface asperity which instead of being removed has been smoothed down. It is a property of the flint microstructure that some asperities have particular shapes (e.g. cigar shaped rather than spherical) which might be misleading. In this case, in fact, the direction of the "striation" was not in the direction of use.

In another instance a feature very similar optically to a comet tail pit on a Neolithic sickle blade, revealed itself by SEM to be a cigar-shaped natural hollow in the flint surface whose edges had been polished during use (figures 223 and 224).

Comet tail pits can be seen on two separate artefacts buffed by dust-free hide. They appear to owe their shape to the asperities from the flint surface being plucked from one spot forming the head and then dragged against the surface forming the tail. As no abrasives were used in this experiment in which the hide was kept as free of ambient dust as possible (section 6.4.10), it is suggested that this illustrates one of the mechanisms of polish formation. In figure 70 the smoothed surface of the tail is quite apparent suggesting that the asperities removed from the head of the pit had rubbed against the tail area flattening it. Figure 66 is at a much higher magnification clearly showing where the crystal(s) have been removed leaving behind the unpolished flint which forms the head of the pit. In spite of the higher magnification the tail area does not seem to have been smoothed as much as in figure 70. This may be due to the fact that there was a difference in the pressure when the removed asperities had rubbed against the area of the tail.

7.1.7 Polish from Multiple Uses

Vaughan (1985:40) and others mention that they recognise two use-wear "polishes" superimposed on each other on the same edge. This seems to contrast with my observations that polish from working soft materials has always been polished over by harder materials. In the particular case of artefacts from the experiments in section 6.4.7, the polish produced by hard hammer retouch (figure 94) was totally repolished by use on hide with ochre. The quartzite hammer is a very hard material leaving a very well linked polish but the use on hide smoothed the area in such a way that the hard hammer traces were no longer visible. The same occurred in the experiments described in section 6.5.2 [a] where polish from cutting bracken and incising birch were polished over by the gravelly sand (figures 30 and 133). As polish seems to be a smoothing of the flint surface the only way of distinguishing between one polishing episode and another would be through the distribution of the polish on the edge. When the two distribution patterns are superimposed, it is generally no longer possible to distinguish one from the other.

7.1.8 Residues

The experimental programme has shown that residues which are only "deposited" (see Glossary) on the surface of the flint are unlikely to survive archaeologically since they can be washed off with thorough cleaning. When they were "adhering" to the surface more drastic steps such as movement in sand or immersion in warm NaOH w/v 20 to 30 % as advocated by Keeley were needed to remove them. The experiments in sections 6.5.1 and 6.5.2 have shown that the more stubborn residues still "adhering" to the surface of the flint even after cleaning with sodium hydroxide were removed by movement in the sediment. It is suggested that under very exceptional conditions where organic matter survives such residues may also be preserved. No recognisable residues have been observed "embedded" **in** or **under** the polish produced by any of the worked materials whether by optical microscopy or by the SEM. The bar on micrographs taken by SEM act as a scale for what is being observed. From the micrographs of Neolithic sickle blades (figure 216, 220 and 222) and experimental blades polished without water to a macroscopic gloss (EXP/103 figure 217) one can measure that the deformed layer in each case is a well under a micron. This more or less excludes the possibility that phytoliths from plants or other large particles from meat, or hides can be "trapped" in this layer.

7.1.9 Use-Wear Polish as Diagnostic of Worked Material

The arguments for and against use-wear polish being distinctive of a specific worked material have been discussed at length in Chapters 1 and 3. The experience of several microwear analysts, inside and outside our Institute, has already suggested that the characteristics which were supposed to be diagnostic of one worked material only were also being observed when other materials were being worked. The experiments described in Chapter 6 both on use-wear and on polish formation are, I hope, convincing enough to clearly demonstrate that there is no direct correlation between polish and the worked material. The degree of polish development is dependent on several variables: the hardness of the worked material, the presence or absence of abrasives, the presence or absence of a fluid medium, the motion, angle and duration of work.

Blades macroscopically as glossy as Neolithic "sickle blades" with polish appearing by optical microscopy identical to the one observed on the Neolithic implements, have been produced by polishing the flint surface with 1/4 micron diamond grit. Examination by SEM of the experimentally produced "sickle blades" and the Neolithic ones showed that the surfaces of both appear extremely smoothed (figures 216, 217 and 222). There are two visible differences between the two surfaces:

a) the direction of the striations which in the Neolithic sickles are all parallel to the edge indicating that they were used in a longitudinal motion such as cutting and/or reaping while the mechanically polished artefact does not show evidence for motion in a repeated direction;

b) the mechanically polished surface appears smoother than the archaeological one. This is most probably due to the fact that the sickle may have been used on grasses or reeds in which the silica particles are of a size in the order of at least several microns while the experimental blade was polished with a much smaller particle size (¼μ grit) in a hydrocarbon oil base.

Comet tail pits were supposed to be diagnostic of use on plant material as they were first observed on archaeological sickle blades by Witthoft (1967:384) who called them "comet shaped depressions" and went on to state that "Most of the gloss is added to the flint rather than derived from it." (*ibid.*:385).

Keeley (1980:60–61) has used them as a diagnostic feature of use on plants and as a distinguishing feature between "wood polish" and "plant polish".

In this work comet tail pits have been observed on experimental implements used to scrape hide with ochre on which macroscopic polish developed. They were also observed on the blades mechanically polished with 1/4 micron diamond grit. None of these implements had come in contact with silica which could be deposited on the surface.

I suggest that comet tail pits occur whenever flint is being polished by whatever material since they are the result of removal of one or more crystals and their subsequent movement along the flint surface (see also sections 7.1.6 and 7.2.7 and figures 64, 66, 70). They are more noticeable when the surrounding area is well polished and smooth. In an area of little polish development their shape would be unnoticed among the various cracks and crevices of unpolished flint. Some of these crevices might be caused when a flake slides against the core and one or more surficial crystals are removed in much the same way as occurs during use (see also Newcomer 1976). On artefacts used until macroscopic polish has developed the direction of the comet tail pits is indicative of the tool's movement against the worked material.

7.2 Post-depositional surface modification

7.2.1 Mechanical Soil Processes

Weakly developed use-wear polish such as that produced by cutting bracken, and probably by cutting meat and scraping fresh hide, can disappear after movement in wet gravelly sediment in a gentle tumble polisher. The surface of the flint is probably re-polished by the matrix, causing an increase in the extent and distribution of the polished areas.

Well developed polish such as that produced by cutting reeds and rushes is not affected in the same way. It is only slightly dulled. If polish is the result of smoothing of the flint surface by removal of all micro asperities and perhaps by infilling the micro depressions (whether by plastic flow, or solution and precipitation, or a combination of both), no more smoothing or polishing can be done by the sediment, but some of its more abrasive agents may produce sub-optical fine scratches which would be responsible for the duller appearance of tumbled polish.

Experiments in section 6.5.2(a) suggest that water as a fluid enhanced the development of mechanically produced soil sheen (EXP/12 figure 133). The resulting sheen polished over the previous use-wear polish so that a more linked distribution was observed. The directional distribution of the sheen resulting from tumble polishing was contrary to that of the use-wear polish. This is presumably due to the effect of the tumbling motion while the direction of polish would tend to be more random in the case of movement in the sediment in post-depositional circumstances.

There is no doubt that movement of artefacts on archaeological sites is quite common. Movement is caused not only by periglacial soil processes, including cryoturbation, gelifluction, soil creep or other disturbances which leave stratigraphic evidence, but other movements occur which are quite minor and are therefore stratigraphically invisible. They are probably caused by trampling, bioturbation, and the settling of the sediment under pressure. Only now that refitting of artefacts is practised more frequently has it become obvious how common this phenomenon is. The question is whether such non-violent movement over thousands of years can affect flint surfaces to cause **PDSM**. The relatively gentle experiments in section 6.5 1 and 6.5.2 have produced sheen and affected flint surfaces which suggests that mechanically induced PDSM does not require violent processes in order to occur although this does not preclude PDSM being primarily chemical in a given natural situation (see section 7.2.4 below).

7.2.2 Bright Spots from Mechanical Processes

Flat bright spots, very similar to those observed on archaeological artefacts, can be produced by friction of flint on flint with water as a medium. They are easier to produce on edges and ridges which are more vulnerable and where more pressure can be exerted i.e. a given force can be brought to bear over a smaller area. They occur only very randomly during tumble polishing where little pressure is exerted. Their appearance and distribution is affected by the microtopography of the flint surface. In the soil, flat bright spots could occur during a brief but intense movement while the implement is rubbing against another implement or stone in wet conditions. Flat bright spots **with striations** could be diagnostic of trampling, movement in the sediment or similar post-depositional factors. Any polish present on the edges of artefacts with these spots should be viewed with suspicion, and only attributed to use after careful consideration. Stapert (1976:38) refused to accept the hypothesis that bright spots could be the result of friction "between stones in the soil since the latter process has been in operation for more often on Middle Palaeolithic than on Upper Palaeolithic or later implements". These experiments would however suggest that flat bright spots with striations are related to the particular taphonomic circumstances occurring during burial, rather than being directly related to the duration of burial.

7.2.3 Bright Spots from Chemical Processes

In the experiments described in section 6.5.12, two types of bright spots developed (figure 202 and 204). The flat variety developed without the striations observed on mechanically produced bright spots. It formed on a flat surface near the edge though not **on** it while the bright spots having a distribution more similar to use-wear polish developed on the more rounded areas of the artefact near the bulb of percussion. The same was observed on the Hengistbury Head scraper discussed in section 4.1 figure 47. No obvious explanation for this phenomenon has yet been established though one is naturally led to suspect that the distribution of these bright spots on the microtopography may be influenced by some aspect of the constitution of the flint at and around their place of occurrence. It is important to note however that the same process (distilled water, heat, and calcium carbonate in solution) produced both the bright spots, the sheen, and the incipient patination. The presence of striations in bright spots caused by mechanical friction and their absence in striations formed by chemical processes deserves further investigation. If a direct correlation could be demonstrated, the presence/absence of striations could be a good indicator of the chemical or mechanical predominance of the PDSM mechanism. Non-striated bright spots would be indicative of dissolution suggesting that other polish present on the artefact might also have been formed in the same way (see section 6.5.12).

7.2.4 Sheen from Mechanical Processes

Macroscopic sheen appears only after prolonged movement of the artefacts in the sediment and it develops faster when water and/or gravel are present. Coarse flint develops sheen more slowly than fine flint (cf. **EXP/E5** section 6.5.2). Shorter tumbling times did not develop as much soil sheen as is observed on certain archaeological implements. Hence chemical processes either alone or in conjunction with mechanical ones may be producing sheen on archaeological implements even after relatively brief movement.

7.2.5 Sheen from Chemical Processes

It was observed (see section 6.5.11b etc.) that sheen, whether produced by chemical or mechanical processes, seems to occur earlier on Brandon than on the coarser Israeli flint. This is compatible with the fact that use polish also forms faster on finer flint than on coarser varieties. It strongly suggests that sheen like use-wear is a smoothing of the flint surface whether by the removal of asperities, plastic deformation, solution and redeposition of silica or a combination of these processes. The fact that sheen is slower to form on coarse flint was surprising at first, since most of the flint artefacts from Israel which I have sampled and/or examined (which are generally coarser than Brandon flint) were affected by sheen. Semenov has also noted (1964:11) that sheen appears more on flint of "chalk origin" than on "limestone flint". In fact, after careful consideration, it seems reasonable that any smoothing of a flint surface is achieved faster on fine flint than on coarse flint with widely spaced particles requiring much more removal or infilling or both. The explanation for the higher proportion of flint from the Near East being affected by soil sheen may be due to the particular soil processes specific to that area, such as the presence of some water combined with relative high mean temperatures. Further consideration also needs to be given to the types and proportions of silica phases present in different flints (Clayton 1983).

7.2.6 Post-Depositional Processes and the Vulnerability of Edges

Even when the whole artefact is affected by either mechanical or chemical sheen, the distribution of this sheen seems to be more concentrated on the edges and becomes progressively less marked as one moves away from the edges. This distribution of "polish" was formerly considered a diagnostic feature of use-wear polish but the present work suggests that this idea should now be abandoned. Intensity of polish being a function of proximity to flint edge actually seems to apply whenever the flint surface is being polished by whatever mechanism. As the particles of the edge are more exposed than those inland, they will consequently be preferentially removed whether by chemical or mechanical means. This has been observed in experiments for mechanical (sections 6.5.1, 6.5.2, 6.5.3) and chemical PDSM (sections 6.5.5 to 6.5.12) as well as from the experiments described under polish formation (sections 6.4.5 to 6.4.12). Polish can be used as a good indicator of use-wear when it presents a discrete distribution on edges that may plausibly have been used while being absent on other edges of a piece. This pattern of distribution is disturbed when post-depositional processes occur after use either degrading the polish completely or smoothing the surface still further making it very difficult, if not impossible, to isolate the use-wear polish from the subsequent smoothing.

7.2.7 Edge Damage

Gentle tumbling in fine sediment without gravel does not cause serious edge damage (see Glossary) as seen from experiments in section 6.5.2 and 6.5.4. Predictably, a certain amount of edge damage is produced by the movement of artefacts in gravelly sediment. Its amount and distribution can mislead if it is the only criterion used to reconstruct tool use. Furthermore, the removal of microflakes may remove the polish from the edge, hindering a correct reading of the traces.

7.2.8 Micropitting

What are optically observed as pits in the polished surface of the flint are sometimes referred to as micropitting in the microwear literature. Keeley (1980) mentions pitting on tools used to work bone ("tiny pits smaller than 1 micron" *ibid.*:43) or antler ("pockmarks" *ibid.*:56). He also noticed some pitting (circular 5 microns and less *ibid.*:49–50) on tools used to work dry hide without lubricants. He suggests the latter might be "micropotlids" from thermal stress.

According to my observations micropitting appears during work on all the harder materials. Under the SEM pits appear as hollows. They may result from the removal of asperities as the worked material and the flint slide against each other or they may be the remains of deeper depressions in the natural microtopography which have not been obliterated by the smoothing process. One is often able to observe the irregular crystals at their bottom. It may be possible to distinguish pits formed by plucking out of particles later in the working process from those present earlier (whether originally present or due to plucking out during earlier work) by the degree of rounding of the pit edges. Pits due to plucking out may show comet tails due to scoring by the removed asperity, although a chance collision of a free particle with a natural microtopographical pit might also give this feature.

In the experiments on post-depositional processes (6.5.2 a and b) pits have been observed after contact with gravelly sand. It may be that the contact of the artefacts with the gravel in the sand during a rotative motion has caused large asperities or a group of crystals to be removed hence leaving the pits.

SEM micrographs of EXP/103 (figure 218 and 217) show pits in a variety of sizes. Some of the larger ones are clearly suggestive of a hollow whose edges have been smoothed during use. The association of pits with work on hard materials and in tumbling experiments suggest that the hypothesis proposed for them by Diamond (1979:165), namely that they are the result of surface fatigue, may also be correct. This author would suggest that the nature of this "fatigue" may be high energy impact on asperities or high pressure contact (perhaps as a result of the load being brought to bear on a small area by a relatively undeformable, i.e. hard, material) causing subsurface cracking rather than simply elastic and/or plastic deformation.

7.2.9 Patination

Patination seems to be a complex process which is very sensitive to small changes in the composition of the surrounding solution. For patination to occur the pH of the solution seems to be a critical variable. This was observed in the experiment described in section 6.5.10 where it was not possible to replace the evaporated water for two days. During the first 9 days no patina whatsoever was visible while two days only at the increased concentration patinated the artefacts to the core. The effects of other ions [i.e. other than H^+] in solution requires further careful consideration, however. These experiments have not unfortunately clarified whether the process of patination involves removal of interstitial matter, precipitation of new crystals or a combination of both. Glare was observed to precede patination in the experiments described in section 6.5.12. In experiments described in sections 6.5.6 and 6.5.7 and 6.5.8 no patination occurred due to the slow processes and lack of time. However glare started to form on the surface of the flint artefacts. This glare may be due to a small scale dissolution of the more prominent crystals within the unpolished spaces. The dissolved silica may also be redeposited in the small fissures of the flint surface producing a smoothing and an increase in the reflectivity of the surface observed as glare.

In the initial stages of patination it appears, at least subjectively, that there is an element of crystallisation of new material from the solution. Much more detailed work needs to be carried out to understand the processes better. It is clear from these experiments, however, that polish on patinated flint artefacts can probably **not** be used to reconstruct tool function as it either does not survive or it will be too greatly distorted by the processes that have caused patination.

In conclusion it seems that the disappearance of use-wear traces on artefacts subjected to simulated post-depositional processes and the appearance of new ones in the form of soil sheen has serious implications for microwear analysis undertaken on such collections. An example is discussed in Chapter 4.1 regarding the site of Meer (Van Noten, 1978). Keeley (1980: 34–35), though warning of the traces caused by natural agencies, concluded "most natural processes, then leave traces which are unlikely to cause much confusion for the microwear analyst". I beg to differ: post-depositional processes cause many problems to the microwear analyst and some confusion too as some post-depositional traces can be very similar to use wear. A very critical approach is required to distinguish one from the other. Recent microwear reports do not seem to take enough notice of the problem, which is rarely mentioned. In the presence of **PDSM**, such as sheen, bright spots and/or patination, flint surfaces are generally so altered that no comparison can be made between them and the use-wear polish produced experimentally. It is not possible to recognise the pattern of polish distribution (surface smoothing) caused by use when subsequent smoothing has disturbed the previous pattern.

7.3 Future research

The specific points requiring further research have been noted in the text and it is sought here only to make a general observation.

The "death sentence" imposed by the present work on such expressions as "bone polish", "hide polish" and "wood polish" should not mean the end of microwear analysis. As mentioned earlier, other more personal methods have now been developed by many competent microwear analysts to incorporate all the criteria necessary to reconstruct tool use when polish is present. Grace (1989), for example, has now simplified and improved his method to make it accessible to many aspiring microwear analysts. On artefacts affected by PDSM on which use-wear polish has been destroyed or is not identifiable, edge-wear and edge characteristics can be made to give a lower level of interpretation as was done in Chapter 4 of this work.

Commenting on Semenov's work which had preceded his, Keeley said in his book *Experimental Determination of Stone Tool Uses* (1980:108:) "Disappointment and disillusionment followed as one investigator after another found Semenov's results impossible to substantiate".

Science progresses by such leaps when one stage brings a group of researchers to find fault with previous work and to take it one step further. It is hoped that this work has played a part in such a transition and that the flaws in this work may spur others to continue the investigation of polish and post-depositional processes on flint artefacts to develop the usefulness of such studies still further.

The criticism of the Keeley method made in this thesis should in no way diminish the importance of Keeley's contribution in recognising that microscopic polish observed at around x200 is a valuable indicator in the interpretation of function on flint tools and in greatly stimulating further research in this field.

References

Adam, N.K. (1927). The Polishing of Surfaces. *Nature* 119: 162

Akoshima, K. (1981). An experimental study of microflaking. Report of Tohoku University Microflaking Research team (1). *Journal of the Archaeological Society of Nippon* LXVI(4):1–5.

Anderson, J.C., K.D. Leaver, R.D. Rawlings and J.M. Alexander (1985). *Materials Science* Third Edition. Van Nostrand Reinhold (UK).

Anderson, P.C. (1980). A Testimony of Prehistoric Tasks: Diagnostic Residues on Stone Tool Working Edges. *World Archaeology* 12 (2) :181-194.

Anderson-Gerfaud, P. (1981). *Contribution Methodologigue a l'Analyse des Microtraces d'Utilisation sur les Outils Prehistoriques*. These de 3e cycle, Universite de Bordeaux I, Talence

(1982). Comment preciser l'utilisation agricole des outils pre-historiques? *Cahiers de L'Euphrate* 3:149–164.

(1986). A Few Comments Concerning Residue Analysis of Stone Plant-Processing Tools. Owen L.R. & G. Unrath (eds.) *Early Man News* **9/10/11**:69–81.

Andersen, H. H. & H. J. Whitlow (1983). Wear Traces and Patination on Danish Flint Artefacts. *Nuclear Instruments and Methods in Physics Research* **218**:468–474.

Ataman, K. (1989). *The Chipped Stone Assemblage from Can Hasan III A Study in Typology, Technology and Function*. Ph.D. Thesis submitted at the Institute of Archaeology, University College, London.

Bamforth, D.B. (1988). Investigating Microwear Polishes with Blind Tests: The Institute Results in Context. *Journal of Archaeological Science* **15**:11–23.

Bar-Yosef, O., B. Vandermeersch, B. Arensburg, P. Goldberg, H. Laville, L. Meignen, Y. Rak, E. Tchernov, & A. M. Tillier (1986). New Data on the Origin of Modern Man in the Levant. *Current Anthropology* **27(1)**: 63–64.

Barton, R.N.E. (1981). Some Conjoined Artifacts from New Mesolithic Site at Hengistbury Head, Dorset. *Proceedings of theDorset Natural History and Archaeological Society* **103**:13–20.

Barton, R.N.E. and C.A. Bergman (1982). Hunters at Hengistbury: some evidence from experimental archaeology. *World Archaeology* **14(2)**:237–248.

Beilby, G.T. (1921). *Aggregation and flow of solids*. Macmillan and Co. Ltd. London.

Beyries, S. (1987). *Variabilite de l'industrie lithique au Mousterien: approche fonctionnelle sur quelques gisements Francais*. British Archaeological Reports; International Series 328; Oxford.

Beyries, S. & F. Delamare & F. Quantin (1988). Traceologie et rugosimetrie tridimensionnelle. S. Beyries (ed.) Industries Lithiques: Traceologie et Technologie. *British Archaeological Reports*; International Series 411 (ii).

Bocquet, A. (1980). Le Microdenticule un outil mal connu. *Bulletin de la Societe Prehistorique Francaise* 77:76–85.

Boekelman, K (1986). Rast Unter Baumen: Ein Ephemerer Mesolitischer Lagerplatz Aus Dem Duvenseer Moor. *Offa Buecher* **43**:149–163.

Bordes, F. (1950). Du poli particulier de certains silex tailles. *L'Anthropologie* **54**:161–3.

Bradley, R. & C. Clayton (1987). The Influence of Flint Microstructure on the Formation of Microwear Polishes. Sieveking G. de G. & M.H. Newcomer (eds.) *The human uses of flint and chert*: 1–10 Cambridge University Press Cambridge.

Broadbent, N. (1979). Coastal Resources and Settlement Stability : A critical Study of a Mesolithic Site Complex in Northern Sweden. AUN 3 *Archaeological Studies Uppsala University of North European Archaeology*.

Broadbent, N. & K. Knutsson (1975). An experimental analysis of quartz scrapers. Results and applications. *Fornvannen* **70**:113–128.

Broderick, M. (1988). Residue Analysis of Foxie Otter Site Chipped Stone. Christopher C. Hanks *In The Foxie Otter Site*. The Regents of the University of Michigan:100–203.

(1979). Ascending Paper Chromotographic Technique in Archaeology. B. Hayden (ed.) *Lithic Use-Wear Analysis*:375–384 Academic Press New York.

Bueller, J. (1984). A Pilot Microwear Analysis of Sampled Tools of Layer 7 at Sefunim. Sefunim Prehistoric Sites Mount Carmel, Israel by A. Ronen. *British Archaeological Reports International Series* **230 (1)**:201–214.

(1983). Methodologial problems in the microwear analysis of tools selected from the Natufian sites of El Wad and Ain Mallaha. *Travaux de la Maison de l'Orient* **5**:107–125.

Cahen, D. & J. P. Caspar (1984). Les Traces d'Utilisations des Outils Prehistoriques. *L'Anthropologie* **88(3)**:277–308.

Cahen, D & L.H. Keeley (1980). Not less than two, not more than three. *World Archaeology* **12(2)**:167–179.

Cahen, D., L.H. Keeley & F.L. Van Noten (1979). Stone Tools, Toolkits, and Human Behavior in Prehistory. *Current Anthropology* **20(4)**:166–180.

Cahen, D. & J. Moeyersons (1977). Subsurface movements of stone artefacts and their implications for the prehistory of Central Africa. *Nature* **266**: 812–815.

Charles, R.J. (1967). The Nature of Glasses. *Scientific American* **217(3)**:127–136.

Clark, J.G.D. (1954). *Excavations at Star Carr*. Cambridge University Press.

Clayton, C.J. (1983). *Geochemistry of Chert Formation in Upper Cretaceous Chalks*. Unpublished Ph.D. Thesis University of London.

Cornish, D.C. (1961). *The Mechanism of Glass Polishing*. British Scientific Instrument Research Association South Hill Chiselhurst Kent. B.S.I.R.A. Research Report R 267

Cook, J. & J Dumont (1987). The development and application of microwear analysis since 1964. Sieveking G. de G. & M.H. Newcomer (eds.). *The Human uses of Flint and Chert*: 53–61. Cambridge University Press.

Cotterell, B. & J. Kamminga (1979). The Mechanics of Flaking. B. Hayden (ed.) *Lithic Use-Wear Analysis*: 97–112. Academic Press New York.

Crabtree, Don E. (1972). An introduction to flintworking. E.H. Swanson, Jr. & B.R. Butler (eds.) *Occasional Papers of the Idaho State University Museum* **28**:1–98.

Curwen, E.C. (1941). Some food-gathering implements: Study in Mesolithic tradition. *Antiquity* **15**:320–37.

(1940). Notes and News. *Antiquity* **XIV**:435–437.

(1936). Non-Crescentic Sickle-Flints from Sussex. *The Antiquaries Journal* **XVI**:85–90.

(1935). Agriculture and the flint sickle in Palestine. *Antiquity* **9**:62–66

(1930). Prehistoric flint sickles. *Antiquity* **4**:179–186.

Del Bene, T. A. (1980). An X-Ray Analysis of Depositional Polish. *Lithic Technology* **IX(2)**:36–37.

(1979). Once Upon a Striation: Current Models. B. Hayden (ed.) *Lithic Use-wear Analysis*: 167–177 Academic Press New York.

Diamond, G. (1979). The Nature of So-Called Polished Surfaces on Stone Artefacts. B. Hayden (ed.) *Lithic Use-Wear Analysis*: 159–66. Academic Press, New York.

Dimbleby J.W. (1965). The Mesolithic site on Iping Common, Sussex in Keef P.A.M., J.J. Wymer & J.W. Dimbleby. *Proceedings of the Prehistoric Society* **XXXI**:85–94.

d'Errico, F. (1985). Traces D'Usure sur l'Industrie Lithique: Approche Methodologique et Proposition d'une Technique. *L'Anthropologie* **89(4)**:439–456.

d'Errico, F. and L. Varetto (1985). Scanning electron microscope study of microwear on original stone implements. *Antropologia Contemporanea* **8(2)**:135–142.

Dowson, D. (1979). *History of Tribology*. Longman, London and New York.

Dumont, J.V. (1982). The quantification of microwear traces: a new use for interferometry. *Word Archaeology* **14**:206–217.

(1983). An Interim report of the Star Carr Microwear Study. *Oxford Journal of Archaeology* **2(2)**:127–145.

Evans, A.G. (1979). Abrasive Wear in Ceramics: An Assessment. Hockey B.J. & R.W. Rice (eds.) *The Science of Ceramic Machining and Surface Finishing II*: 1–14 U.S. National Bureau of Standards.

Evans, J. (1897). *Ancient Stone Implements, Weapons and Ornaments of Great Britain* (2nd edition). Longmans Green, London.

Frere, J. (1800). Account of flint weapons discovered at Hoxne in Suffolk. *Archaeologia* **13**:204–205.

Frison, G.C. (1968). A Functional Analysis of Certain Chipped Stone Tools. *American Antiquity* **33(2)**:149–155.

Gendel, P.A. and L. Pirnay (1982). Microwear Analysis of Experimental Stone Tools: Further test results. *Studia Praehistorica Belgica* **2**:251–65.

Grace, R. (1981). *Attribute Analysis as a Method of studying Functional Variability in Palaeolithic Assemblages*. Unpublished B.A. Dissertation, University of London, Institute of Archaeology.

(1988). *Interpreting the Function of Stone Tools*. Ph.D. Thesis submitted at the Institute of Archaeology University College London.

(1989). *Interpreting the Function of Stone Tools: The Quantification and Computerisation of Microwear Analysis*. British Archaeological Reports International Series 474.

Grace, R., K. Ataman R. Fabregas and C.M.B. Haggren (1988). A multi-variate approach to the functional analysis of flint tools. Beyries (ed.) Industries Lithiques Traceologie et Technologie. *British Archaeological Reports International Series* 411 (ii):217–230

Grace, R., I.D.G. Graham and M.H. Newcomer (1985). The Quantification of Microwear Polishes. *World Archaeology*. **17(2)**:112–120.

Griffiths, D. R., C.A. Bergman, C.J. Clayton, K. Ohnuma, G.V. Robin & J. Seeley (1987). Experimental investigation of the heat treatment of flint. Sieveking G. de G. and M. H. Newcomer (eds.) *The Human Uses of Flint and Chert*: 43–52. Cambridge University Press. Cambridge.

Gysels, J. and D. Cahen (1981). Premiers resultats de l'analyse des traces microscopiques d'usure de quelques outils de Mesvin IV. *Notae Praehistoricae* **1**:75–82.

Hayden, B. (ed.) (1979). *Lithic Use-Wear Analysis*. Academic Press New York.

Hayden, B. and J. Kamminga (1973) Gould, Koster and Sontz on "microwear": a critical review. *Newsletter of Lithic Technology* **11(1–2)**:3–8.

(1979). An Introduction to Use-Wear: The First CLUW. B. Hayden (ed.) *Lithic Use-Wear Analysis*: 1–14. Academic Press New York.

Hockey, B.J. and R.W.Rice (eds.) (1979) *The Science of Ceramic Machining and Surface Finishing II*. U.S. National Bureau of Standards. Washington D.C.

Holley, G. and T. Del Bene (1981) An Evaluation of Keeley's Microwear Approach. *Journal of Archaeological Science*. **8**:337–52.

Holmes, D.L. (1987). Problems Encountered in a High-Power Microwear Study of some Predynastic Lithic Artifacts. Sieveking G. de G. & M.H.Newcomer (eds.) *The Human Uses of Flint and Chert*: 91–96. Cambridge University Press.

(1989). The Predynastic Lithic Industries of Upper Egypt. *British Archaeological Reports International Series* 469 (i & ii).

Hurcombe, L. (1985). The Potential of Functional Analysis of Obsidian Tools: a closer view. C. Malone, & S. Stoddart, (eds.) Papers in Italian Archaeology IV. *British Archaeological Reports International Series* 244:50–60.

(1988). Some Criticisms and Suggestions in Response to Newcomer *et al*. *Journal of Archaeological Science*. **15**:1–10.

Iler, R.K. (1979). *The Chemistry of Silica*. John Wiley & Sons Inc. New York.

Juel Jensen, H. (1988a). Functional Analysis of Prehistoric Flint Tools by High-Power Microscopy: A Review of West European Research. *Journal of World Prehistory* **2(1)**:53–88.

(1988b). Microdenticulates in the Danish stone age: a functional puzzle. Industies Lithiques Traceologie et Technologie S. Beyries (ed.). *British Archaeological Reports International Series* 411(i): 231–252 Oxford.

Kamminga, J. (1979). The Nature of Use-Polish and Abrasive Smoothing on Stone tools. B. Hayden (ed.) *Lithic Use-Wear Analysis*: 143–158. Academic Press New York.

Keeley, L.H. (1974 a). The Methodology of Microwear Analysis: A Comment on Nance. *American Antiquity* **39(1)**:126–128.

(1974 b). Technique and methodology in microwear studies: a critical review. *World Archaeology*. **5(3)**:323–336.

(1976).Microwear on Flint: Some Experimental Results. *2nd International Symposium on Flint Staringia* **3**:49–51.

(1977). The Functions of Palaeolithic Flint Tools. *Scientific American* **237(5)**:108–26.

(1978). Preliminary Microwear Analysis of the Meer Assemblage. F.Van Noten (ed.) *Les Chasseurs de Meer*: 78–86 Disertationes Archaeologicae Gandenses 18, Bruges, de Tempel.

(1980). *Experimental Determination of Stone Tool Uses: A Microwear Analysis*. Chicago Press.

(1981). Reply to Holley and Del Bene. *Journal of Archaeological Science* **8**:348–352.

Keeley, L.H. & M.H. Newcomer. (1977). Microwear Analysis of Experimental Flint Tools: A Test Case. *Journal of Archaeological Science* **4**: 29–62

Keeley, L.H. & N. Toth (1981). Microwear polishes on early stone from Koobi Fora Kenya. *Nature* **293**:464–465.

Kirchner, H.P., R.M. Gruver, and D.M. Richard. (1979). Fragmentation and Damage Penetration During Abrasive Machining of Ceramics. Hockey B.J. and R.W. Rice (eds.) *The Science of Ceramic Machining and Surface Finishing II*: 23–42. U.S. National Bureau of Standards Washington D.C.

Knutsson, K. (1986) SEM-analysis of Wear Features on Experimental Quartz Tools. Owen L.R. & G. Unrath (eds.) *Early Man News* **9/10/11**:35–46.

Landingham, R.L., A.W. Casey, and R.O. Lindhal. (1979). Effects of Various Polishing Media and Techniques on the surface Finish and Behavior of Laser Glasses. Hockey, B.J.,and R.W. Rice (eds.) *The Science of Ceramic Machinng and Surface Finishing II*. U.S. National Bureau of Standards Washington D.C.

Levi Sala, I. (1982) *Carinated Tools or Cores: a Microwear Study of Material from Ksar Akil*. Unpublished B.A. Dissertation, University of London, Institute of Archaeology.

(1986a). Experimental Replication of Post-Depositional Surface Modifications on flint. *Early Man News* **9/10/11** Owen L.R. & G. Unrath (eds): 103–108.

(1986b). Use Wear and Post-depositional Surface Modification: A Word of Caution. *Journal of Archaeological Science* 13 (3):229-244.

(1987). Microscopic Analysis of Some Mousterian Pieces from Kebara: The effect of Post-Depositional Processes on Functional Interpretation. *Mitekufat Haeven New series Journal of the Israel Prehistoric Society* **20**:143–150.

(1988). Processes of Polish Formation on Flint Tool Surfaces. S. Beyries (ed.) Industries Lithiques : Traceologie et Technologie. *British Archaeological Reports International Series* 411(ii):83–98 Oxford.

(1989). Functional Analysis and Post-depositional Alterations of Microdenticulates. R.N.E. Barton (ed.) *The Mesolithic Excavation at Hengistbury Head*. Monograph Series, Dorset Natural History and Archaeological Society.

Loy, T. (1983) Prehistoric blood residues: Detection on tool surfaces and identification of species of origin. *Science* 220.

Maher, M. (1983). *Knapping areas at the Mesolithic site, West Heath Hampstead*. Unpublished B.A. Dissertation, University of London, Institute of Archaeology.

Malkin, S and M. Huerta. (1979). Glass Grinding Mechanisms and Influence on the Finished Surface and Strength. Hockey B.J. and R.W. Rice (eds.) *The Science of Ceramic Machining and Surface*

Finishing II: 93–106. U.S. National Bureau of Standards Washington D.C.

Mansur-Franchomme, M.E. (1983)a. Scanning Electron Microscopy of Dry Hide Working Tools: The role of abrasives and humidity in microwear polish formation. *Journal of Archaeological Science* **10(3)**:223–230.

(1983)b. Traces d'Utilisation et Technologie Lithique: Exemples de la Patagonie. These 3eme Cycle, Universite de Bordeaux I.

Mansur M.E. (1982). Microwear analysis of natural and use striations: new clues to the mechanisms of striation formation. *Studia Prehistorica Belgica* **2**:213–234

Masson, A. (1982) Silice et Traces d'Usage. La Part du Silex. *Nouvelles Archeologiques Museum Histoire naturelle* Lyon **20**:31–49.

Masson, A., E. Coqueugniot & S. Roy (1981). Silice et traces d'usage: le lustre des faucilles. *Nouvelles Archeologiques Museum Histoire naturelle* Lyon **19**:43–51.

Meeks, N.D., G. de G. Sieveking, M.S. Tite & J. Cook (1982). Gloss and Use-Wear Traces on Flint Sickles and Similar Phenomena. *Journal of Archaeological Science* **9**:317–40.

Micheelsen, H. (1966). The Structure of dark flint from Stevns, Denmark. *Medd fra Dansk Geol. Forening*, **16**:286–368.

Miller, R. (1982). Pseudo-tools created by livestock from Halawa, Syria. *Journal of Field Archaeology* **9**:281–283.

Moss, E.H. (1983)a. Some Comments on Edge Damage as a Factor in Functional Analysis of Stone Artifacts. *Journal of Archaeological Science* **10**:231–242.

(1983)b. *The Functional Analysis of Flint Implements.Pincevent and Pont d'Ambon: two case studies from the French Final Palaeolithic*. British Archaeological Reports International Series 177.

(1986)a Further work on the function of flint tools at Pincevent (Seine et Marne), France: Sections 36 and 27. Studies in the Upper Palaeolithic of Britain and Northwest Europe. D.A. Roe (ed.). *British Archaeological Reports International Series* **296**:175–186

(1986)b Aspects of site comparison: debitage samples, technology and function. *World Archaeology* **18(1)**:116–133

(1986)c. What Microwear Analysts Look at. Owen L. & G. Unrath (eds.) *Early Man News* **9/10/11**:91–96.

(1987). A Review of "Investigating Microwear Polishes with Blind Tests". *Journal of Archaeological Science* **14**:473–481.

Moss, E.H. & M.H. Newcomer (1982). Reconstruction of Tool Use at Pincevent: Microwear Experiments. Cahen D. (ed.) Recent Progress in Microwear Studies. *Studia Praehistorica Belgica* **2**:289–312. Tervuren.

Munday, F.C. (1976) The Avdat/Aqev area: its habitat and geographic setting; in Marks A.E. (ed.). *Prehistory & Palaeo-Environments in the Central Negev, Israel*. Vol. I SMU Press Dallas

Newcomer, M.H. (1976). Spontaneous Retouch. *Staringia* **3**:62–64.

(1987). Review on "Use-Wear Analysis of Flaked Stone Tools" 1985 by P.C. Vaughan, 1985 University of Arizona Press. *Quaternary Science Reviews*:**XViii–XX**.

Newcomer, M. H. & L. H. Keeley (1979). Testing a Method of Microwear Analysis with Experimental Flint Tools. B. Hayden (ed.) *Lithic Use-Wear Analysis*: 195–206 Academic Press New York.

Newcomer, M., R. Grace , and R. Unger-Hamilton (1986). Investigating Microwear Polishes with Blind Tests. *Journal of Archaeological Science* **13(3)**:203–217.

(1987). Microwear Methodology: A Reply to Moss, Hurcombe and Bamforth. *Journal of Archaeological Science* **15**:25–33.

Nilsson, S. (1838-43). *Skandinaviska nordens urinvare*. Berlingska Boktryckeriet, Lund.

Odell, G.H. (1982). Emerging directions in the analysis of prehistoric stone tool use. *Reviews in Anthropology* **9**:17–33.

(1985) On Evaluating "Blind Tests" in Lithic Use-Wear Research. *The Western Canadian Anthropologist* **2**:26–30.

(1986) Book Review on P.C.Vaughan 1985 "Use-wear Analysis of Flaked Stone Tools" University of Arizona Press. *Lithic Technology* **15(3)**:115–120

Odell, G., and F. Odell-Vereecken (1980). Verifying the reliability of lithic use-wear assessments by "blind tests": the low-power approach. *Journal of Field Archaeology* **7**:87–120.

Owen, L., G. Unrath & P. Vaughan, (eds.) (1984). Register of microwear analysts and their research. *Early Man News* **7–8**:30–89.

Owen, L. R. (1986). Register of microwear analysts and their research 2. *Early Man News* **9/10/11**:187–189.

OECD (1969). *Glossary of Terms and Definitions in the Field of Friction, Wear and Lubrication-Tribology*. Research Group on Wear of Engineering Materials, Organisation for Economic Cooperation and Development (OECD), Paris.

Ollendorf, A.L. (1987). Phytolith Study at Tel Miqne, Israel. *Journal of Field Archaeology* **14(4)**:453–463.

Pfeiffer, L. (1912). *Die Steinzeitliche Technik und ihre Beziehungen zur Gegenwart: ein Beitrag zur Gaschichte der Arbeit*. Festschrift zur XLIII Algemeinen Versammlung der Deutschen Anthropologischen-Gesellshaft Vol 1. Gustav Fischer, Jena.

Plisson, H. (1983). De la conservation des micro-polis d'utilisation. *Bulletin de la Societe Prehistorique Francaise* **80(3)**:74–77

(1985). Correspondance scientifique. *Bulletin de la Societe Prehistorique Francaise* **82(4)**:99– 101.

(1986). Alteration des micropolis d'usage: quelques experiences complementaires. Owen L.R. & G. Unrath (eds.) *Early Man News* **9/10/11**:111–116.

Plisson, H. & M. Mauger (1988). Chemical and Mechanical Alteration of Microwear Polishes: An Experimental Approach. *Helinium* **XXVIII(1)**:3–16

Prost, D.C. (1988). Essai d'etude sur les mecanismes d'enlevement produits par les facons agricoles et le pietinement humain sur des silex experimentaux. S. Beyries (ed.) Industries Lithiques Traceologie et Technologie. *British Archaeological Reports International Series* 411(ii):49–64.

Quinn, T.F.J. (1971). *The Application of Modern Physical Techniques to Tribology*. Newnes-Butterworths London.

Rabinowicz, E. (1966). Friction in R.M. Besancon (ed.) *The Encyclopaedia of Physics*. Rheinhold New York.

Rosenfeld, A. (1971). The Examination of Use Marks On Some Magdalenian End Scrapers. *British Museum Quarterly* **35**:176–182.

Rottlaender, R. (1975). The formation of Patina on Flint. *Archaeometry* **17**:106–110.

(1976). Some Aspects of the Patination of Flint. *Staringia* **3**:54–56.

Rowlett, R.M. & M. C. Robbins (1982). Estimating original assemblage content to adjust for post-depositional vertical artefact movement. *World Archaeology* **14**:73–83.

Schick, T. & M. Stekelis (1977). Mousterian Assemblages in Kebara Cave, Mount Carmel. *Eretz Israel* **13**:97–149.

Schmalz, R.F. (1960). Flint and the Patination of Flint Artifacts. *Proceeding of the Prehistoric Society* **XXVI**:44–49.

Schneider Jr., S.J. & R.W. Rice (eds.) (1972). *The Science of Ceramic Machining and Surface Finishing*. National Bureau of Standards Special Publication 348.

Seitzer Olausson, D. (1980). Starting from Scratch: The History of Edge Wear Research from 1838 to 1978. *Lithic Technology* **9(2)**:48–60.

Semenov, S. (1964). *Prehistoric Technology* (translation M.W. Thompson). Moonraker Bradford-on-Avon.

Shafer, H.J. and R.G. Holloway (1979). Organic Residue Analysis in determining stone tool function. In *Lithic Use-Wear Analysis* B. Hayden (ed.): 385–400. Academic Press New York.

Shea, J. J. (1987). On Accuracy and Relevance in Lithic Use-wear Analysis. *Lithic Technology* **16**:2–3.

Shepherd, W. (1972). *Flint*. London. Faber & Faber.

Shipman, P. & J.J. Rose (1983). Bonemark Criteria in Early Hominid Behaviour. *Journal of Anthropological Archaeology* **2**:57–98.

Spurrell, F.C.J. (1884). On some Palaeolithic Knapping Tools and modes of using them. *Journal of the Royal Anthropological Institute* **13**:109–118.

(1892). "Notes on Early Sickles". *Archaeological Journal* **49**:53–68

Stapert, D. (1976). Some Natural Surface Modifications on Flint in the Netherlands. *Palaeohistoria* **18**:8–41.

Stordeur, D. (ed.) (1987). *La Main et L'Outil: Manches et Emmanchements Prehistoriques*. Travaux de la Maison de L'Orient Lyon. No. 15.

Sussman, C. (1988). *A Microscopic Analysis of Use-Wear and Polish Formation on Experimental Quartz Tools*. British Archaeological Reports International Series 395.

Tixier, J. (1963). *Typologie de L'epipaleolithique du Mahgreb*. Alger: Centre de Recherches Anthropologiques Prehistoriques et Ethnographiques, Memoires II.

Tringham, R., G. Cooper, G. Odell, B. Voytek & A. Whitman (1974). Experimentation in the formation of Edge Damage: A new Approach to Lithic Analysis. *Journal of Field Archaeology* **1**:171–96.

Unger-Hamilton, R. (1983). An investigation into the variables affecting the development and the appearance of plant polish in flint blades. M.-C. Cauvin (ed.) Traces d'utilisation sur les outils neolithiques du Proche Orient. *Travaux de la Maison de l'Orient* Lyon No.5: 243–250.

(1984). The Formation of Use-Wear Polish on Flint: Beyond the "deposit versus abrasion" controversy. *Journal of Archaeological Science* **11**:91–98.

(1988). *Method in Microwear Analysis*. British Archaeological Reports International Series 435.

Unger Hamilton, R., R. Grace, R. Miller & C. Bergman (1987). Drill bits from Abu Salabikh, Iraq. D. Stordeur (ed.). *La Main et l'Outil Manches et Emmenchemants Prehistoriques*: 269–285.

Unrath, G., L.R. Owen, A. Van Gijn, E.H. Moss, H. Plisson & P. Vaughan. (1986). An Evaluation of Microwear Studies: A Multi-Analyst approach. Owen L.R. & G. Unrath (eds.) *Early Man News* **9/10/11**:117–176.

Van Gijn, A. (1986). Fish Polish, Fact or Fiction. Owen L.R. and G. Unrath (eds.) *Early Man News* **9/10/11**:13–28.

Van Noten, F. (1978). *Les Chasseurs de Meer* Dissertationes Archaeologicae Gandenses 18, De Tempel Bruges.

Vaughan, P.C. (1981). *Lithic Microwear Experimentation and theFunctional Analysis of a Lower Magdalanian Stone Tool Assemblage*. Ph.D.dissertation, University of Pennsylvania.

(1985). *Use-Wear Analysis of Flaked Stone Tools*. University of Arizona Press, Tuscon.

(1987). Wear Analysis of a Lower Magdalenian flint assemblage from southwestern France in G. de G.

Sieveking and M. H. Newcomer (eds.) *The Human Uses of Flint and Chert*: 111–114. Cambridge University Press.

Vayson, A. (1920). La plus ancienne industrie de Saint-Acheul. *L'Anthropologie* **30**:441–496.

(1922). L'etude des outillages en pierre. *L'Anthropologie* **31**:1–38.

Villa, P. (1977). Sols et Niveaux D'Habitat du Paleolithique Inferieur en Europe et au Proche Orient. *Quaternaria* **19**:107–134.

Villa, P. (1982). Conjoinable Pieces and Site Formation Processes. *American Antiquity* **47**: 276–290.

Villa, P. & J. Courtin (1983). The Interpretation of Stratified Sites: A View from Underground. *Journal of Archaeological Science* **10**:267–281.

Wilmsen, E.N. (1968). Functional Analysis of Flaked Stone Artifacts. *American Antiquity* **33(2)**:156–161.

Witthoft, J. (1967). Glazed Polish on Flint Tools. *American Antiquity* **32(3)**:383–388.

Yamada, S. (1986). The Formation Process of Use-Wear Polishes. *Archaeology and Natural Science* **19**:101–123.

Appendix 1

1. Cutting Hide

a) **Task:** fresh cow hide cut into strips for thongs
Tool: EXP/14 microdenticulate max. length 31mm
Edge Angle 45 degrees
Morphology **Maximum Thickness** 4mm
Serration 20mm convex
Profile very slightly twisted
Motion cutting—mostly unidirectional
Angle perpendicular to worked surface
Duration 5 minutes
Prehension proximal held in right hand
Polish very little
Striations none
Edge Wear none macroscopic
Efficiency and Results
very inefficient—thongs not straight edged
Comparison with Archaeological
not comparable

b) **Task:** raw (untanned) deer hide, soaked overnight cut into strips for thongs
Tool: EXP/15/4 microdenticulate—max. length 79mm
Edge Angle 50 degrees
Morphology **Maximum Thickness** 6 mm
Serration 2 mm concave
Profile curved
Motion mostly unidirectional cutting
Angle perpendicular to worked surface
Duration 15 minutes
Prehension proximal held in right hand
Polish very faint
Striations none
Edge Wear none macroscopic
Efficiency and Results
inefficient edge
Comparison with Archaeological
not comparable

2. De-Hairing Hide

a) **Task:** de-hairing raw (untanned) deer hide soaked overnight
Tool: EXP/15/1 unretouched—max. length 79mm
Edge Angle 50 degrees
Morphology **Maximum Thickness** 6 mm
Serration none, 45 mm, convex edge
Profile curved
Motion unidirectional scraping towards worker
Angle perpendicular to worked surface
Duration 45 minutes
Prehension distal right hand held
Polish faint
Striations none
Edge Wear none macroscopic
Efficiency and Results
75 cm de-haired—efficient sharp edge—higher edge angle more appropriate
Comparison with Archaeological
not comparable

b) **Task:** de-hairing hide as in a)
Tool: EXP/15/2 microdenticulate max.length 79mm
Edge Angle 45 degrees
Morphology **Maximum Thickness** 6 mm
Serration 35 mm almost straight
Profile curved
Motion unidirectional scraping away from worker
Angle perpendicular to worked surface
Duration 5 minutes
Prehension proximal right hand held
Polish none
Striations none
Edge Wear none macroscopic
Efficiency and Results
no de-hairing occurred, experiment suspended, unsuitable edge due to low edge angle and denticulation
Comparison with Archaeological
none

c) **Task:** de-hairing hide as in a) and b)
Tool: EXP/15/3 unretouched edge max.length 79mm
Edge Angle 40 degrees
Morphology **Maximum Thickness** 6 mm
Serration none 31 mm straight edge
Profile curved
Motion unidirectional scraping towards worker
Angle perpendicular to worked surface
Duration 20 minutes
Prehension proximal right hand held
Polish very very faint
Striations none
Edge Wear none macroscopic
Efficiency and Results
12 sq cm de-haired, no denticulation required. high angle edge more appropriate
Comparison with Archaeological
not comparable

3. De-Hairing and Cutting Hide

a) **Task:** de-hairing hide as in 2
Tool: EXP/31/1 unretouched bladelet—max. length 49mm

Edge Angle 38 degrees
Morphology **Maximum Thickness** 6 mm
Serration none - 17 mm convex edge
Profile very curved
Motion scraping, unidirectional, away from worker
Angle perpendicular to worked material
Duration 20 minutes
Prehension proximal in wooden haft up to one-third, right hand held
Polish faint
Striations none
Edge Wear none macroscopic
Efficiency and Results
20 sq cm de-haired, tool efficient but small and curved, improved performance after tip broke making edge less convex.
Comparison with Archaeological
not comparable

b) **Task:** cutting deer hide as in 1 b)
Tool: EXP/31/2 unretouched bladelet max.length 49mm
Edge Angle 40 degrees
Morphology **Maximum Thickness** 6 mm
Serration none, concave, straight edge, 15 mm
Profile very curved
Motion unidirectional cutting towards worker
Angle perpendicular to worked material
Duration 15 minutes
Prehension proximal right hand held
Polish lightly developed
Edge Wear none macroscopic
Efficiency and Results
very sharp and efficient edge, performance improves as hide dries out
Comparison With Archaeological
not comparable

4. Cutting Meat
Task: skinning cow meat and cutting in strips
Tool: EXP/10 microdenticulate max.length 40mm
Edge Angle 40 degrees
Morphology **Maximum Thickness** 4 mm
Serration 20 mm concave
Profile slightly curved
Motion mostly unidirectional cutting, some sawing
Angle mostly perpendicular, some low angle
Duration 15 minutes
Prehension proximal right hand held
Polish hardly any
Striations none
Edge Wear none macroscopic
Efficiency and Results
very efficient for fine cutting on soft material, need for serration not obvious
Comparison with Archaeological
not comparable

5. Cutting Cartilage and Sinew
Task: butchering cow leg, cutting cartilage from bone
Tool: EXP/18 microdenticulate max length 66mm
Edge Angle 40 degrees
Morphology **Maximum Thickness** 4.5 mm
Serration concave 20 mm
Profile slightly curved and twisted
Motion mostly unidirectional cutting, some sawing
Angle mostly perpendicular to worked surface
Duration 60 minutes
Prehension proximal in wooden haft fixed with bees wax, resin and sinew
Polish faint
Striations some parallel to edge
Edge Wear none macroscopic
Efficiency and Results
not very efficient due to curved and twisted profile and to position of serration on concave mesial part of edge
Comparison With Archaeological
not comparable

6. Cutting and Whittling Dry Reed
a) **Task:** deliberately used in two motions as in basketry
Tool: EXP/7 microdenticulate max. length 29mm
Edge Angle 35 degrees
Morphology **Maximum Thickness** 4 mm
Serration 20 mm slightly concave
Profile flat
Motion sawing and whittling towards worker
Angle sawing perpendicular, whittling low angle
Duration 10 minutes
Prehension proximal right hand held
Polish very well developed
Striations parallel to edge from cutting and perpendicular from whittling
Edge Wear a few macroscopic scars
Efficiency and Results
not efficient, angle too low for whittling length insufficient without haft

Comparison with archaeological
not comparable

b) **Task:** cutting and whittling dry reeds as in a)
Tool: EXP/8 microdenticulate max. length 23mm
Edge Angle 38 degrees
Morphology **Maximum Thickness** 4 mm
Serration 16 mm sinuous
Profile flat
Motion towards worker both in whittling and cutting
Angle low angle and high angle
Duration 20 minutes
Prehension proximal in right hand
Polish not so well developed as in a)
Striations some parallel and some perpendicular to edge
Edge Wear denticulation worn down differentially
Efficiency and Results
more concave part of edge more efficient, less polish than 6 a) in spite of longer duration of work due to coarse flint on edge wearing down of serration also removes polished parts of edge
Comparison with Archaeological
not comparable in polish or in wear pattern

c) **Task:** whittling dry reeds as in a) and b) and cutting strips for basketry
Tool: EXP/9 microdenticulate max. length 30mm
Edge Angle 35 degrees
Morphology **Maximum Thickness** 5 mm
Serration 20 mm almost straight
Profile very slightly curved and twisted
Motion mostly unidirectional both in cutting and whittling
Angle cutting perpendicular, whittling low angle
Duration 10 minutes
Prehension proximal right hand held
Polish well developed, micropitted, smooth and highly reflective but restricted to the very edge
Striations perpendicular and parallel to edge, none in the polish
Edge Wear serration worn differentially and markedly, macro scars
Efficiency and Results
implement efficiently used but need for serration was not obvious
Comparison With Archaeological
not comparable in polish, striation or edge damage

7. Reaping Fresh Phragmites
Task: cutting 300 reeds
Tool: EXP/3 microdenticulate max. length 57mm
Edge Angle 38 degrees
Morphology **Maximum Thickness** 4 mm
Serration 22 mm concave
Profile slightly curved and twisted
Motion unidirectional cutting motion towards worker some sawing
Angle perpendicular to worked material
Duration 60 minutes
Prehension resin covering proximal, right hand held
Polish very well developed, very smooth and reflective, up to 2–3mm from edge, micropitting
Striations some parallel to edge
Edge Wear some new scars, no differential wear of serration
Efficiency and Results
task efficiently carried out, need for serration not obvious
Comparison With Archaeological
wear of serration comparable but polish not comparable

8. Reaping Bracken Stems
a) **Task:** reaping bracken stems (June)
Tool: EXP/12 microdenticulate max.length 55mm
Edge Angle 40 degrees
Morphology **Maximum Thickness** 6 mm
Serration 35 mm convex
Profile slightly twisted
Motion unidirectional towards worker and some back and forth sawing
Angle perpendicular to stems
Duration 10 minutes
Prehension proximal right hand held
Polish not well developed, more on dorsal
Striations some parallel to edge
Edge Wear no macroscopic scars, serration not macroscopically worn
Efficiency and Results
task carried out satisfactorily but need for serration not obvious - hafting advisable
Comparison With Archaeological
state of serration comparable, polish appearance not comparable

b) **Task:** reaping bracken stems (July)

Tool: EXP/19 microdenticulate max length 56mm
Edge Angle 45 degrees
Morphology **Maximum Thickness** 6 mm
Serration 27 mm concave
Profile very slightly curved, almost flat
Motion unidirectional towards operator
Angle perpendicular to stems
Duration 12 minutes
Prehension proximal hand held
Polish essentially similar to EXP/12 (8a), better developed
Striations none
Edge Wear no macroscopic wear on serration
Efficiency and Results
edge satisfactory for task, no obvious need for serration, uncomfortable without hafting
Comparison with Archaeological
polish not comparable, state of serration comparable

c) **Task:** reaping bracken stems (September)
Tool: EXP/23 microdenticulate max.length 67mm
Edge Angle 37 degrees
Morphology **Maximum Thickness** 6 mm
Serration 15 mm concave
Profile very slightly curved
Motion unidirectional cutting towards worker
Angle perpendicular to stems
Duration 20 minutes
Prehension distal left hand held
Polish better developed than a) similar to b)
Striations rare
Edge Damage no macroscopic scars or wear on serration
Efficiency and Results
edge satisfactory for task, better with haft, quantity reaped unsatisfactory
Comparison with Archaeological
polish not comparable, state of serration comparable

9. Whittling Seasoned Soft Wood

a) **Task:** shaping wood splinter into a point
Tool: EXP/5 microdenticulate max. length 57mm
Edge Angle 45 degrees
Morphology **Maximum Thickness** 4 mm
Serration 34 mm convex to straight
Profile very slightly curved
Motion unidirectional shaving motion
Angle low angle
Duration 5 minutes
Prehension left lateral in wooden haft right hand held
Polish almost none formed
Striations none
Edge Wear some scars removed serration differentially worn
Efficiency and Results
tool inefficient, unretouched high angle edge would be more appropriate
Comparison with Archaeological
not comparable

b) **Task:** shaping wooden point
Tool: EXP/6 microdenticulate max. length 44mm
Edge Angle 50 degrees
Morphology **Maximum Thickness** 4 mm
Serration 22 mm slightly concave
Profile flat
Motion shaving like pencil-sharpening away from worker
Angle low
Duration 40 minutes
Prehension proximal covered with resin held in right hand
Polish bright restricted to 1 mm from edge little developed
Striations very few
Edge Wear some scars removed, serration differentially worn
Efficiency and Results
task accomplished but unretouched high angle edge more appropriate
Comparison with Archaeological
not comparable

10. Incising Fresh Birch

Task: incising grooves into birch pole
Tool: EXP/27 microdenticulate max.length 63mm
Edge Angle 45 degrees
Morphology **Maximum Thickness** 6 mm
Serration 30 mm concave
Profile twisted and slightly curved
Motion sawing back and forth
Angle perpendicular to worked surface
Duration 10 minutes
Prehension right lateral in wooden haft fixed with bees wax and sinew, right hand held
Polish not well developed
Striations parallel to edge
Edge Wear some scars added serration differentially worn
Efficiency and Results

tool very appropriate for task

Comparison with Archaeological
polish not comparable, state of serration comparable

11. Sawing Shell

Task: slicing a cowrie shell
Tool: EXP/11 microdenticulate max. length 45mm
Edge Angle 42 degrees
Morphology **Maximum Thickness** 6 mm
Serration 30 mm concave
Profile almost flat
Motion sawing back and forth
Angle perpendicular to worked surface
Duration 15 minutes
Prehension proximal right hand held
Polish very intense accompanied by macroscopic gloss
Striations profuse parallel to edge
Edge Wear extreme, serration totally worn out after 5 minutes

Efficiency and Results
denticulation too fine for such hard material

Comparison With Archaeological
no archaeological implement used on such hard material - neither polish nor edge damage comparable

12. Sawing Bone

Task: grooving cow bone to observe effect on undenticulated edge
Tool: EXP/52 unretouched bladelet max. length 56mm
Edge Angle 30 degrees
Morphology **Maximum Thickness** 5 mm
Serration none, concave to straight
Profile very slightly twisted
Motion sawing back and forth
Angle perpendicular to worked surface
Duration 10 minutes
Prehension proximal held in right hand
Polish macroscopic on both aspects of edge, well developed, bright, restricted to edge
Striations some parallel to edge
Edge Wear macroscopic, 2 half moon scars 5 mm across and several smaller removed from edge

Efficiency and Results
task carried out, no regular denticulation occurred

Comparison With Archaeological
not comparable in edge wear or in polish

NB: this experiment was repeated on all the other materials used in this programme but no regular denticulation as in the microdenticulates resulted.

13. Cutting Straw

a) **Task:** cutting dry rushes
Tool: EXP/1
Edge Angle 55 degrees
Morphology **Maximum Thickness** 6 mm
Serration 25 mm concave
Profile twisted and curved
Motion back and forth cutting
Angle perpendicular to worked surface
Duration 30 minutes
Prehension beeswax and resin on proximal
Polish none
Striations none
Edge Wear none macroscopic

Efficiency and Results
tool quite adequate for task

Comparison With Archaeological
no comparison possible on polish and/or edge damage

b) **Task:** cutting soaked rushes
Tool: EXP/1
Edge Angle 55 degrees
Morphology **Maximum Thickness** 6 mm
Serration 25 mm concave
Profile twisted and curved
Motion back and forth cutting motion
Angle perpendicular to worked material
Duration 30 minutes
Prehension beeswax and resin on proximal end
Polish well developed with macroscopic gloss
Striations none
Edge Wear none macroscopic

Efficiency and Results
tool quite adequate for task

Comparison with Archaeological
no similar development and quantity of polish on archaeological tools

Appendix 2
Tabulation of Microdenticulates on Basis of

1. **Shape of Denticulated Edge:** Measured for its implications regarding function according to the Bordes system of holding a pencil against the edge—Bocquet suggests a more precise, detailed and time consuming method useful in typology.

   26 concave
   2 convex
   4 straight
   1 sinuous consisting of two adjoining concave lines
   1 sinuous consisting of two adjoining convex lines

2. **Wear of Denticulation**
   (Implications regarding hardness of worked material)

   24 regularly worn
   10 differentially worn

3. **Breakage of distal and/or proximal ends**
   Breakage was taken into consideration only when near enough to denticulation to be caused by use or possible hafting—therefore small distal end breaks which could be post-depositional were ignored.
   Implications on a) hafting
   b) on hardness of used materials

   24 both distal and proximal present
   4 both distal and proximal broken
   4 only proximal broken
   3 only distal broken

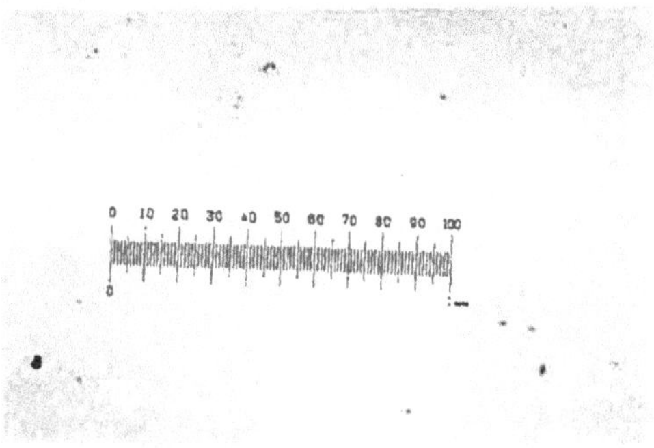

Figure 1a: Microscope stage scale shown at 50 magnifications

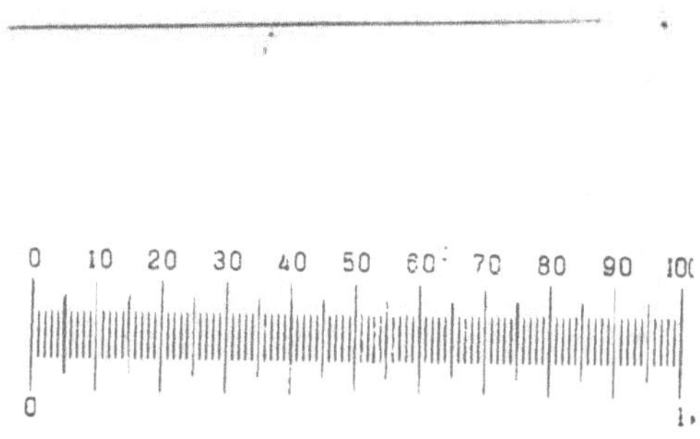

Figure 1b: Microscope stage scale shown at 100 magnifications

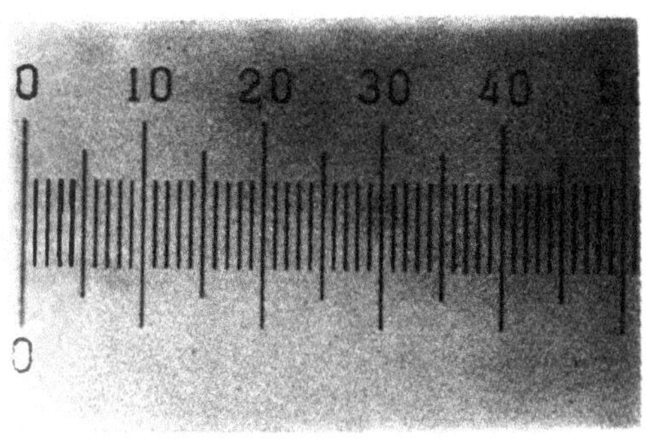

Figure 1c: Microscope stage scale shown at 200 magnifications

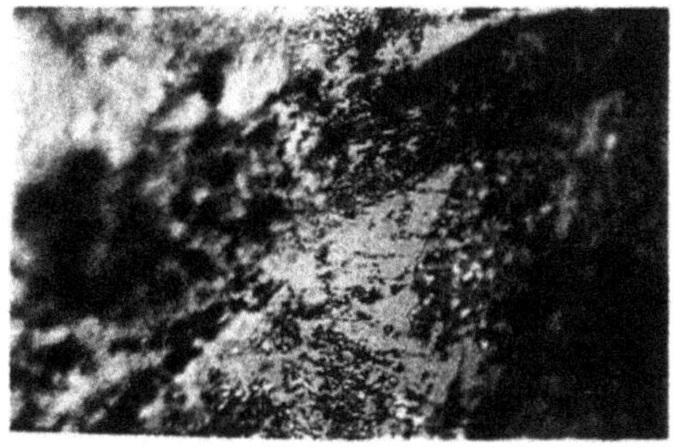

Figure 2: Grooving stone crushed crystals appearing as luminescent specks x200

Figure 3: Initial stages of polish grooving wood x200

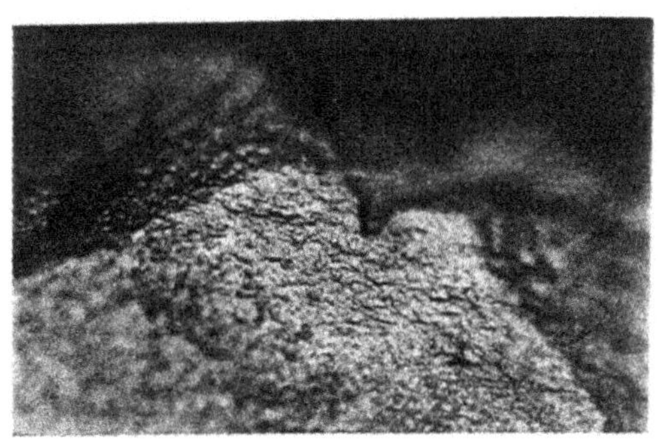

Figure 4: Initial stages

Figure 5: Line of Polish caused by retouch sometimes called streak x100

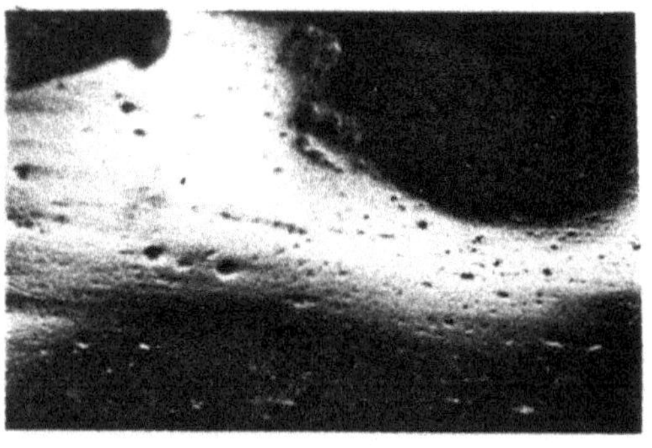

Figure 6: Archaeological "sickle gloss" x100

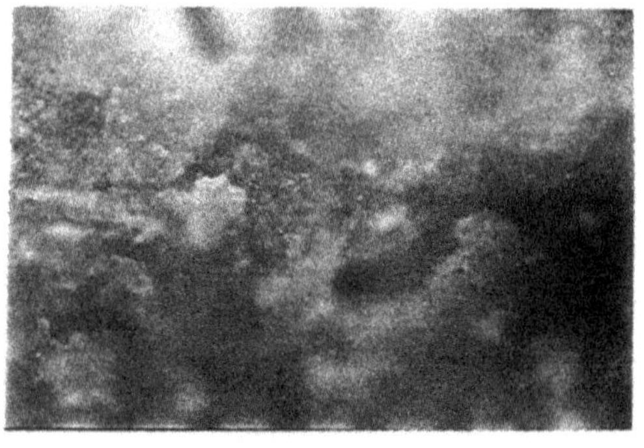

Figure 7: Coarse ISR/5b Flint appearing pedestalled at x200

Figure 8: Macroscopic gloss after whittling green wood
EXP/E5 x100

Figure 9: Same area as in figure 8 after cleaning with NaOH 20% (w/v)
EXP/E5 x100

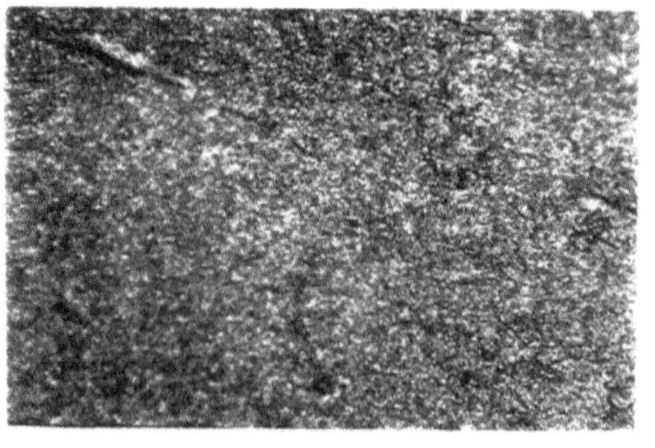

Figure 10: All traces disappear after tumbling in wet gravelly sand x100

Figure 11: Traces from. whittling reeds after washing EXP/9 x100

Figure 12: Same area as figure 11 after shaking in sand for 10 minutes x100

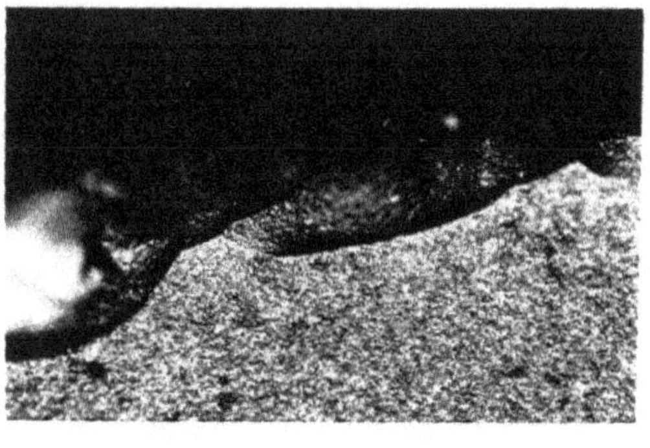

Figure 13: After tumbling for 50 hours in wet sand no traces remain x100

Figure 14: Traces after whittling dry reeds and cleaning EXP/9 x100

Figure 15: The edge in figure 14 after 10 minutes shaking in sand some "striations" removed x100

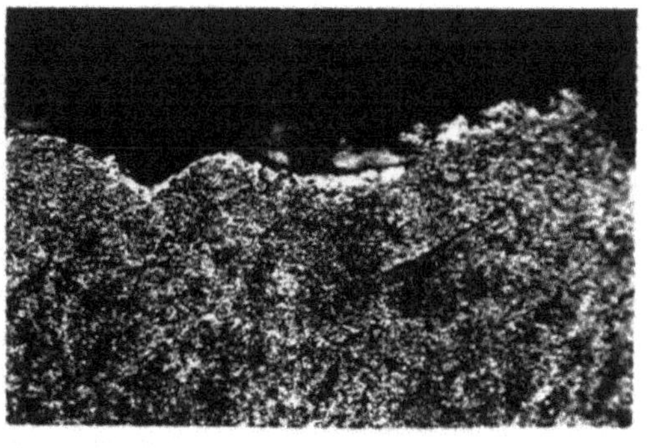

Figure 16: Edge in figures 14 and 15 after 50 hours tumbling in sand only polish survives x100

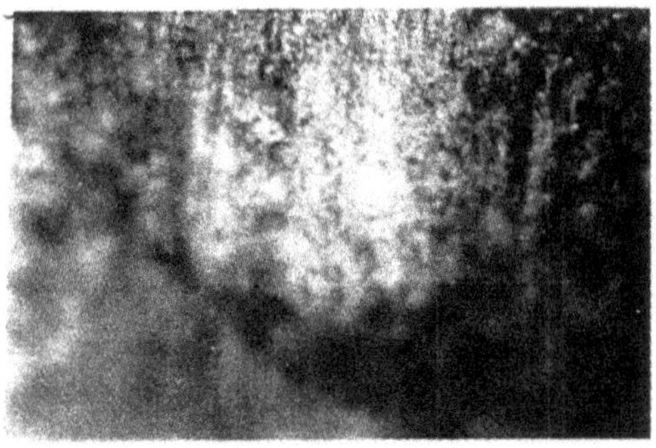

Figure 17: Macroscopic residue from debarking green alder before cleaning
EXP/83 x200

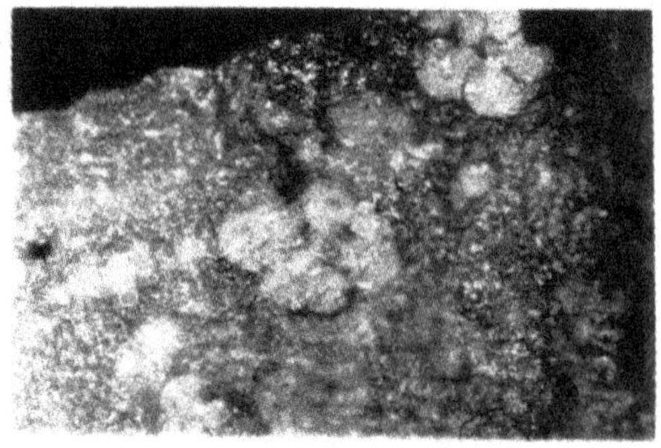

Figure 18: Clean surface of coarse flint, no polish
EXP/83 x200

Figure 19: Traces before chemical cleaning on archaeological artefact x200

Figure 20: Same edge as in figure 19 after 30 minutes each in NaOH 15% and HCL 15% no traces left x100

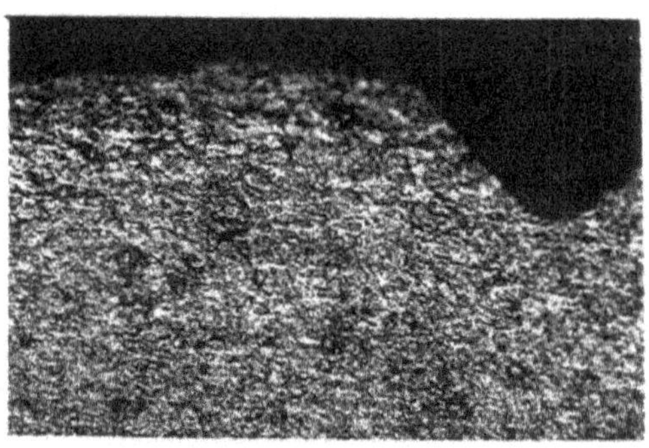

Figure 21: Traces after cutting meat for 60 minutes
EXP/18 x200

Figure 22: Traces after cutting fresh hide for 25 minutes
EXP/31
x200

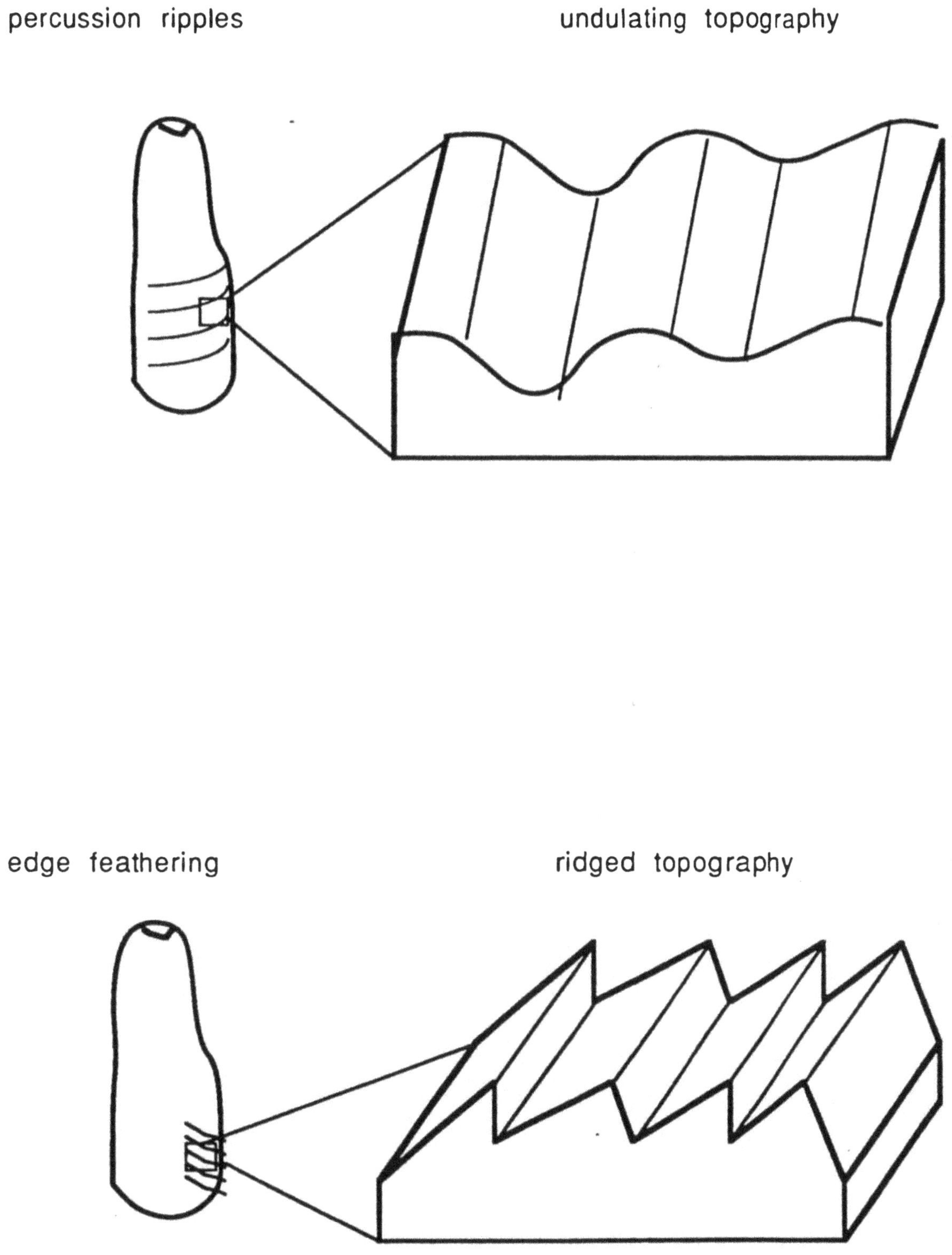

Figure 23: Features of the flint topography. After Grace (1989: Figures 38 and 39)

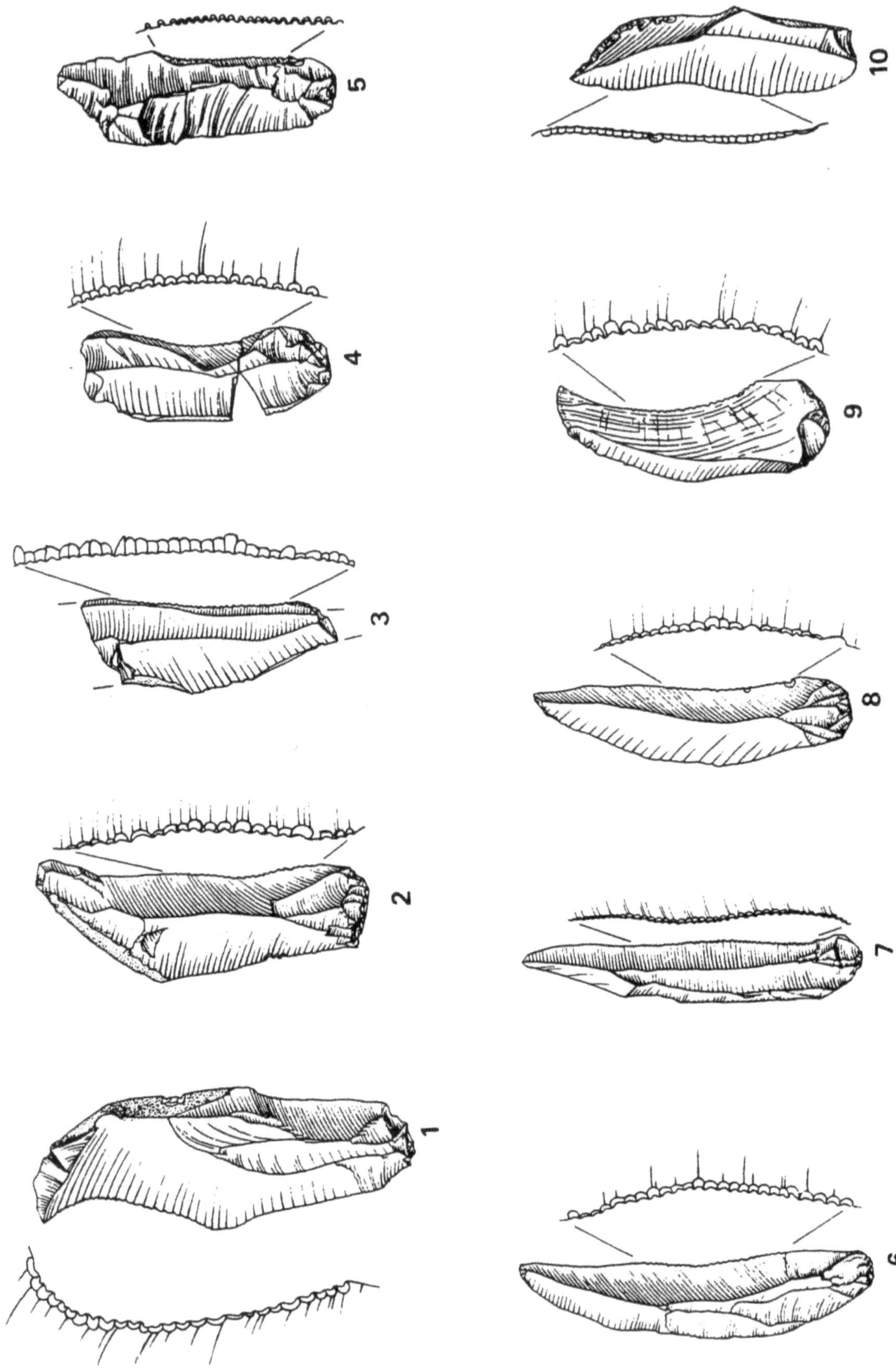

Figure 24: Microdenticulates (actual size) (saw-edge enlarged 2x)

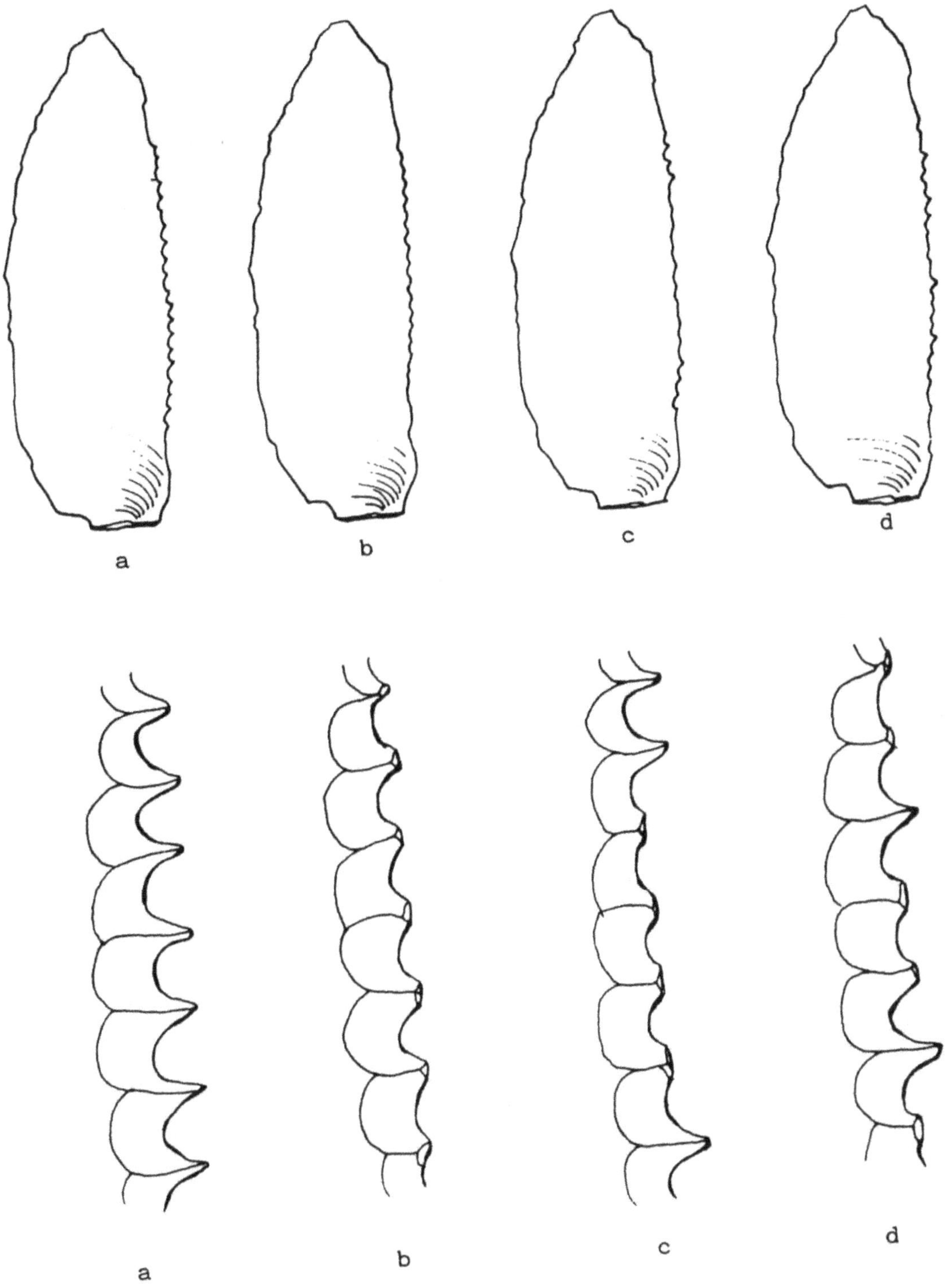

a) Unused edge
b) Edge used on materials of medium hardness
c) Edge used on hard materials like shell
d) Edges of archaeological micro-denticulates described as differentially worn

Figure 25: Patterns of wear on edges of experimental and archaeological microdenticulates

Figure 26: Polish on archaeological micro-denticulate.
x200

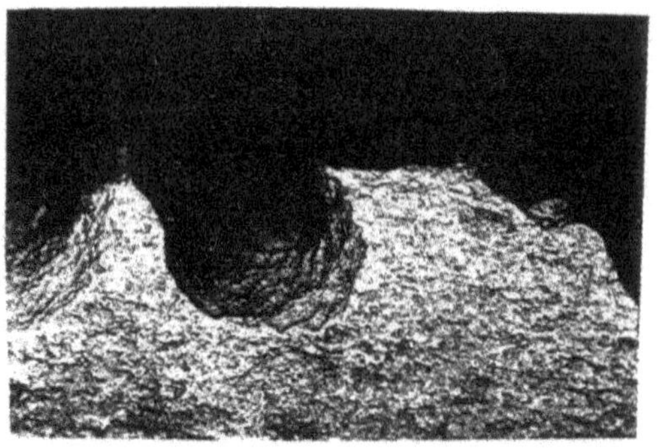

Figure 27: Polish on archaeological micro-denticulate
x200

Figure 28: Polish produced by tumbling in soil 24 hours
EXP/59 x200

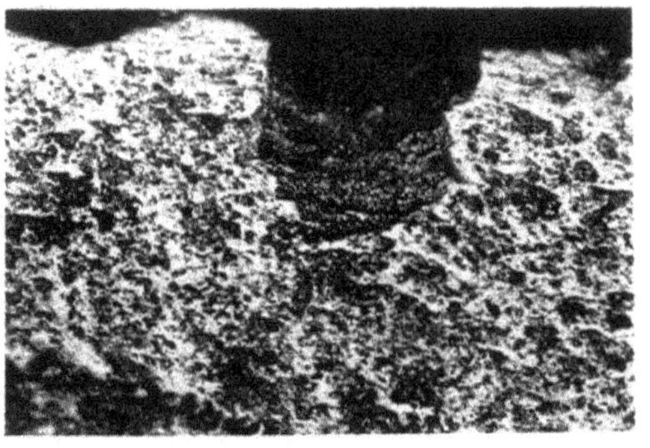

Figure 29: Macroscopic polish from cutting reeds, 60 minutes, no polish in scar
EXP/3 x200

Figure 30: Polish from cutting bracken in June for 10 minutes
EXP/12 x200

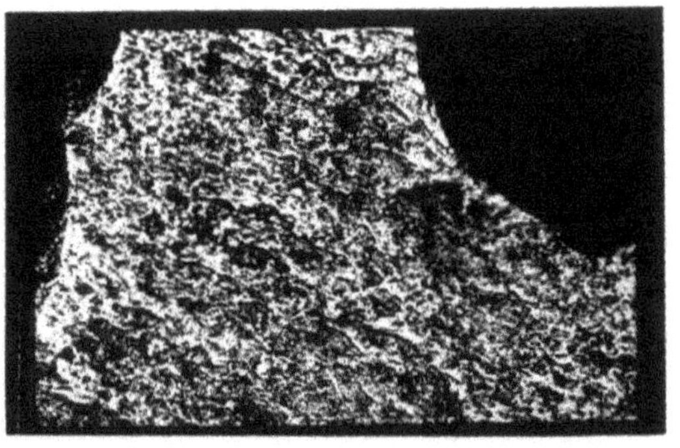

Figure 31: Polish from cutting bracken in September for20 minutes
EXP/19 x200

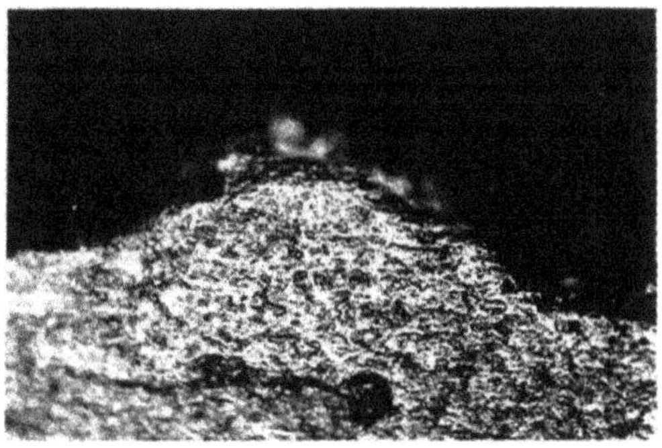

Figure 32: Polish from incising green birch for10 minutes
EXP/27 x200

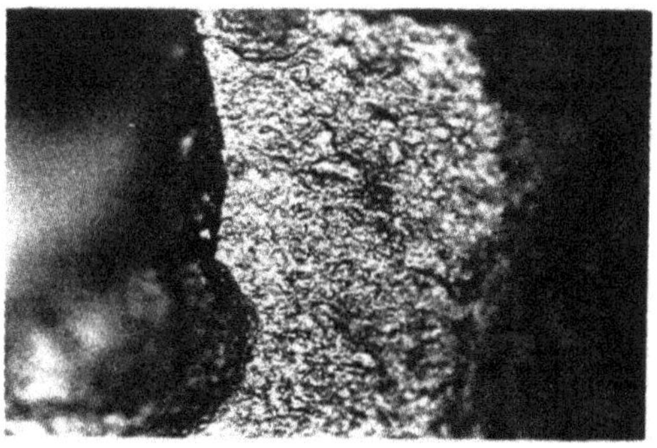

Figure 33: No polish after 30 minutes cutting dry rushes
EXP/1 x200

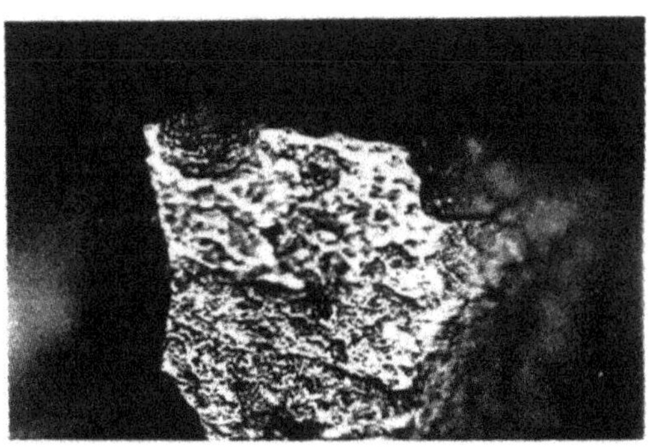

Figure 34: Well developed polish after cutting wet rushes for 15 minutes
EXP/1 x200

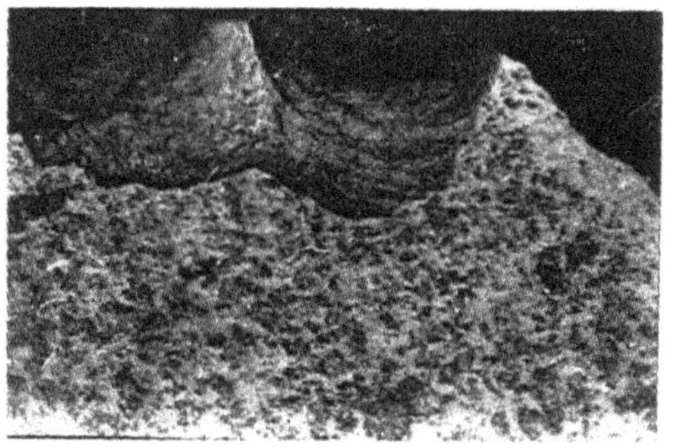

Figure 35: Polish rounding the edge from cutting green reeds for 60 minutes
EXP/3 x200

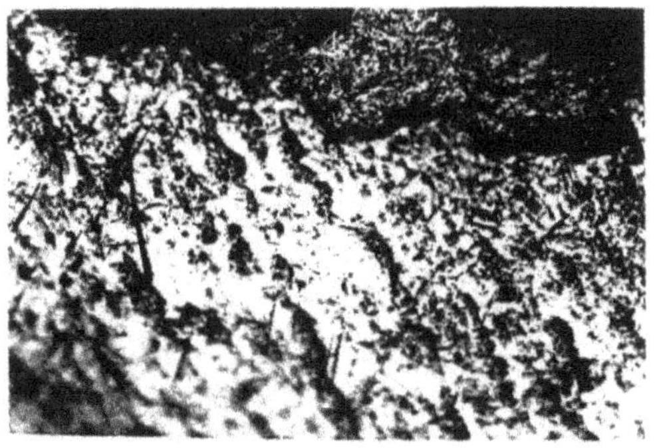

Figure 36: Cluster of bright spots with striations on edge of archaeological artefact x200

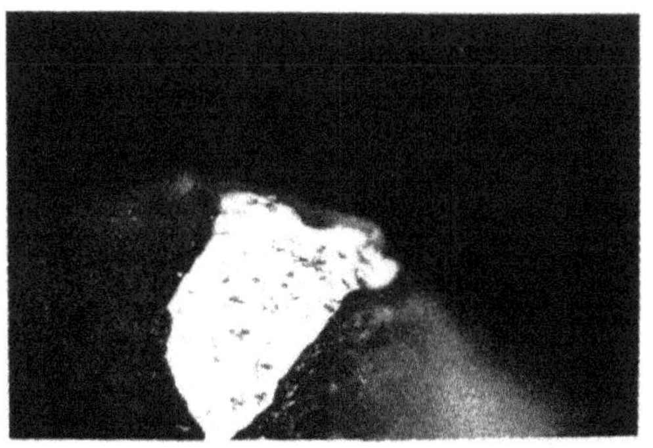

Figure 37: Totally linked polish after 30 minutes on wet rushes
EXP/1 x200

Figure 38: Soil sheen on unused edge
EXP/9 x200

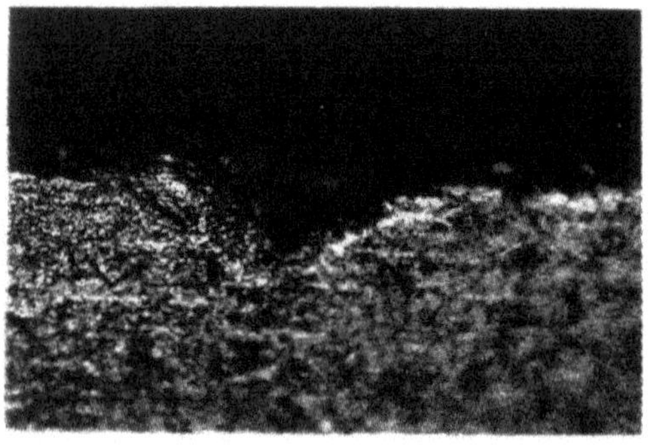

Figure 39: Incising birch for 10 minutes
EXP/27 x100

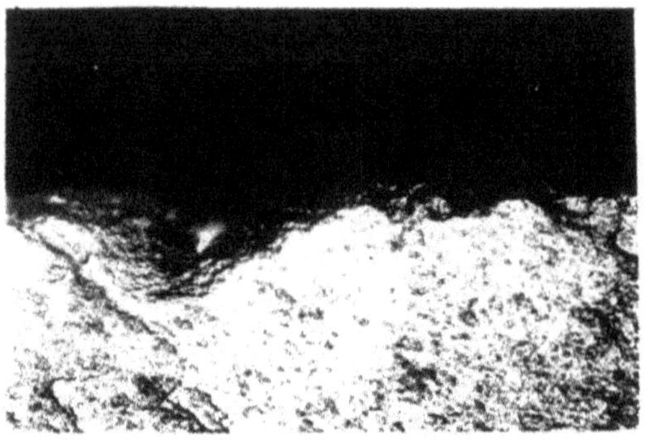

Figure 40: New polish on edge after tumbling see figure 39
EXP/27 x200

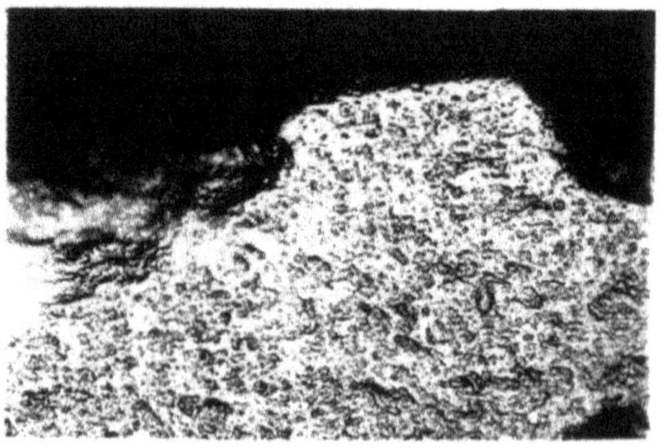

Figure 41: Polish from tumbling better developed than figure 40 and rounding edge
EXP/27 x100

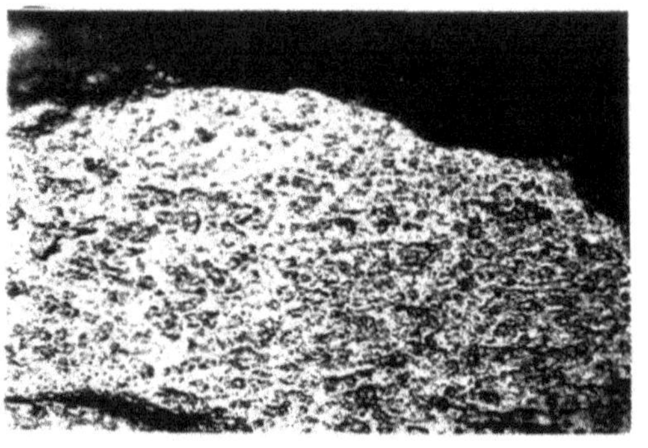

Figure 42: Polish from tumbling similar to presumed "hafting" traces on archaeological artefacts
EXP/27 x200

Figure 43: Bright spots with striations on edge of archaeological artefact x100

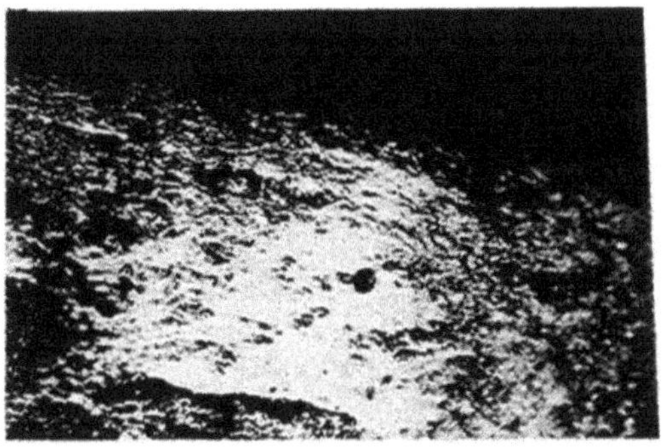

Figure 44: Bright spot from rubbing flint on flint with water for 2 minutes
EXP/77B x200

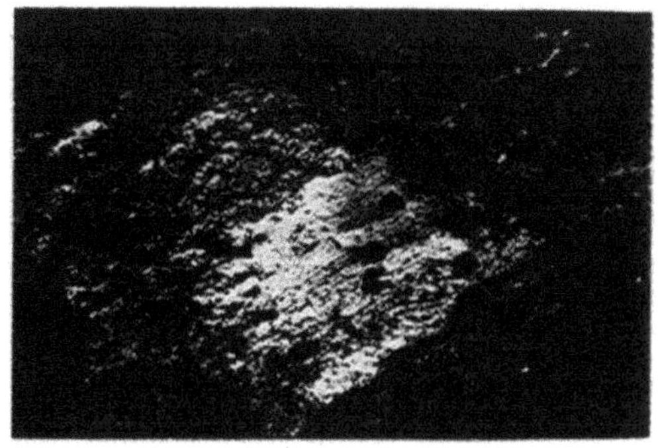

Figure 45: Archaeological bright spot similar to figure 44. x50

Figure 46: Bright spots spots on HH artefact with distribution similar to use-wear x100

Figure 47: Bright spots similar on same artefact as figure46 to figures 35 & 29 used on reeds x100

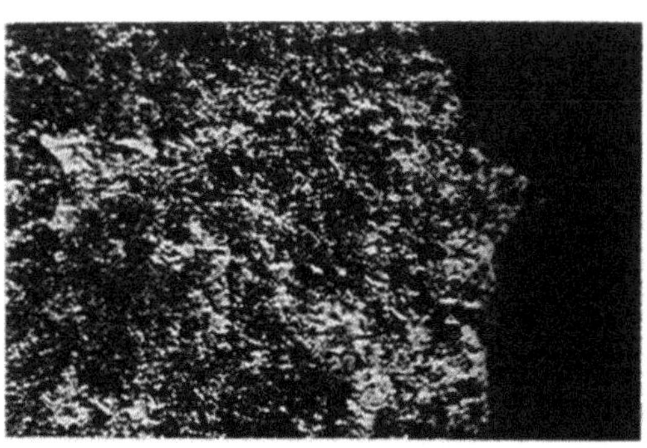

Figure 48: Polish ***before*** brushing experiment
EXP/106 x100

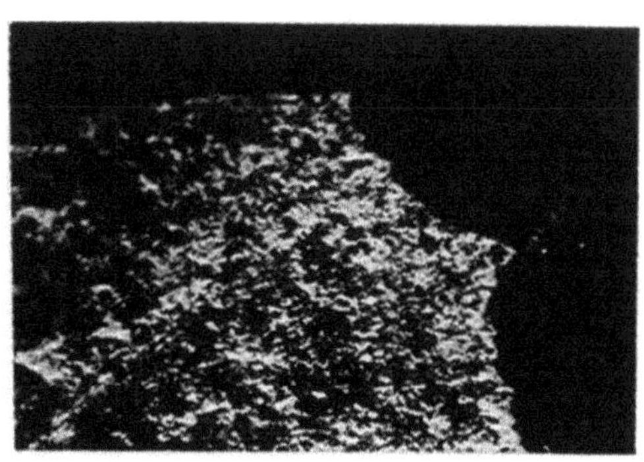

Figure 49: Polish ***after*** brushing with nylon brush
EXP/106 100

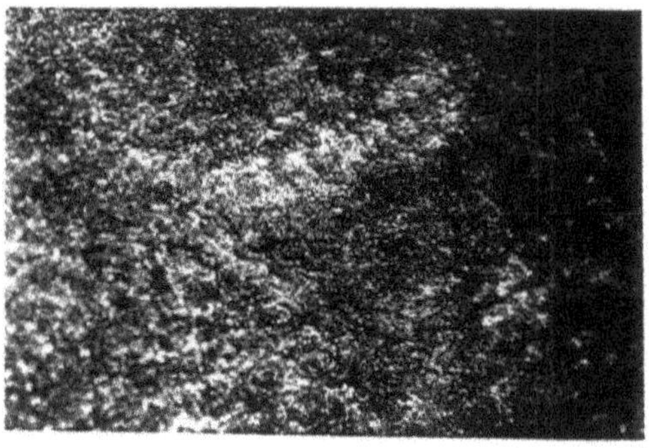

Figure 50: Weakly developed polish ***before*** brushing experiment
EXP/106 x100

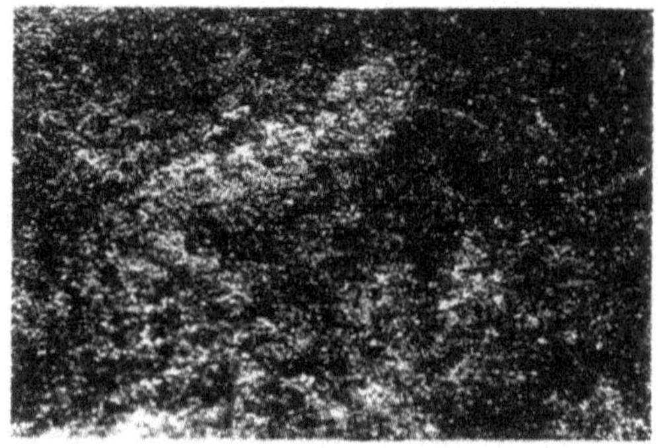

Figure 51: Weakly developed polish ***after*** brushing with nylon brush
EXP/106 x100

Figure 52: Unpolished flint ***before*** brushing experiment
EXP/106 x100

Figure 53: Unpolished flint ***after*** brushing with nylon brush
EXP/106 x100

Figure 54: Weakly developed polish on archaeological artefact from Kebara x200

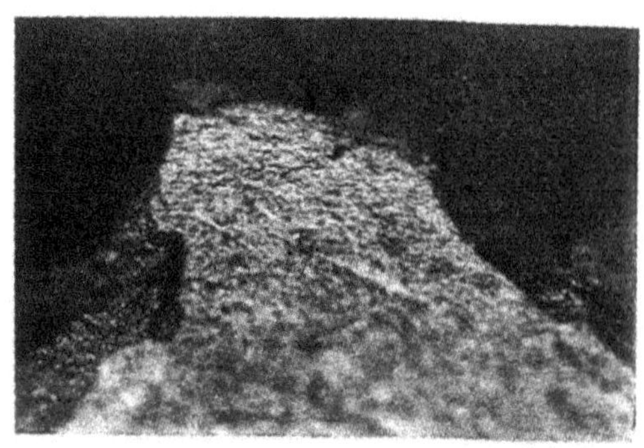

Figure 55: Polish after cutting meat for 60 minutes
EXP/18 x200

Figure 56: Polish from butchering (meat and bone)
EXP/88 x200

Figure 57: Weakly developed polish from butchering
EXP/18 x200

Figure 58: Linear features parallel to edge on archaeological scraper x100

Figure 59: Polish after scraping wet hide for5 minutes.
EXP/128 x200

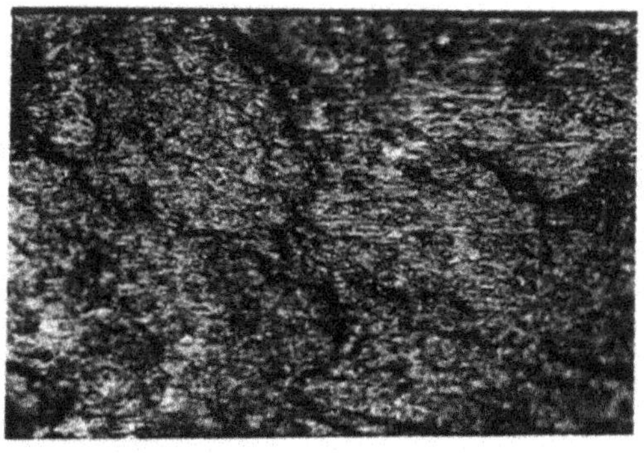

Figure 60: Polish from grooving wood with soft abrasive
EXP/93 x200

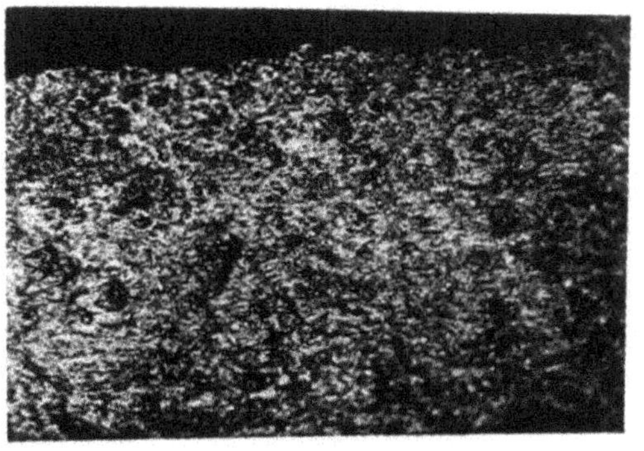

Figure 61: Polish without striations grooving wood with high angled edge
EXP/96 x200

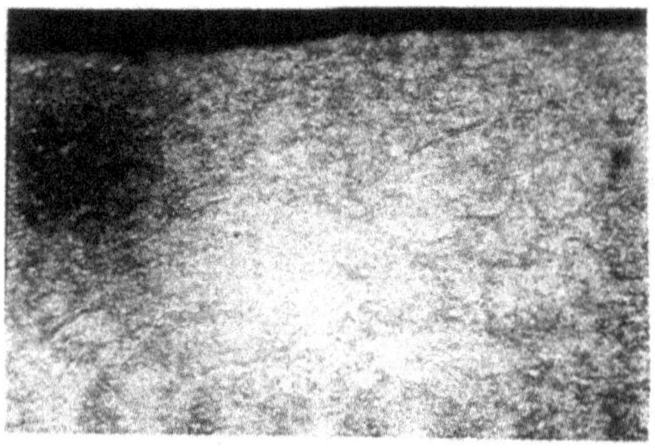

ISR/1: Fine grained does not refract light as BRND x200

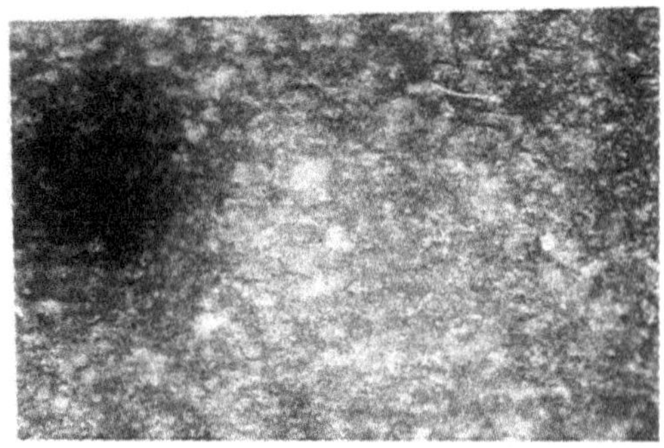

ISR/4a: Coarse grained, many inclusions x100

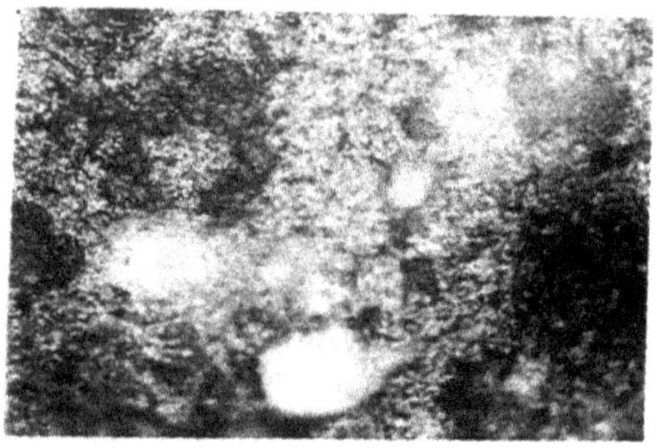

ISR/2: Medium to coarse small and large inclusions large shallow hollows x200

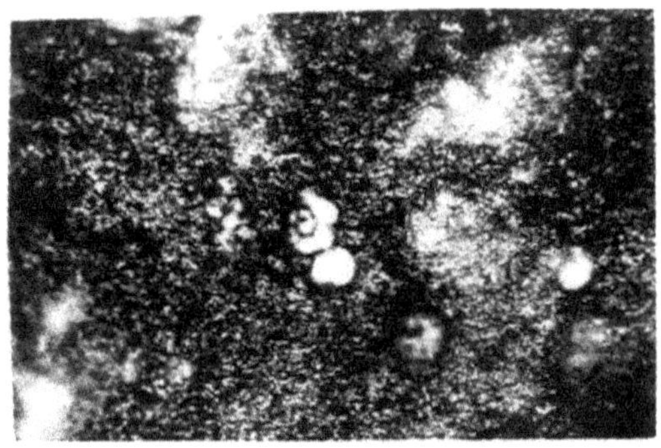

ISR/4b: Medium grained with many ridges and inclusions x100

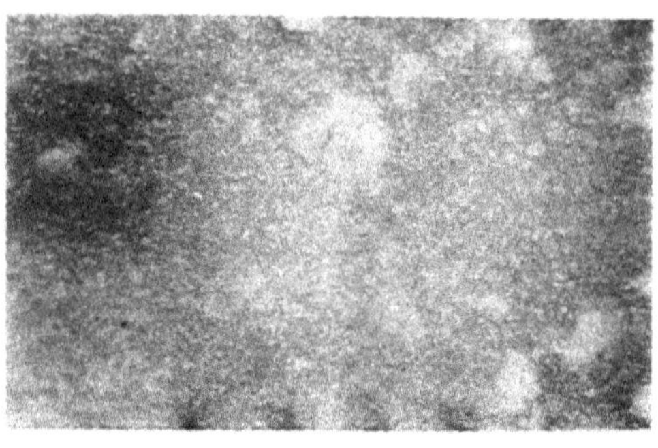

ISR/3: Medium grained weathered nodule with large inclusions x100

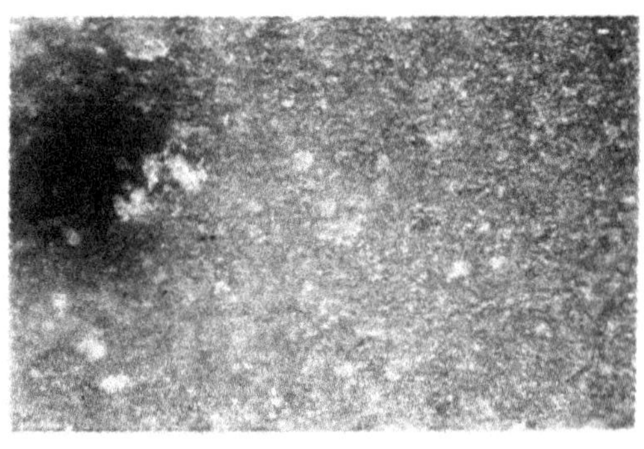

ISR/5a: Not very tightly packed even in this area of finer grain x200

ISR/5b: Flint see figure 7 for Brandon flint see figure 98 For Potters Bar flint see figures 99 and 225

Figure 62: Various Israeli flint types

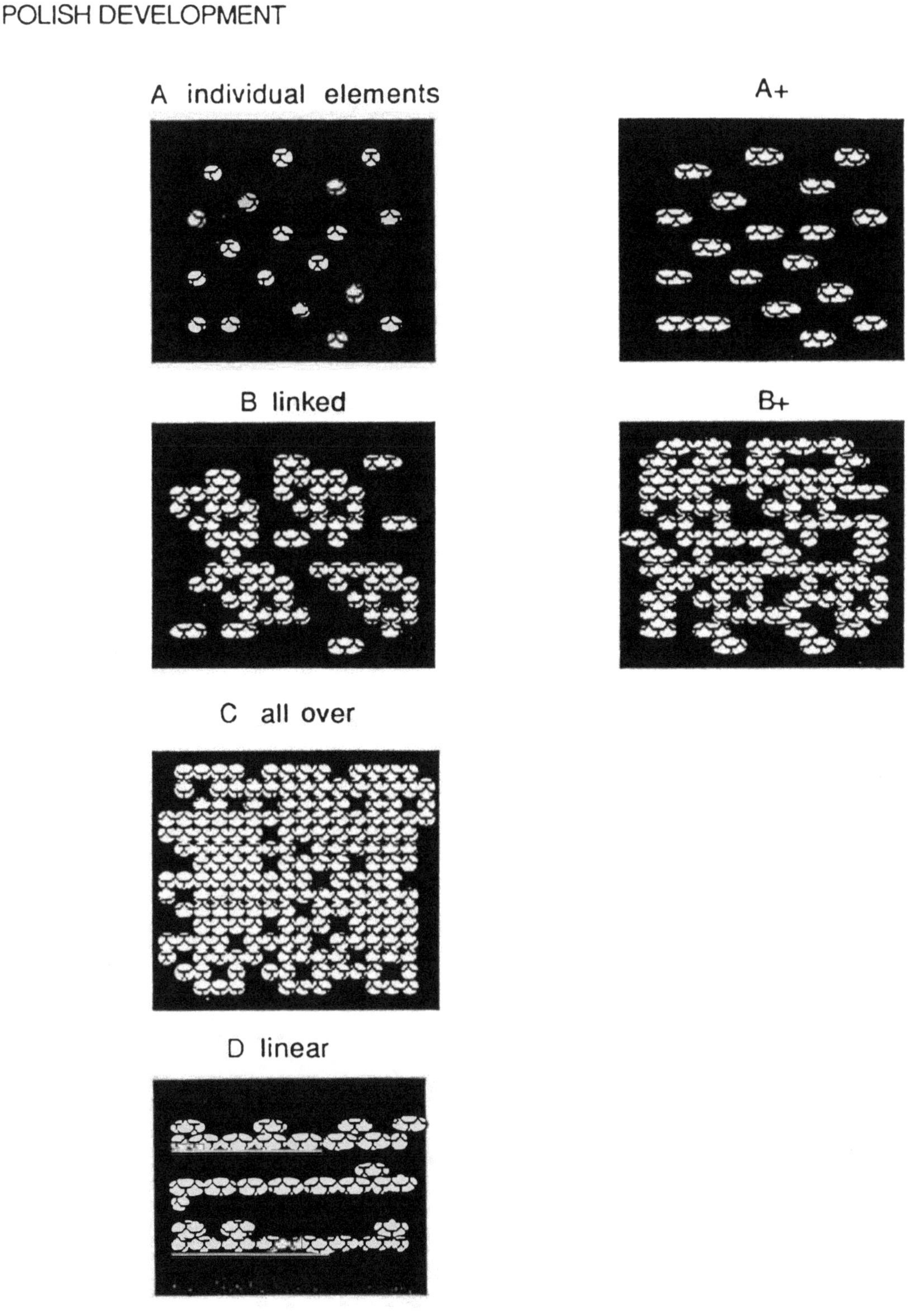

Figure 63: Polish Development. After Grace (1989:Figure 55)

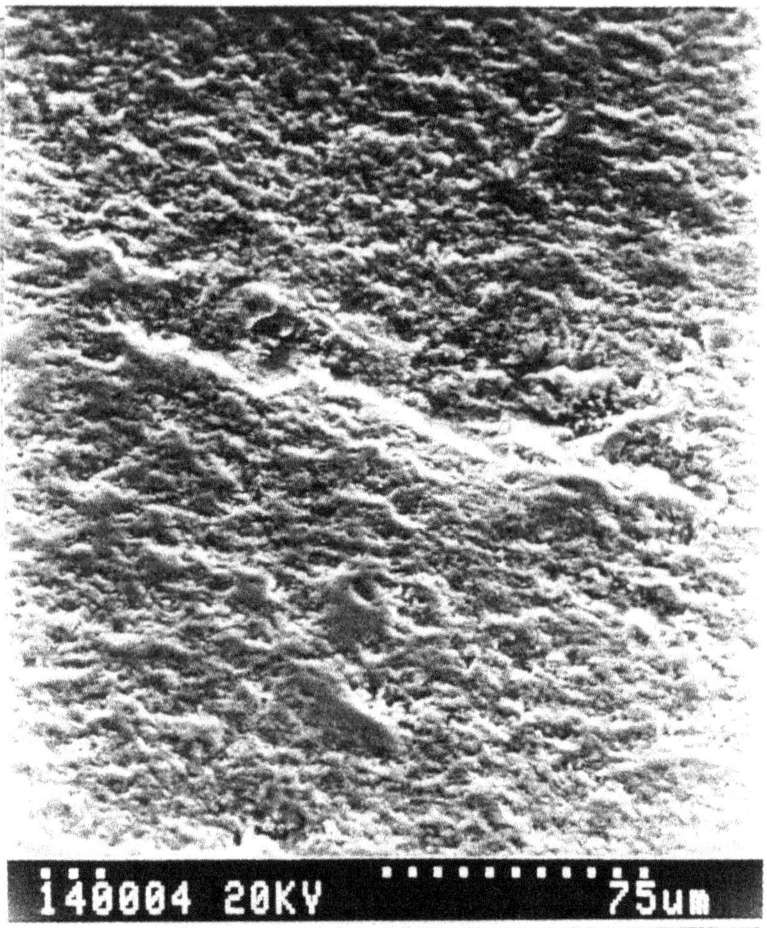

Figure 64: Striation as a groove in the surface smoothed by wet goat hide
EXP/140 SEM

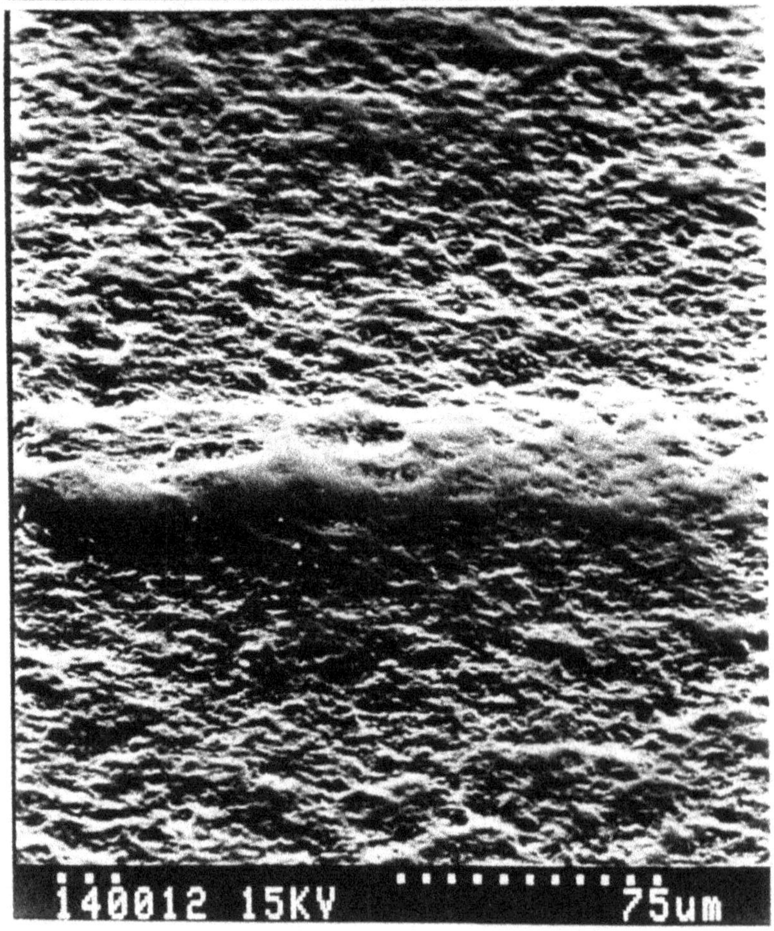

Figure 65: Large asperity plastically deformed as linear feature
EXP/141 SEM

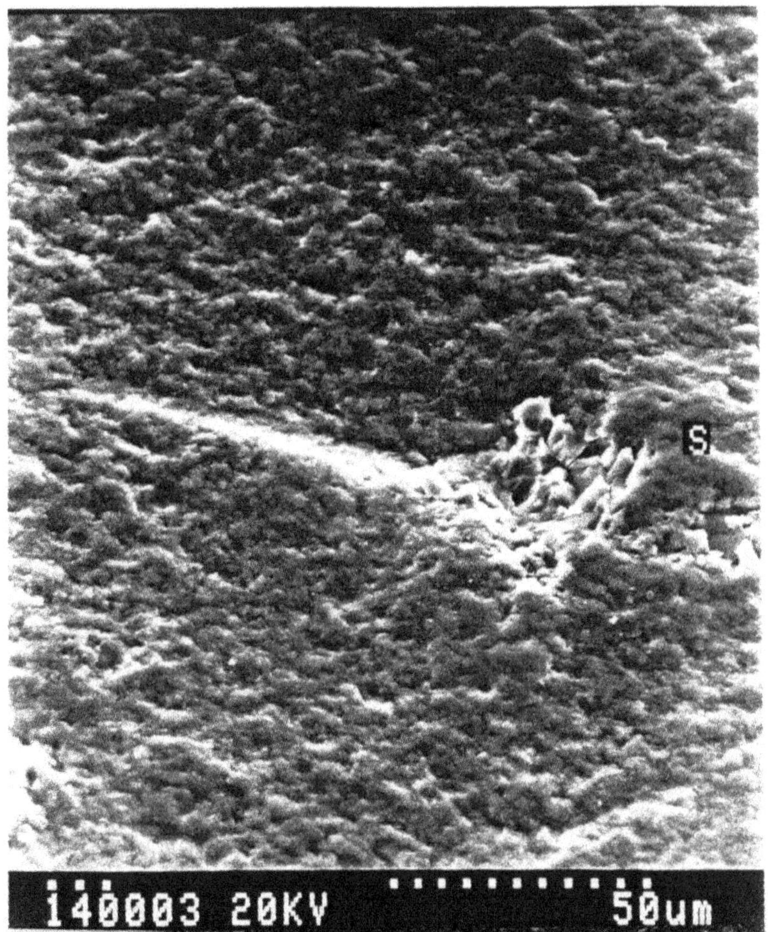

Figure 66: Comet shaped striation in surface smoothed by wet goat hide
EXP/140 SEM

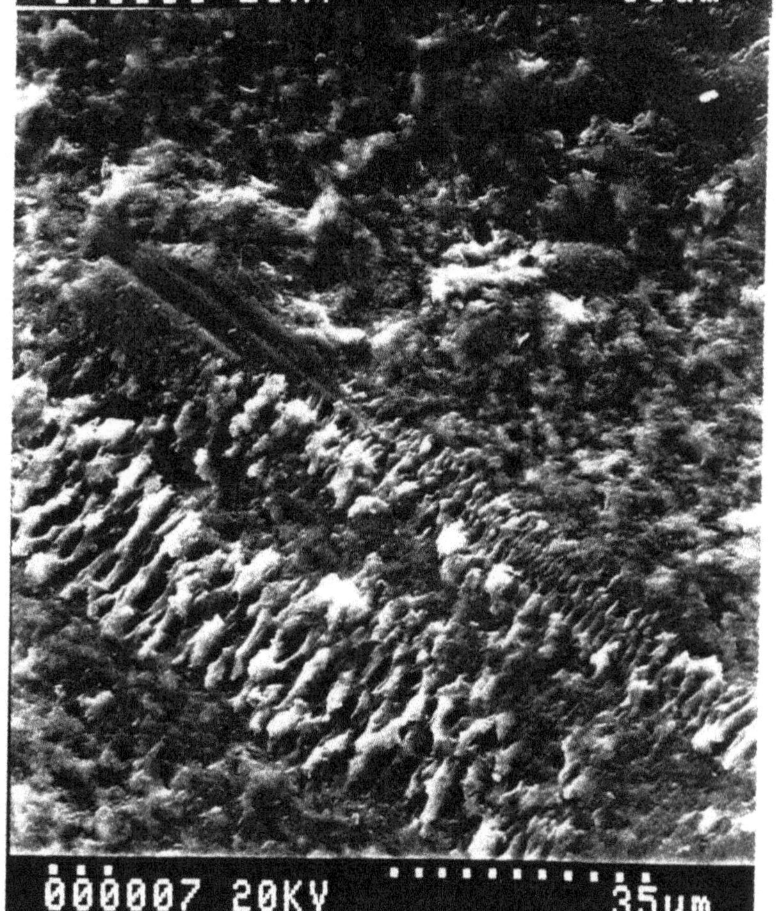

Figure 67: Fern-like striation from scraping hide with ochre
EXP/127 SEM

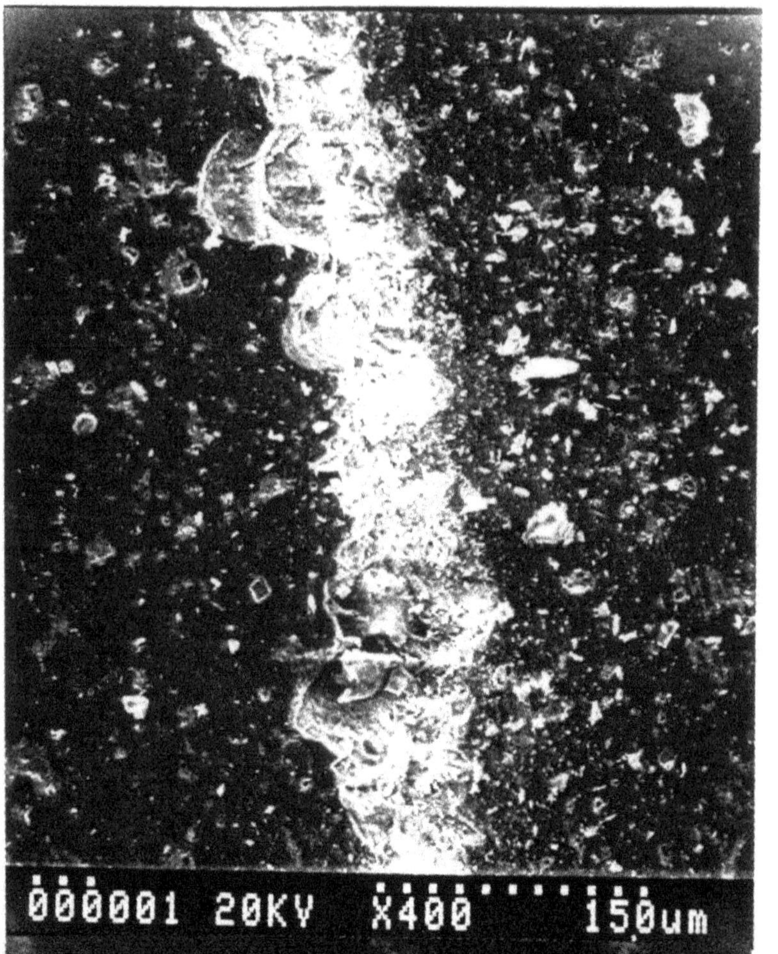

Figure 68: Scratch on surface highly smoothed by ¼μ grit. Note lateral cracks and crushed crystals SEM

Figure 69: Enlarged view of the top left lateral crack in figure 68 above.
EXP/103 SEM

Figure 70: Comet-tail striation with smooth tail on flint polished by dry hide.
EXP/141 SEM

Figure 71: Smooth surface from scraping deer hide with ochre.
EXP/120 SEM

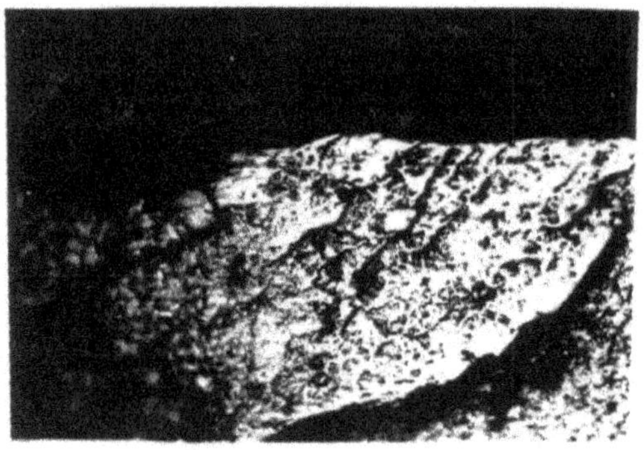

Figure 72: Polish from grooving wood with water.
EXP/92(1) x200

Figure 73: Polish grooving wood with oil.
EXP/92(2) x200

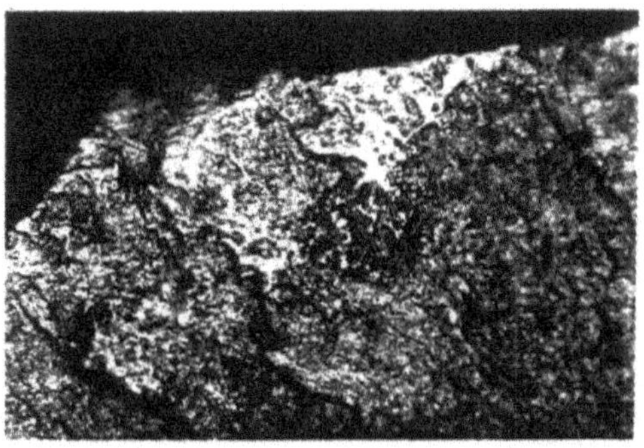

Figure 74: Polish from grooving dry wood
EXP/91 x200

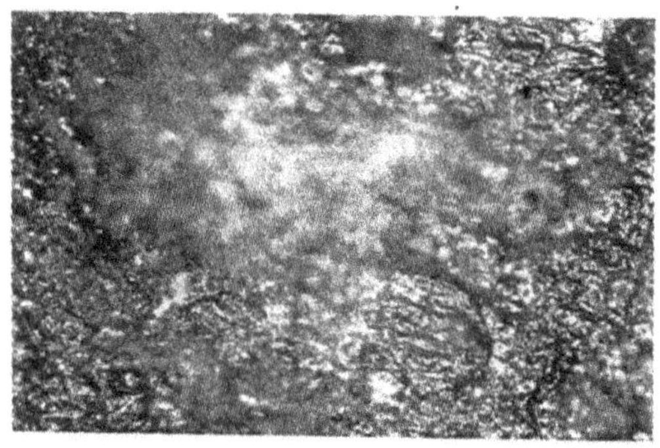

Figure 75: Unused goat hide
EXP/140 and EXP/141 x200

Figure 76: Goat hide after use speckled with flint particles
EXP/140 x200

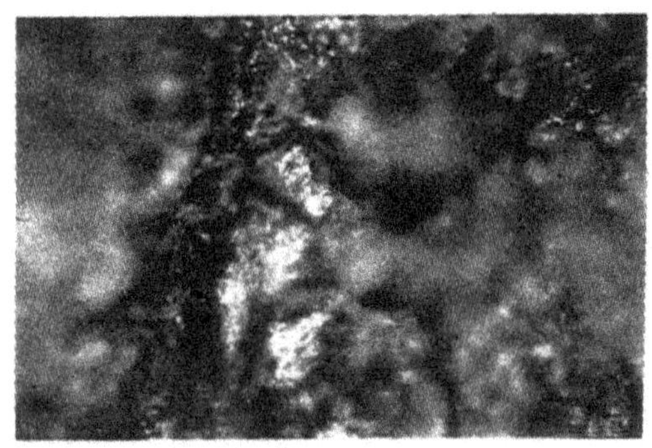

Figure 77: Goat hide after use speckled with flint particles
EXP/141 x200

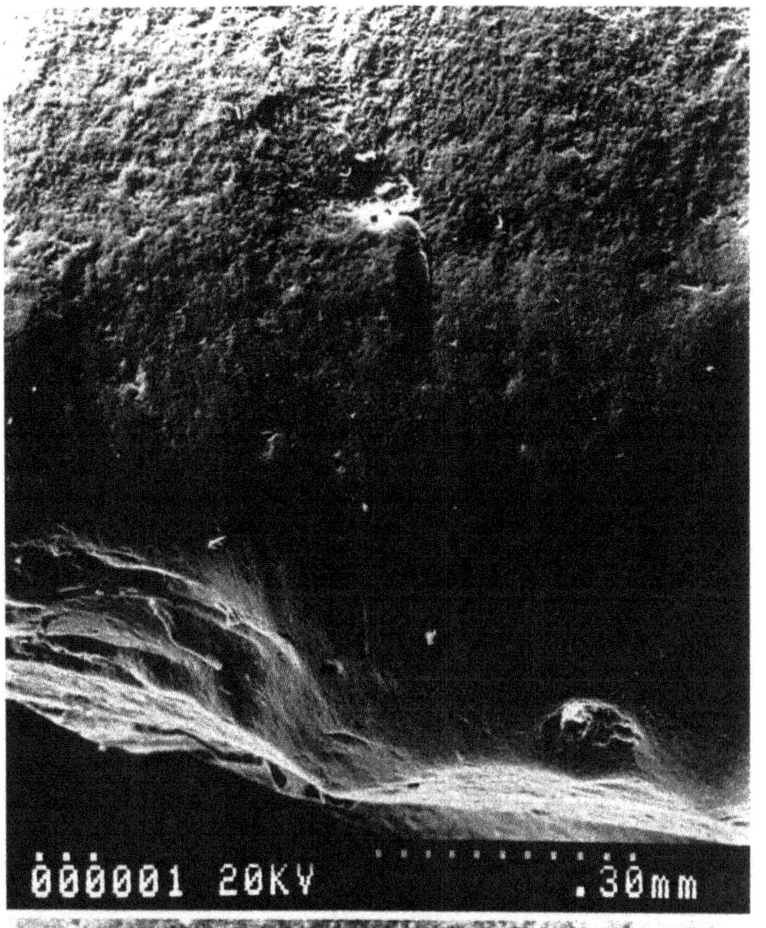

Figure 78: Highly rounded and smoothed edge scraping hide with unsifted ochre and water, cf figure 83.
EXP/118 SEM

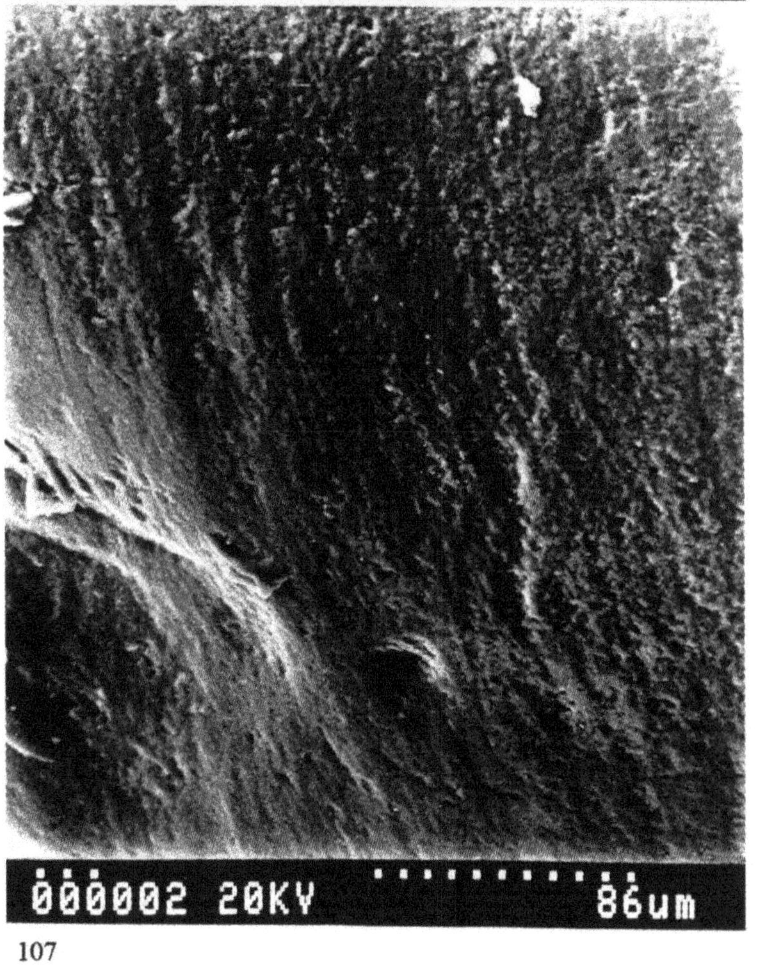

Figure 79: Higher magnification of area in figure 78, note fine striations perpendicular to edge.
EXP/118 SEM

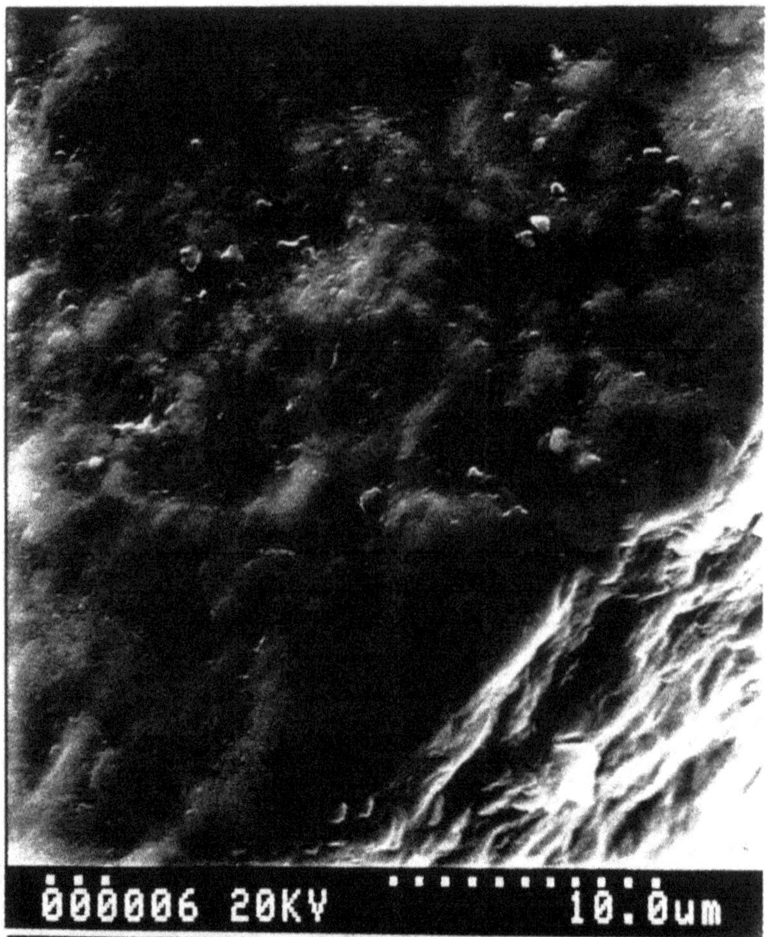

Figure 80: Smooth surface from scraping hide with ochre $<212\mu$ and water.
EXP/127 SEM

Figure 81: Smooth surface with striation from scraping hide with ochre $<212\mu$ and hexane.
EXP/129 SEM

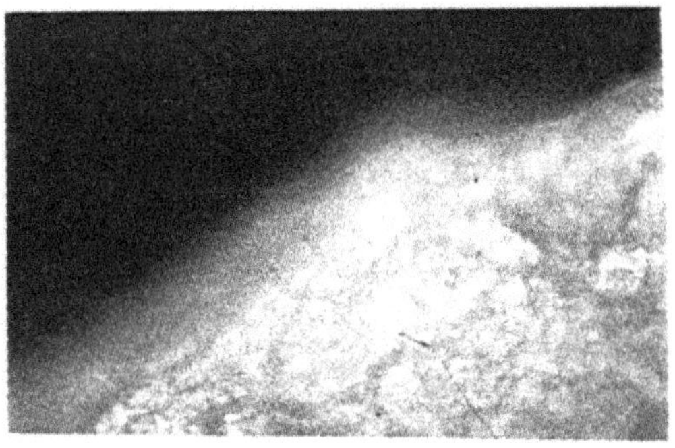

Figure 82: Little developed polish on coarse flint scraping hide 20 minutes.
EXP/117 x100

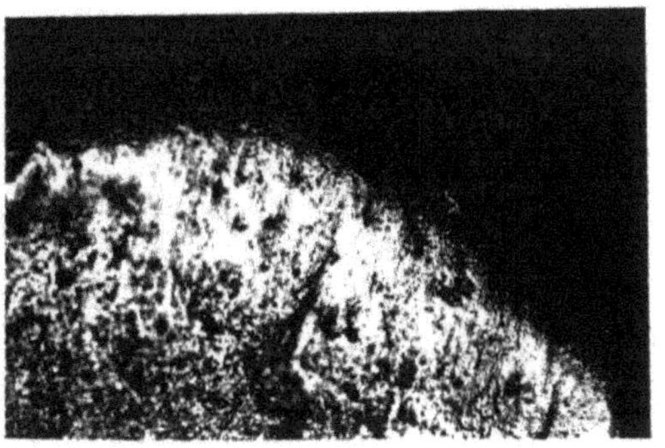

Figure 83: Well developed polish and rounded edge scraping hide with unsifted ochre, cf figures 78 and 79.
EXP/118 x100

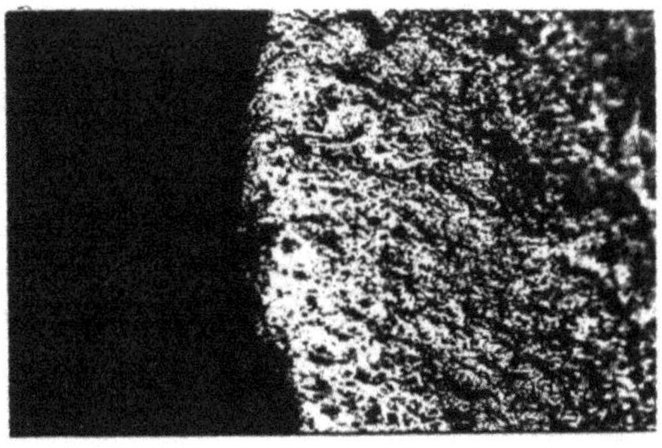

Figure 84: Well developed polish and rounded edge scraping hide with unsifted ochre.
EXP/119 x200

Figure 85: Very well developed polish and rounded edge scraping hide with 212/300μ ochre.
EXP/120 x200

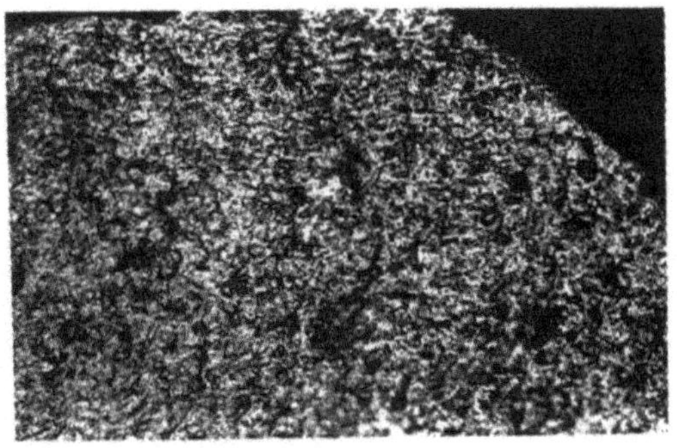

Figure 86: Little developed polish scraping cow hide with 425μ ochre
EXP/121 x200

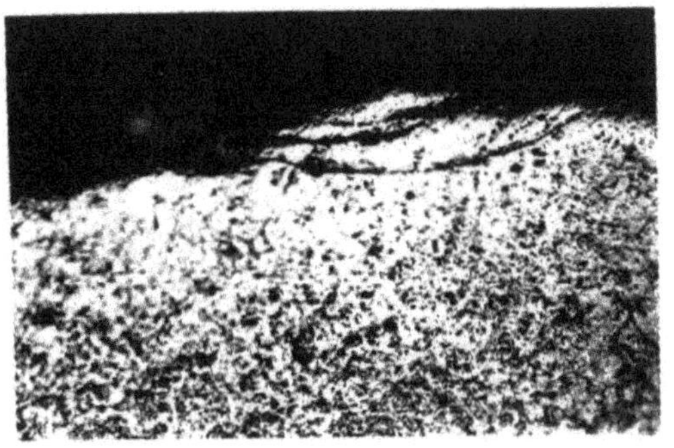

Figure 87: Very well developed polish and rounded edge scraping hide with 212/300μ ochre.
EXP/123 x200

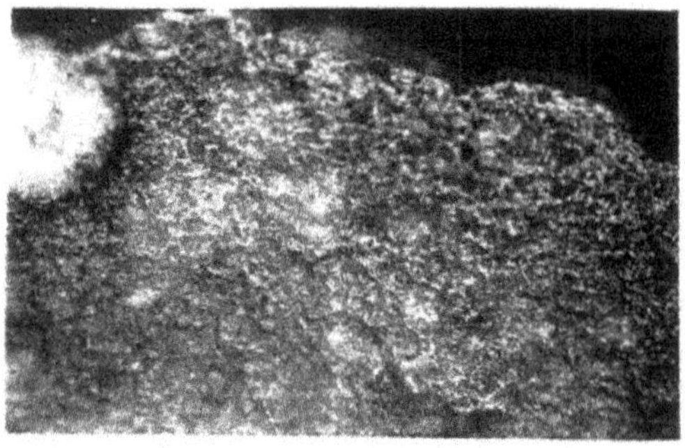

Figure 88: Little developed polish on coarse flint scraping cow hide with <212μ ochre.
EXP/124 x200

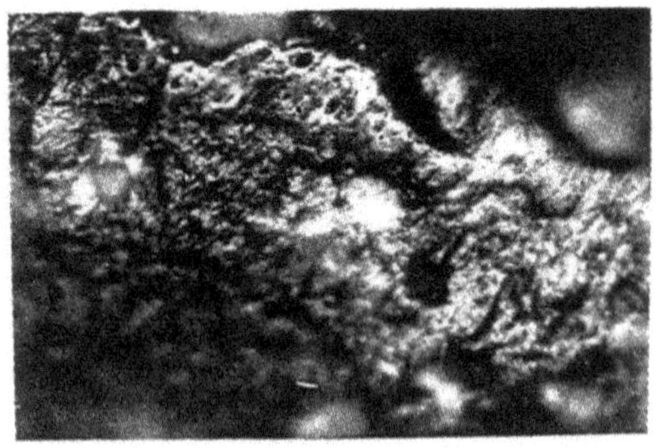

Figure 89: Well developed polish on coarse flint scraping hide with <212m ochre.
EXP/125 x200

Figure 90: Little developed polish scraping cow hide with<212/300μ ochre
EXP/126 x200

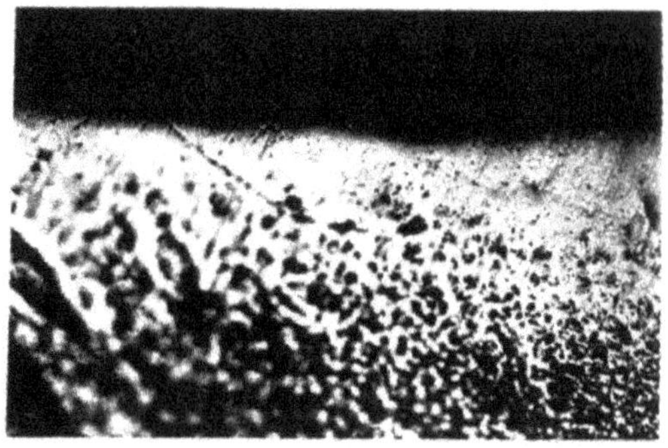

Figure 91: Very well developed polish on fine flint scraping goat hide with <212μ ochre, cf. figure 80.
EXP/127 x200

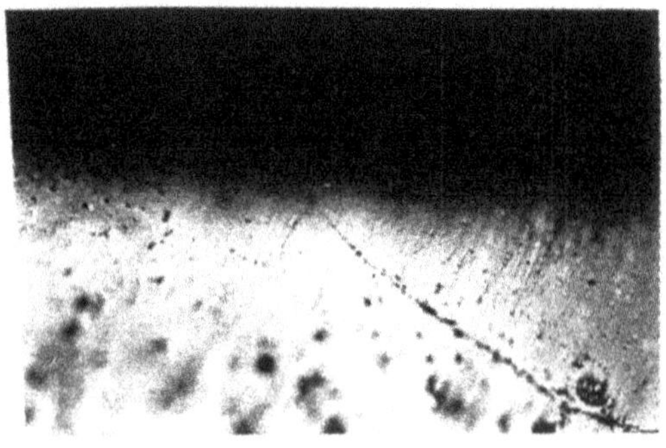

Figure 92: Same area as figure 91 showing curved crack in flint.
EXP/127 x400

Figure 93: Very well developed polish scraping hide with hexane and <212μ ochre, cf figure 81.
EXP/129 x200

Figure 94: Retouch polish from quartzite hammer.
EXP/123 x100

Figure 95: Polish from scraping hide. No retouch traces visible.
EXP/123 x100

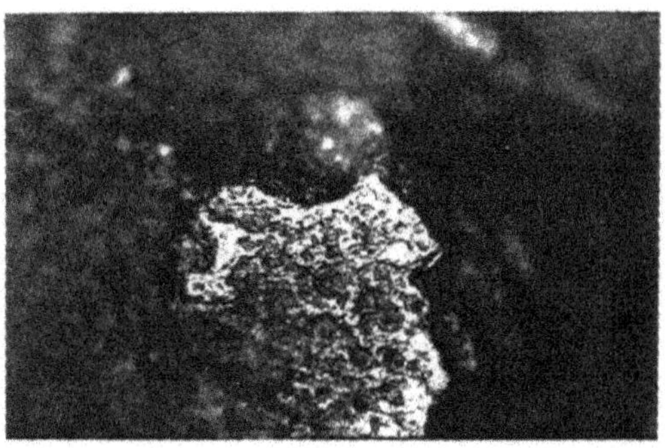

Figure 96: Linked polish from buffing flint with hard wool buff and water.
EXP/108 x200

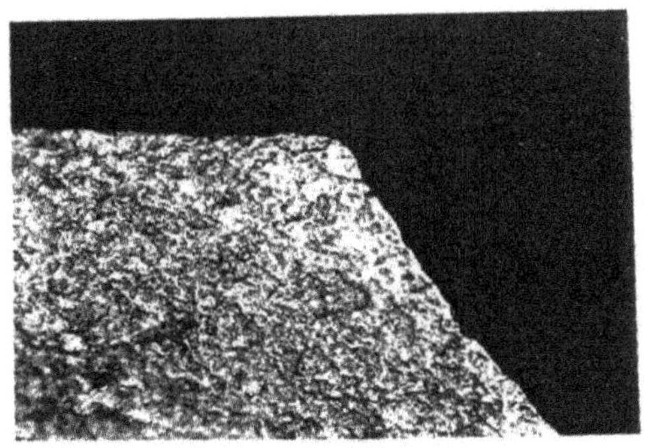

Figure 97: Linked polish from buffing flint with chamois covered buff and water.
EXP/109 x200

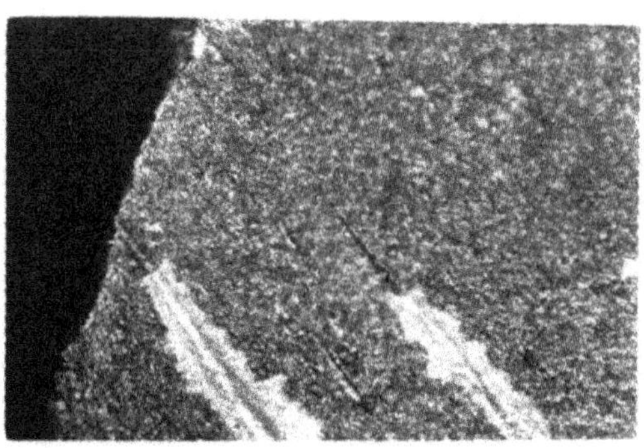

Figure 98: Fine grained Brandon flint.
EXP/129 x100

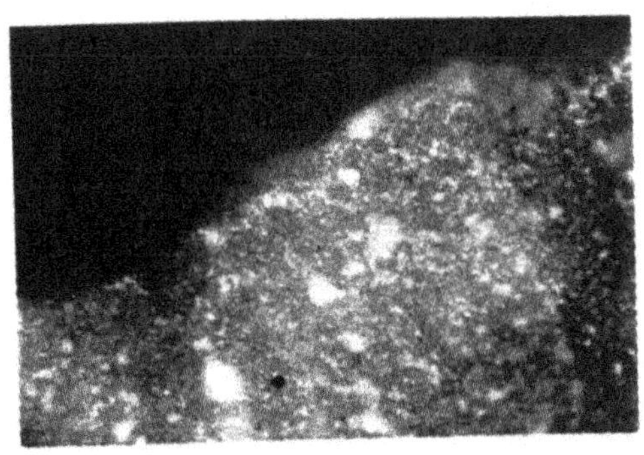

Figure 99: Coarse grained Potters Bar flint.
EXP/145 x100

Figure 100: Flint particles embedded in goat hide used wet to buff flint. EXP/140

Figure 101: Goat hide. as in figure 100 at higher magnification.

Figure 102: X-ray dot mapping of silicon distribution on hide in figure 100.
EXP/140

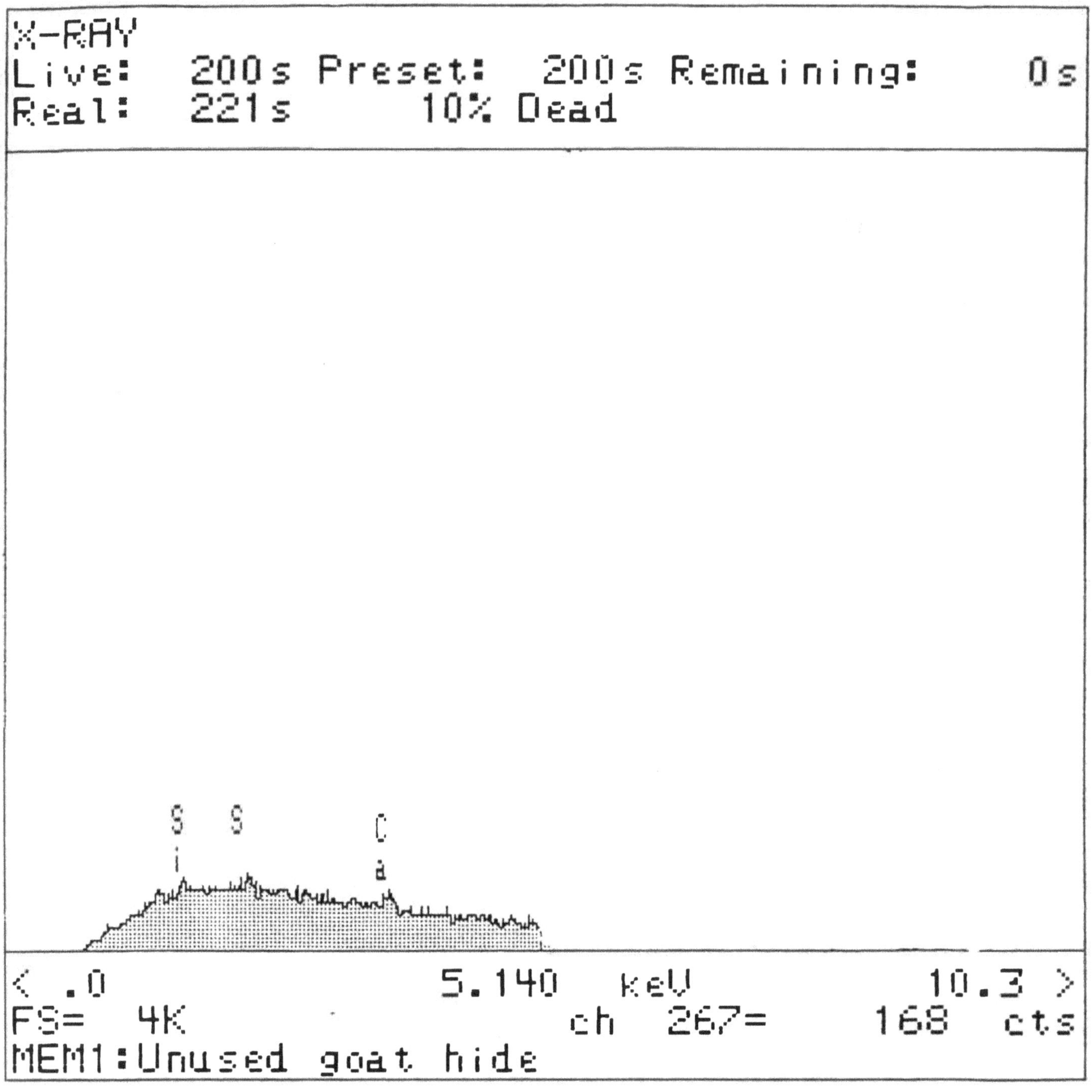

Figure 103: X-ray spectrum of elements on unused goat hide. Note silicon peak. (Signals above 5.140 KeV have been removed as they contained spurious signals probably arising from pole piece of the microscope at the time of observation).

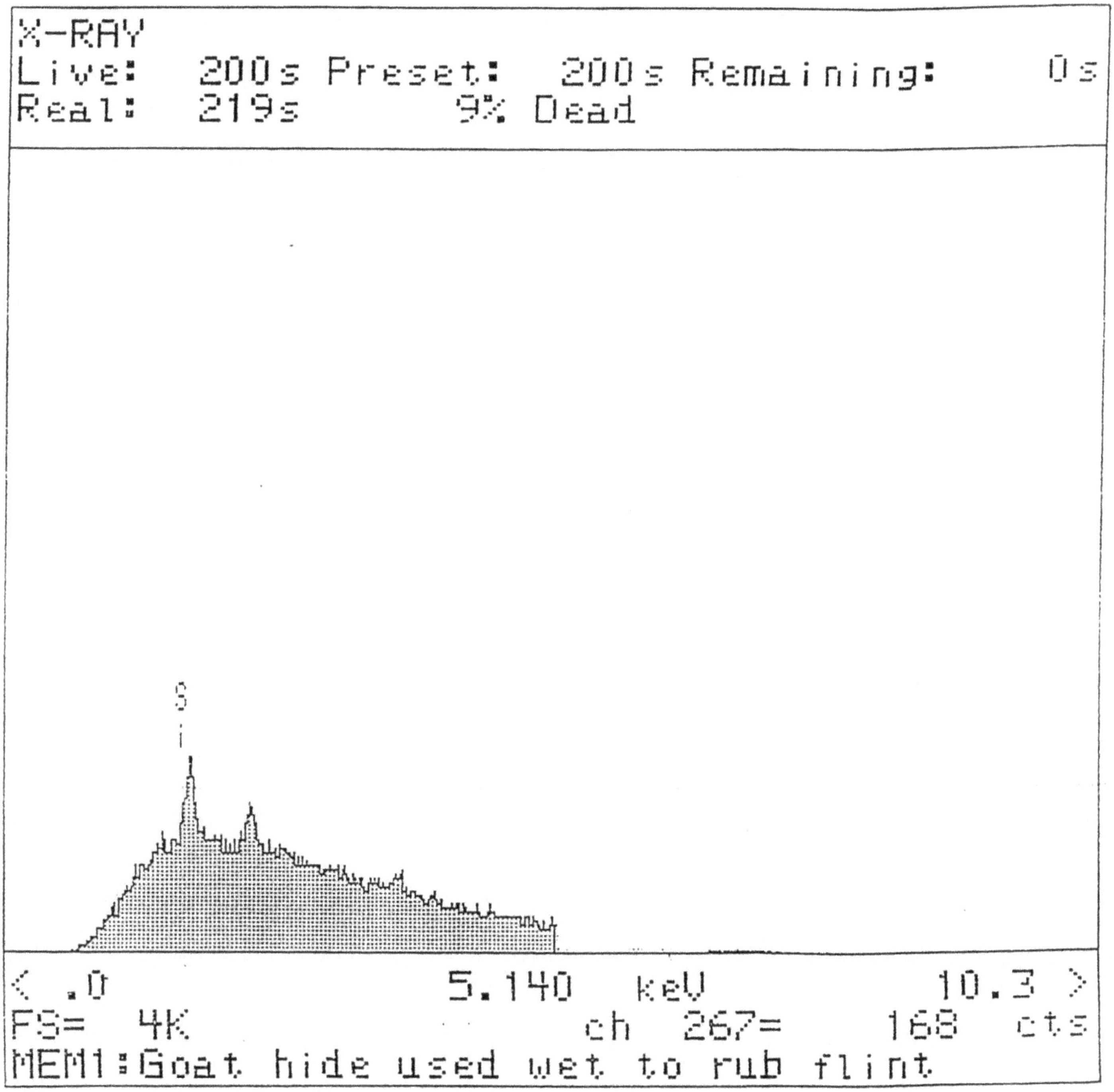

Figure 104: X-ray spectrum of elements on goat hide EXP/140. Note increase in silicon peak. (Signals above 5.140 KeV have been removed as they contained spurious signals probably arising from pole piece of the microscope at the time of observation).

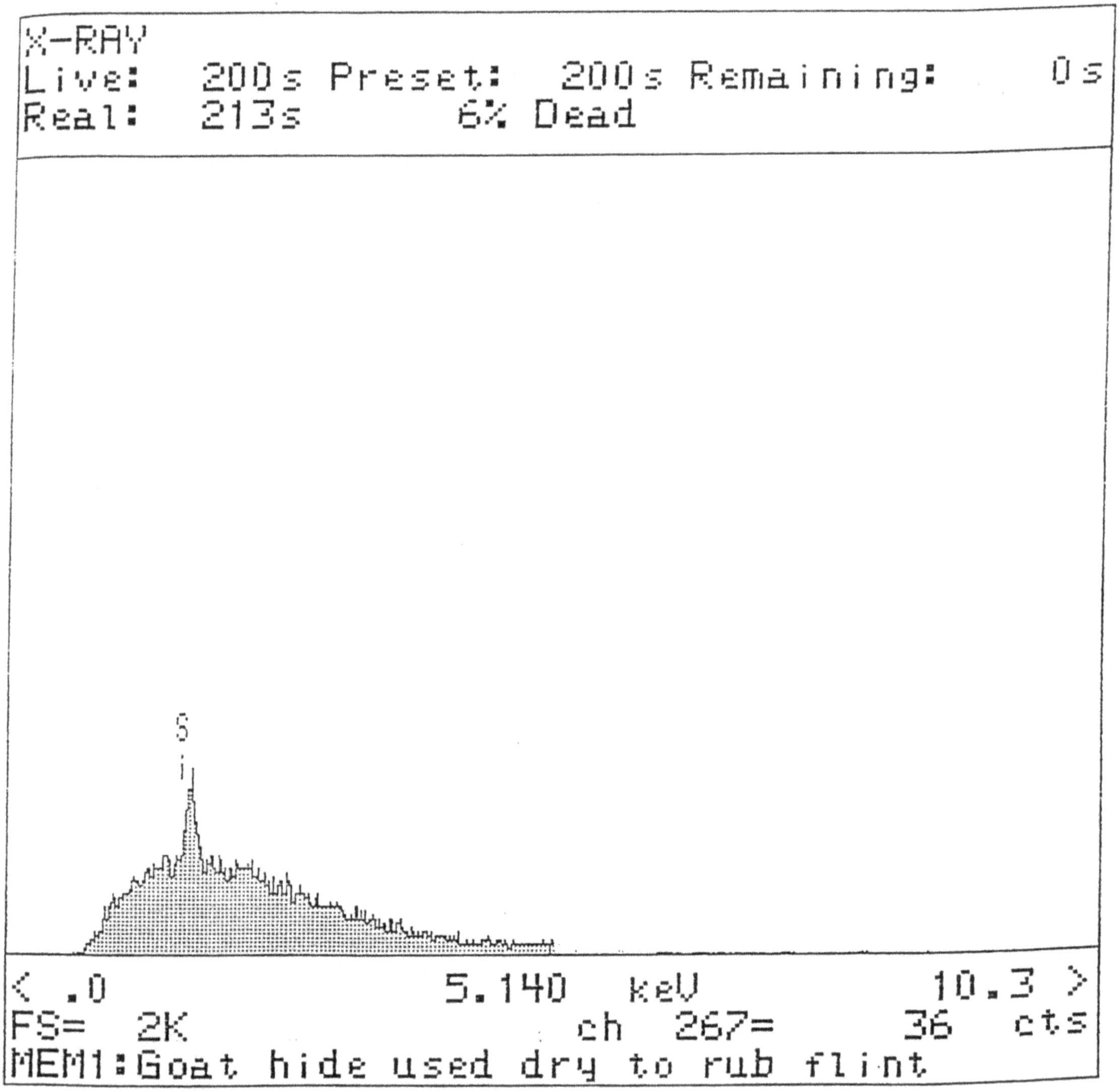

Figure 105: X-ray spectrum of elements on goat hide in EXP/141. Note increase in silicon peak. (Signals above 5.140 KeV have been removed as they contained spurious signals probably arising from pole piece of the microscope at the time of observation).

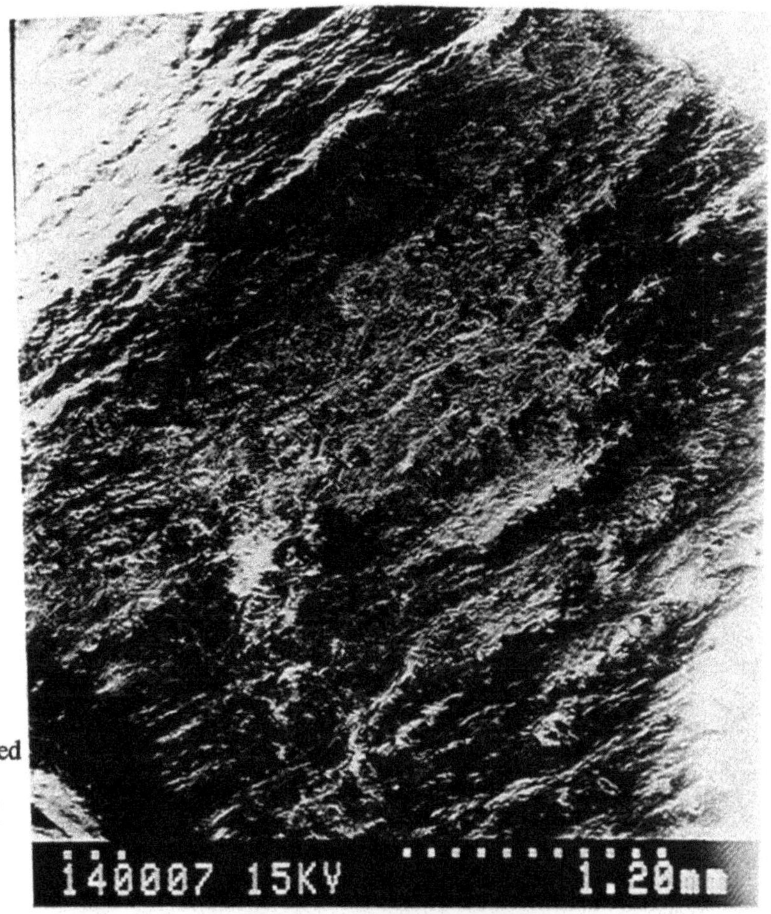

Figure 106: Flint particles embedded in goat hide used dry to buff flint.
EXP/141 SEM

Figure 107: Goat hide as in figure 106 at higher magnification.
EXP/141 SEM

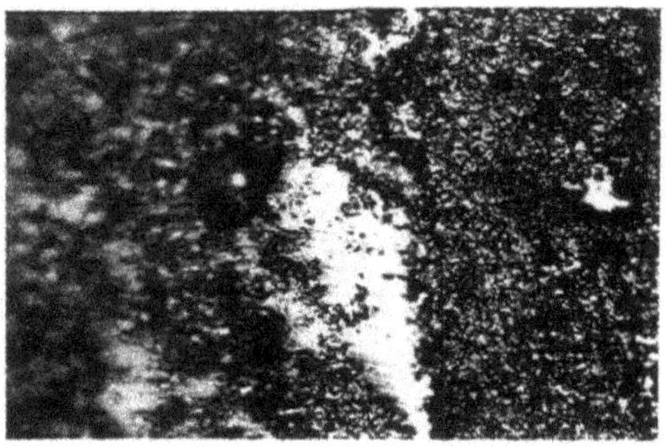

Figure 108: Flat well developed macroscopic polish like bright spots on flint buffed with wet hide.
EXP/140 x400

Figure 109: Diffuse not linked polish on rounded surface of artefact buffed with wet hide.
EXP/140 x100

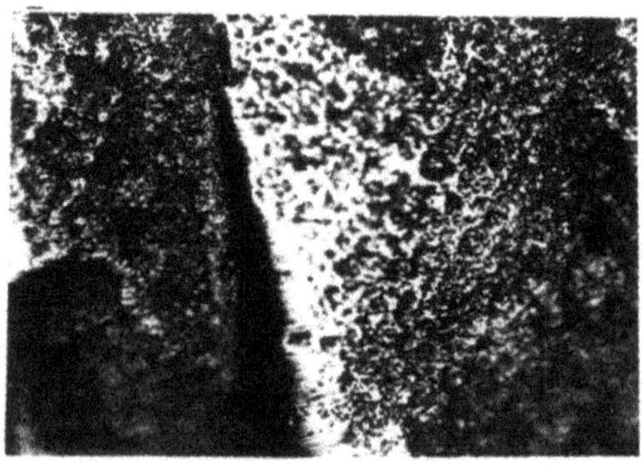

Figure 110: Polish less development than in figure 108 on artefact polished by dry hide.
EXP/141 x200

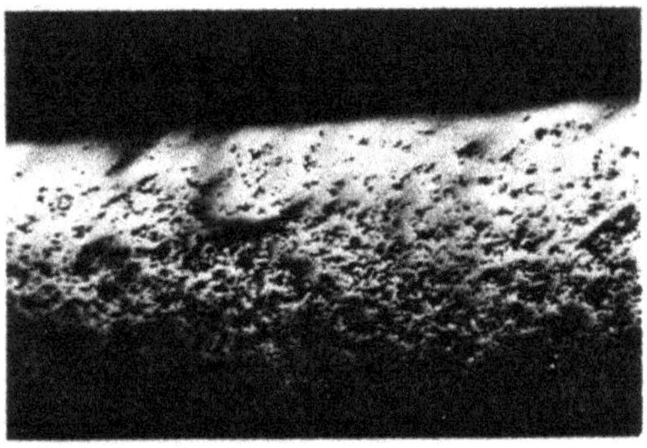

Figure 111: Polishing artefact with ¼μ paste on turntable macroscopic polish.
EXP/85 x100

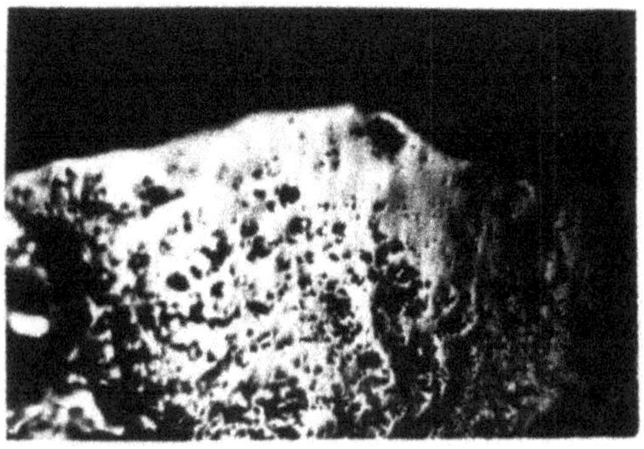

Figure 112: Polishing artefact with ¼μ paste on turntable macroscopic polish.
EXP/97 x100

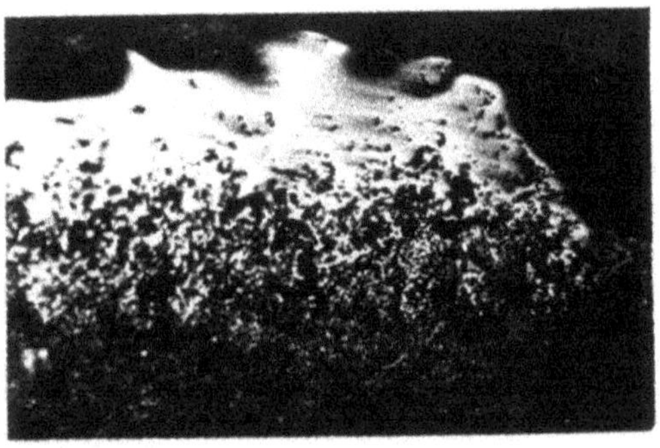

Figure 113: Polishing artefact with ¼μ paste on turntable macroscopic polish.
EXP/84 x100

Figure 114: Polishing artefact from ¼μ paste on turntable.
EXP/94 x100

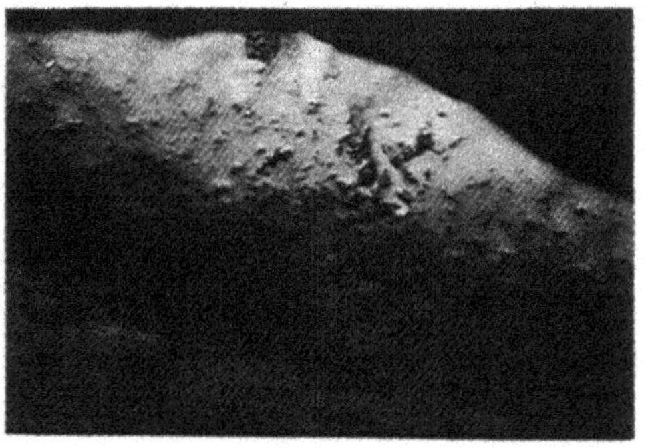

Figure 115: Macroscopic polish from scraping turntable cloth with ¼μ paste.
EXP/89 x200

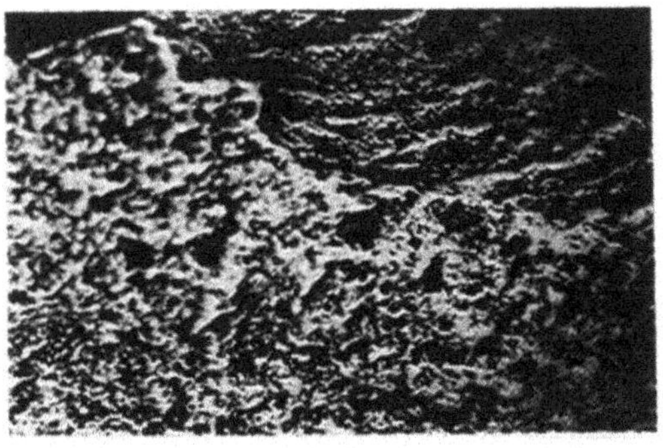

Figure 116: C polish from buffing flint with ¼μ paste for a few minutes.
EXP/103 x100

Figure 117: C+ polish from buffing flint with ¼μ paste for 15 minutes.
EXP/103 x100

Figure 118: C++ polish from buffing flint with ¼μ paste for30 minutes
EXP/103 x100

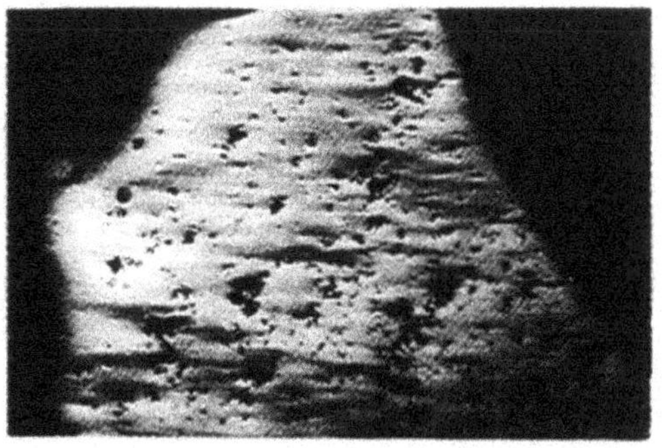

Figure 119: C++ polish from buffing flint with ¼μ paste for 30 minutes.
EXP/101 x100

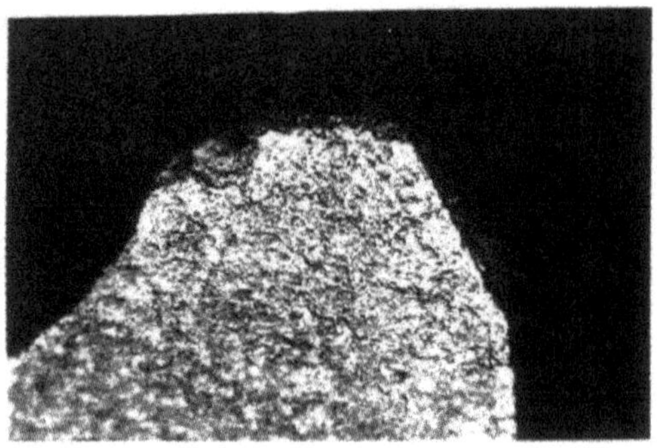

Figure 120: Little developed A polish scraping cow hide for 90 minutes.
EXP/142 x200

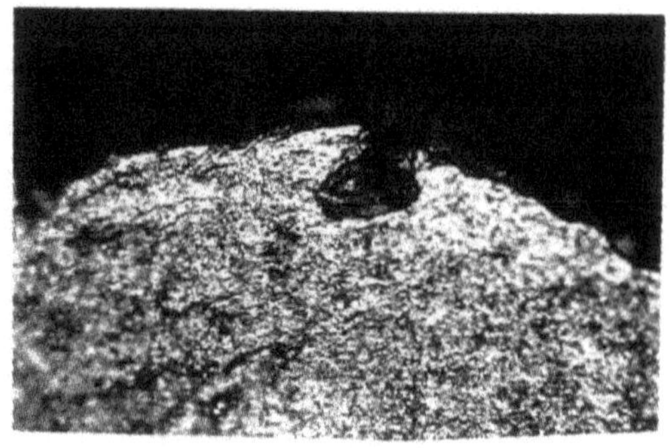

Figure 121: Little developed A+ polish scraping goat hide for 90 minutes.
EXP/143 x200

Figure 122: The same edge as in SEM figure 126.
EXP/143 x400

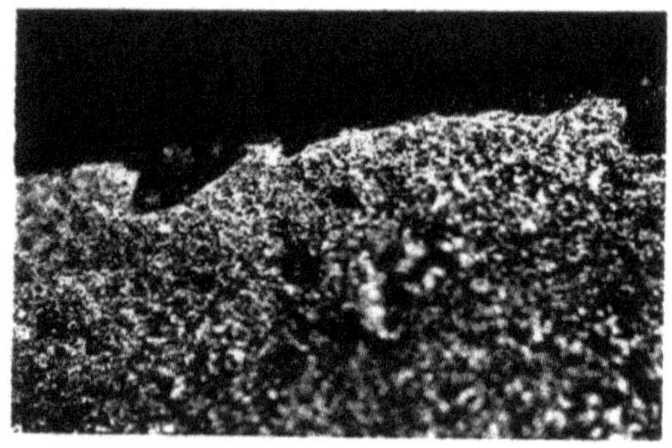

Figure 123: Polish from buffing with cow hide 45 minutes see also SEM figure 127.
EXP/142 x200

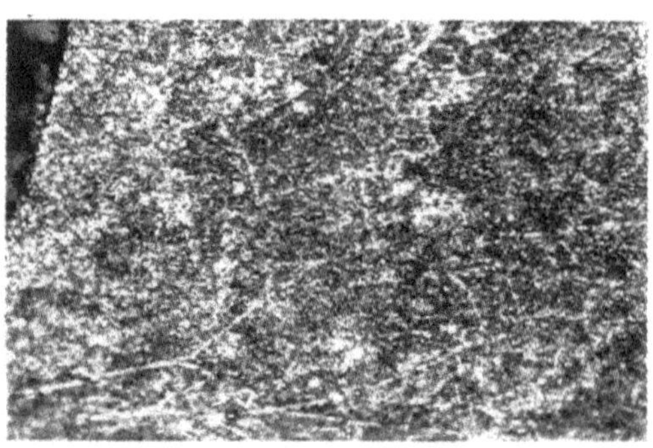

Figure 124: Oddly shaped striations from shaking in gravelly sand for 30 minutes.
EXP/11 x200

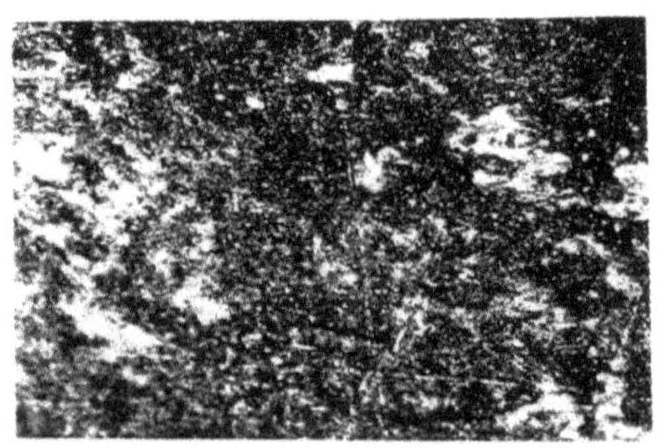

Figure 125: Shaking 30 minutes in gravelly sand. Note asterisk-like striae near use-wear.
EXP/11 x200

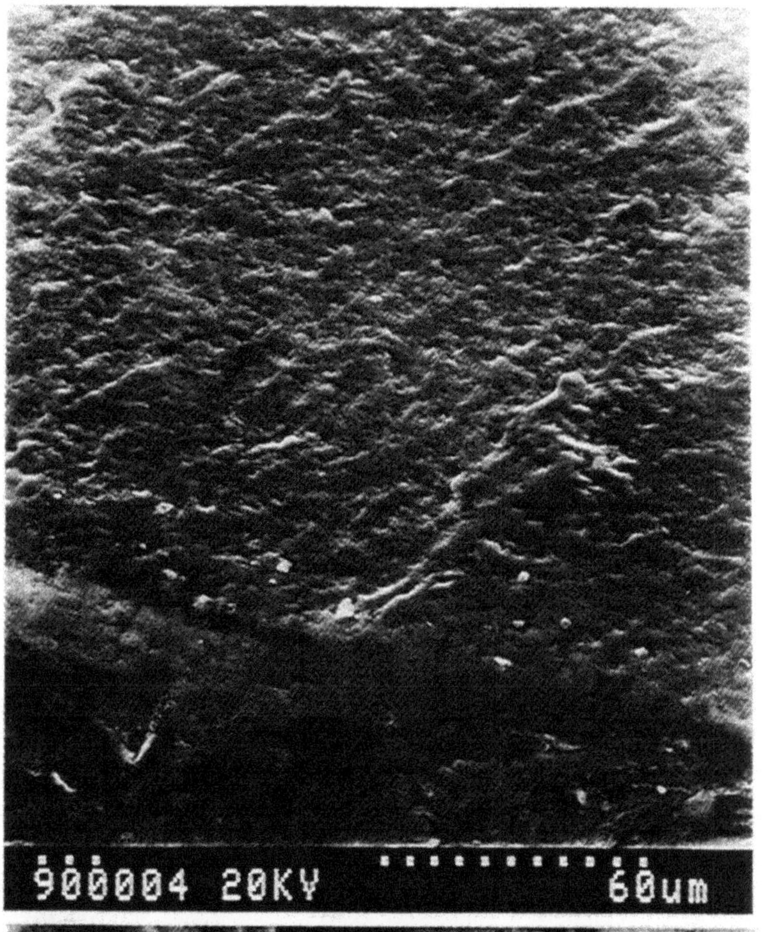

Figure 126: Smooth surface as in figure 122. Note smooth ledge and striation, scraping goat hide.
EXP/143 SEM

Figure 127: Surface from buffing with cow hide smoother than figure 126. Note. however, that this is at much higher magnification.
EXP/142 SEM

Figure 128: Torn cow hide after buffing 45 minutes.
EXP/142 SEM

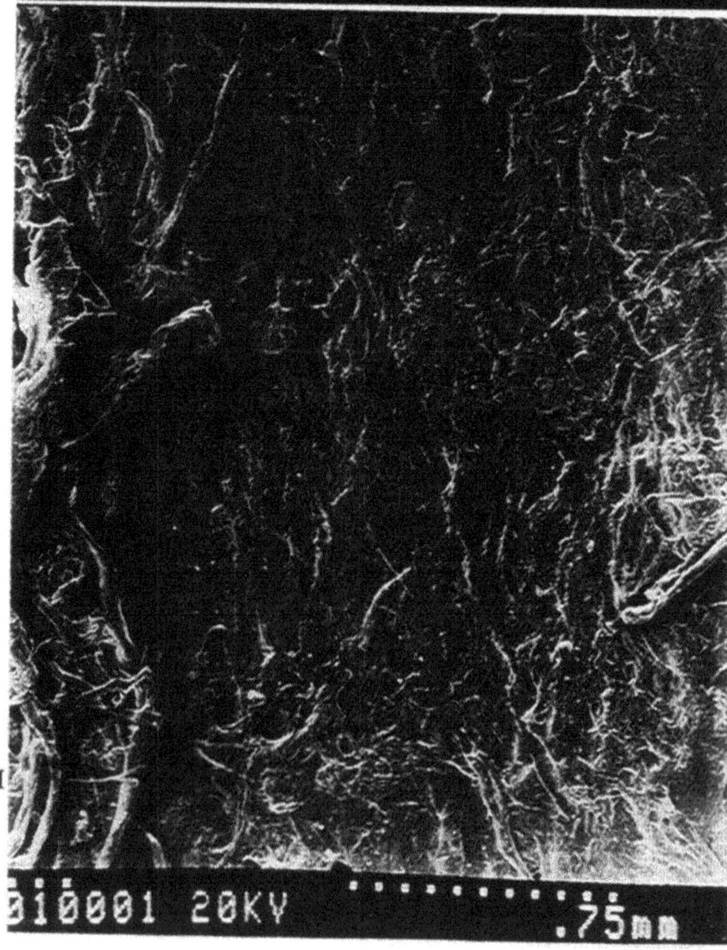

Figure 129: Another area cow hide after buffing 45 minutes, cf. figure 100, goat hide.
EXP/142 SEM

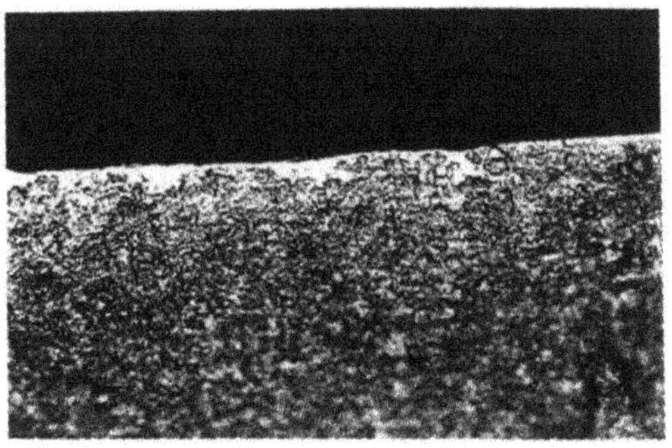

Figure 130: Bright spots on edge of unused artefact shaken for 90 minutes in wet gravelly sand.
EXP/12Δ x200

Figure 131: Abrasion tracks from denticulation before tumbling in fine sand
EXP/12 x100

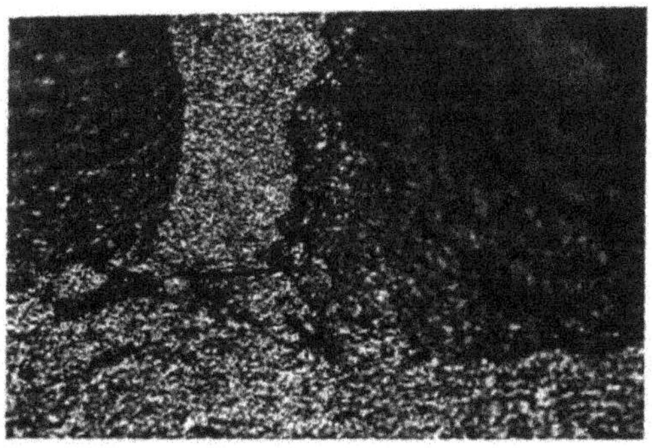

Figure 132: After tumbling flint surface in figure 131 now scoured clean.
EXP/12 x100

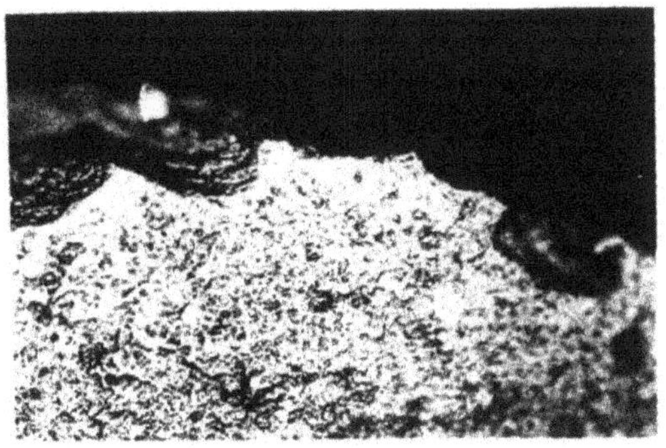

Figure 133: Striations in figure 30 now invisible after tumbling in wet gravelly sand.
EXP/12 x200

Figure 134: Whittling wood. Polish not affected by NaOH cleaning.
EXP/E5 x200

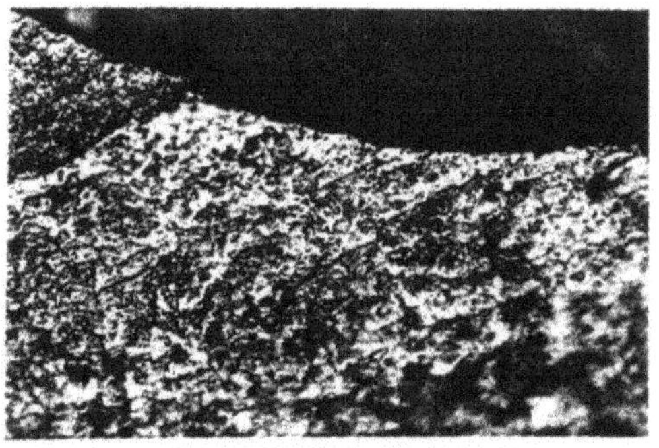

Figure 135: Incising wood polish not affected by NaOH cleaning.
EXP/59 x200

Figure 136: Whittling oak. Polish unaffected by NaOH cleaning.
EXP/75 x100

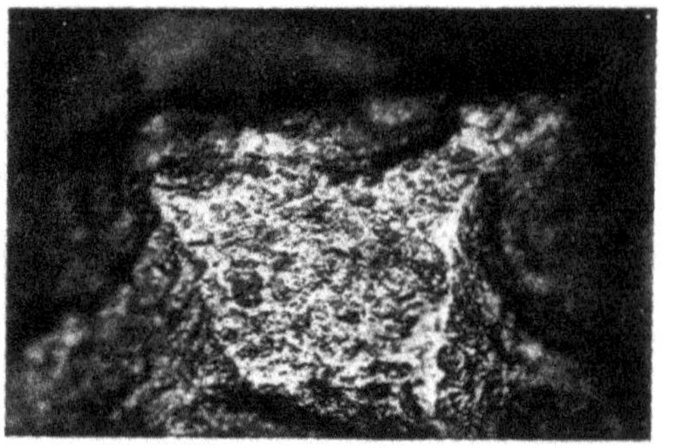

Figure 137: Edge damage from tumbling cf figure 34.
EXP/1 x200

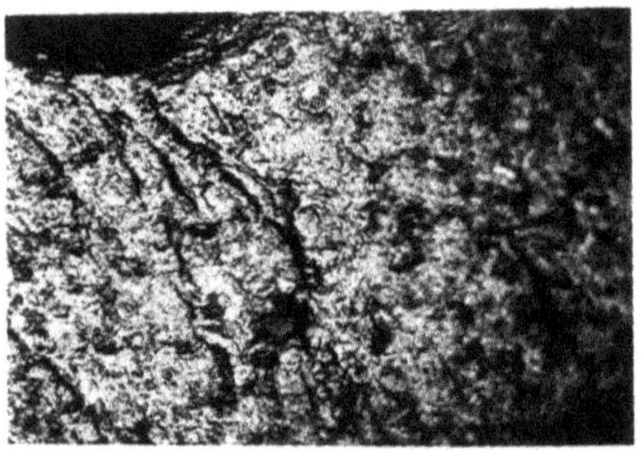

Figure 138: After tumbling appear in similar to PDSM
EXP/75 x200

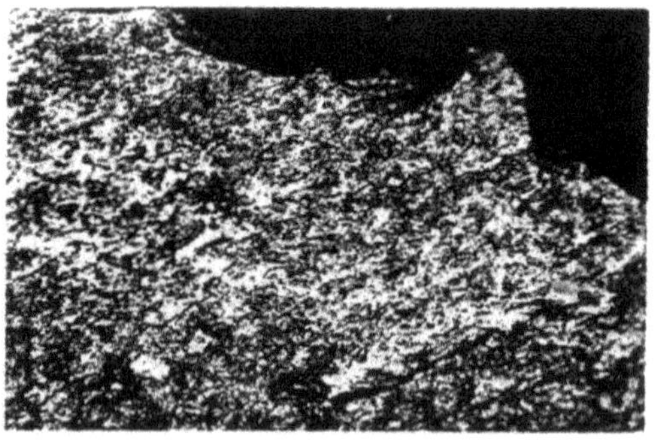

Figure 139: Incising wood. Polish after NaOH cleaning.
EXP/59 x200

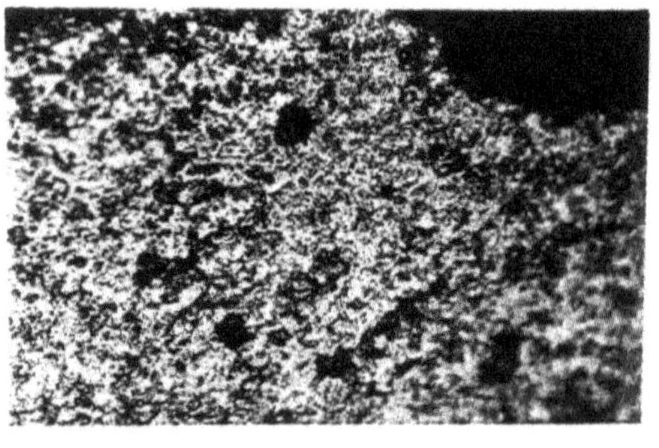

Figure 140: After tumbling polish similar to PDSM pits
EXP/59 x100

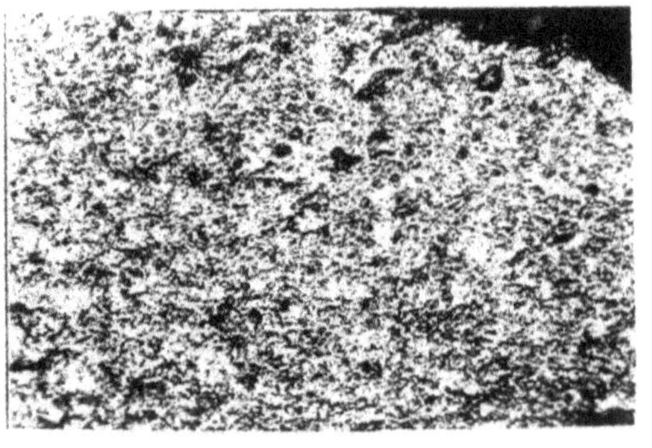

Figure 141: Polish from tumbling appearing similar to PDSM with pits
EXP/59 x100

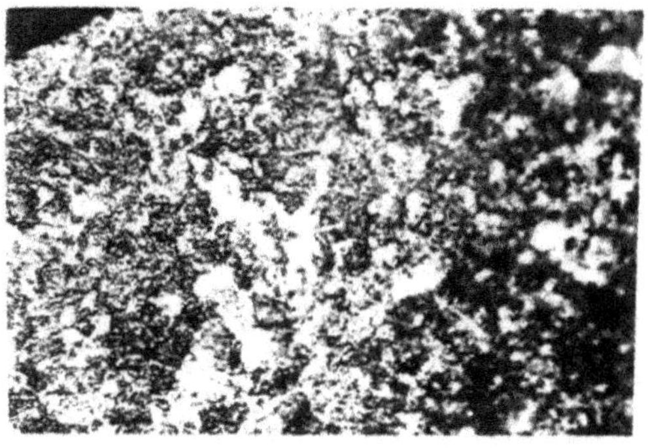

Figure 142: Dry rubbing of flint on flint for two months
EXP/65A x200

Figure 143: Polish from rubbing two wet flint flakes for 5 minutes.
EXP/76B x200

Figure 144: Diffuse polish on bulb rubbing two wet flakes for 2 minutes.
EXP/76B x200

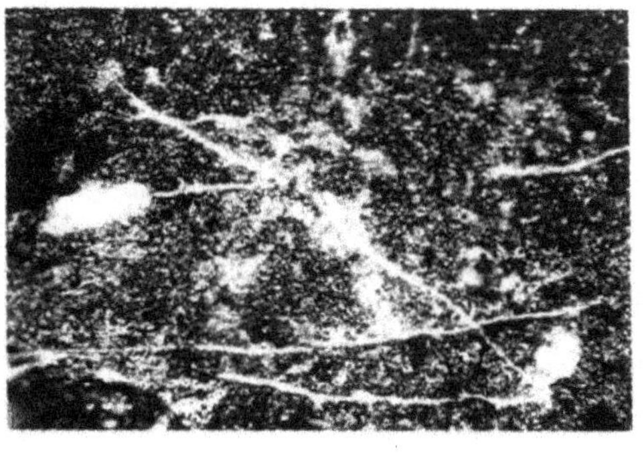

Figure 145: Odd shaped striae after 7 days trampling, cf. figure 124
U166 1985 x200

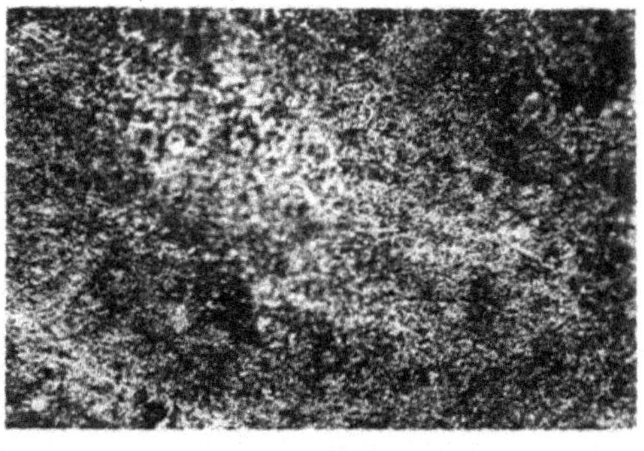

Figure 146: Abrasion after 7 days trampling
L69 1985 x200

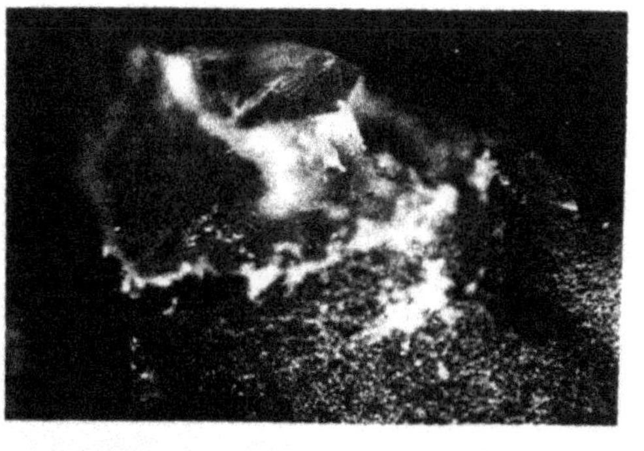

Figure 147: Edge of Klithi experimental artefact crushed before experiment.
U171 1987 x100

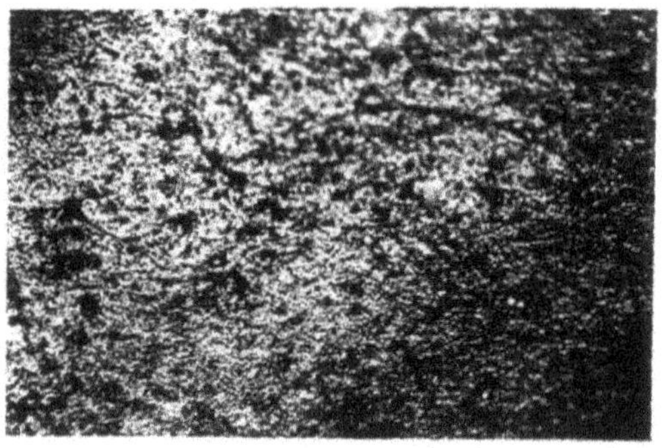

Figure 148: Diffuse polish on Klithi artefact cf. figure 144. Note curved crack.
L78 1987 x100

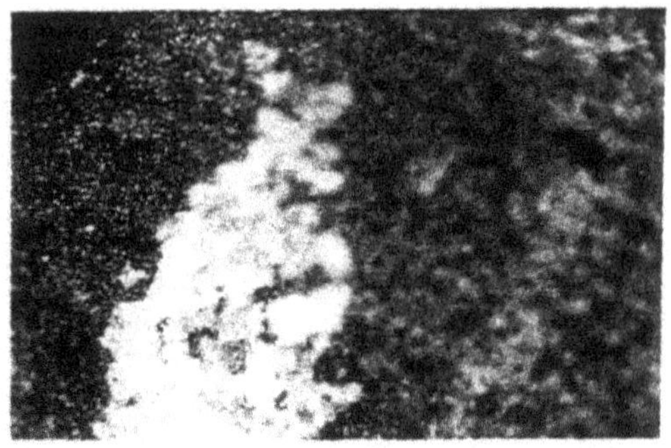

Figure 149: Deposit on Klithi artefact.
L91 1987 x200

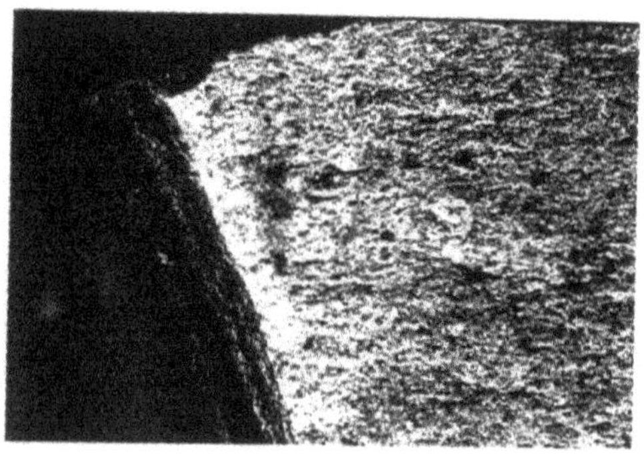

Figure 150: Traces of polish on edge similar to use-wear.
L83 1985 x100

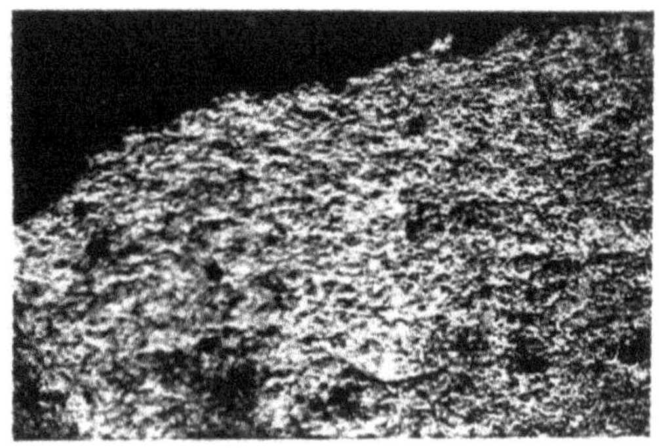

Figure 151: Polish similar to figure 61 grooving wood. Klithi trampling experiment.
L83 1985 x200

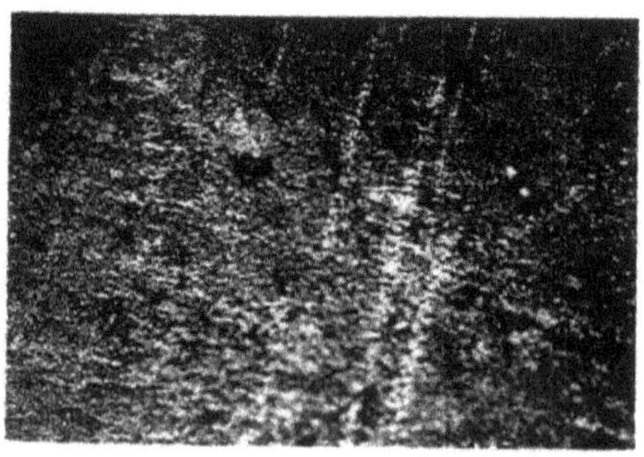

Figure 152: Macroscopic abrasion tracks on Klithi artefact crossing each other
L83 1985 x200

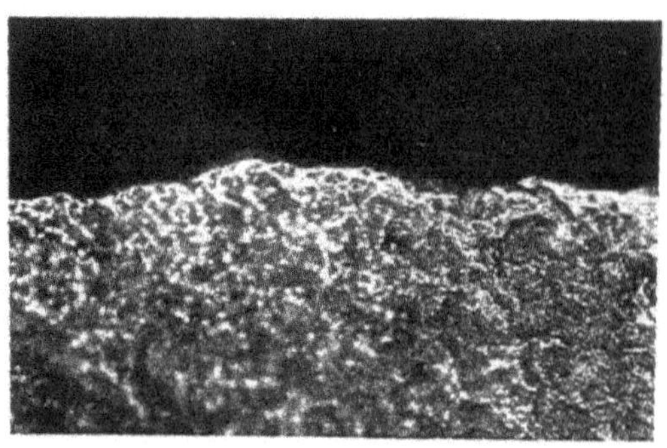

Figure 153: Polish unaffected after 400 days burial in Kebara, cf. figure59, EXP/128
EXP/67 x200

Figure 154: Abrasion after 400 days burial in Kebara
EXP/67 x200

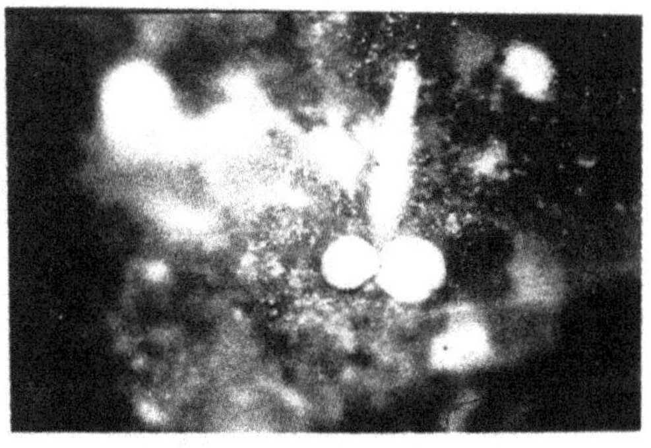

Figure 155: Flint surface insert experiment
EXP/116 x100

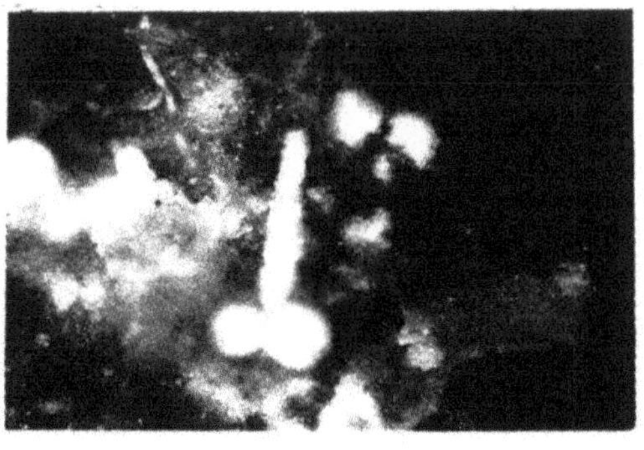

Figure 156: After 2 years in peat and decaying leaves flint surface appears coarser.
EXP/116 x100

Figure 157: Flint surface before experiment.
EXP/99 x200

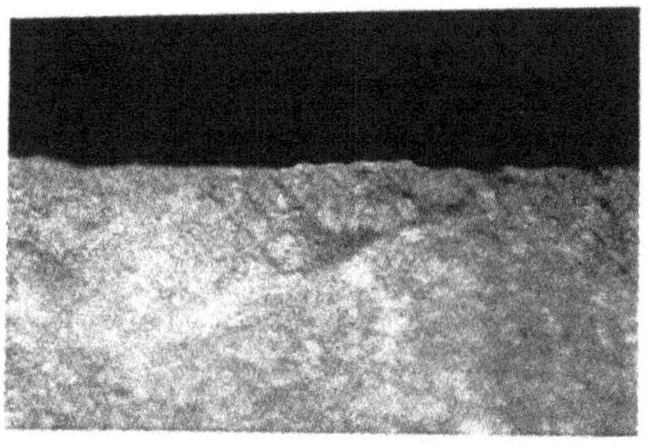

Figure 158: After 4 months in heated peat altered surface extreme glare
EXP/99 x200

Figure 159: Artefact polished by wet hide before experiment
EXP/139 x200

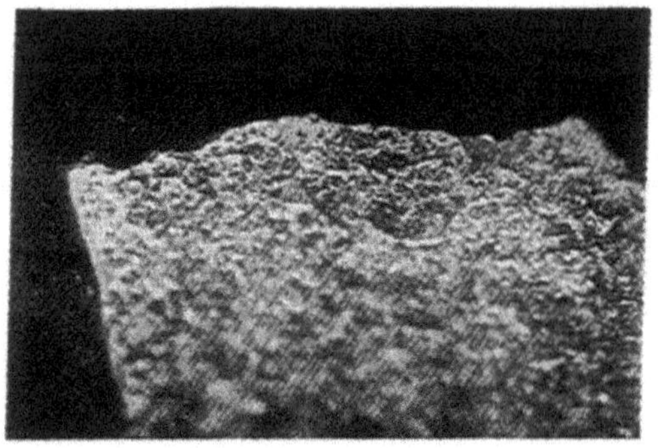

Figure 160: Polish after4 weeks in heated decaying organic matter.
EXP/139 x200

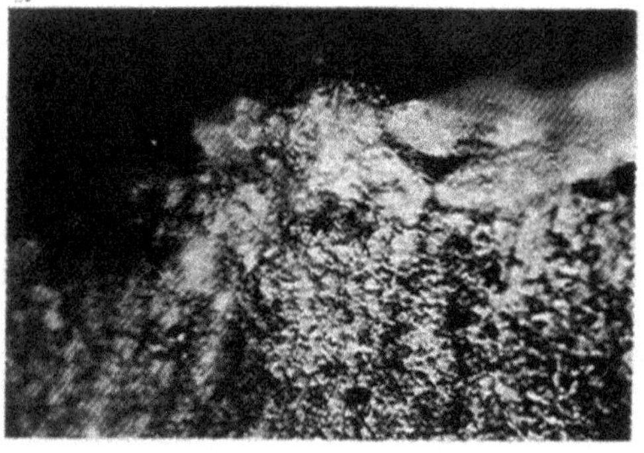

Figure 161: Tracks of little developed polish from hard hammer retouch.
EXP/148 x200

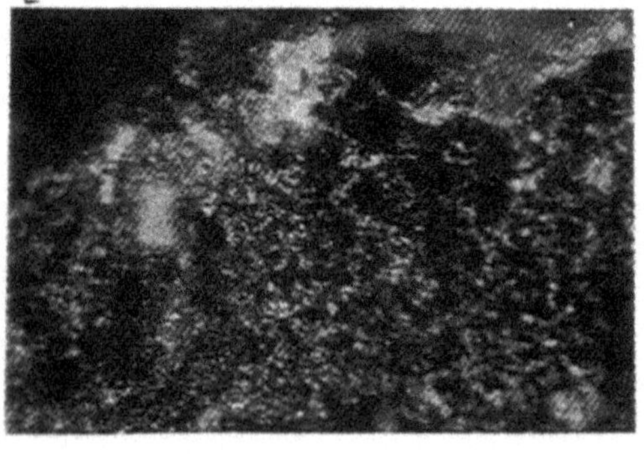

Figure 162: After 14 weeks in heated decaying organic matter. All traces removed.
EXP/148 x200

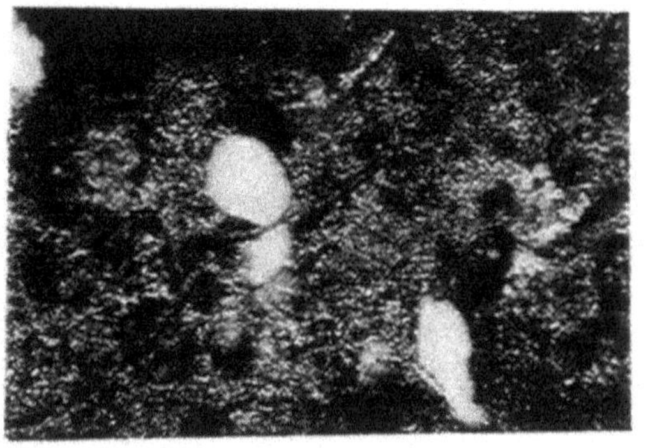

Figure 163: White crystals in hollows of flint surface after 14 weeks in experiment.
EXP/102 x100

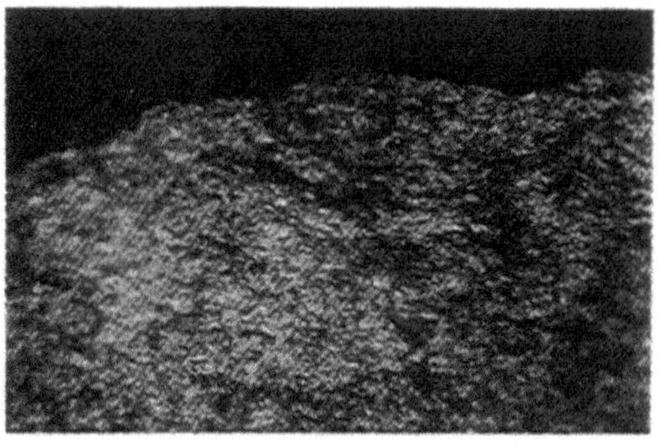

Figure 164: Flint surface seems "polished" after 14 weeks in experiment.
EXP/102 x200

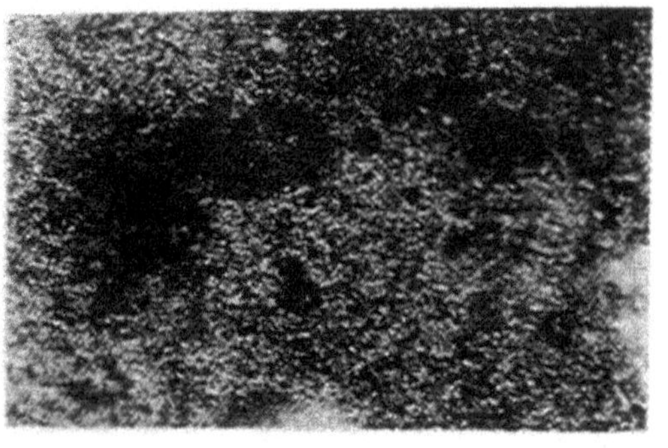

Figure 165: Surface of artefact before experiment
EXP/105 x200

Figure 166: View of the ventral aspect on EXP/148 with encrustations.

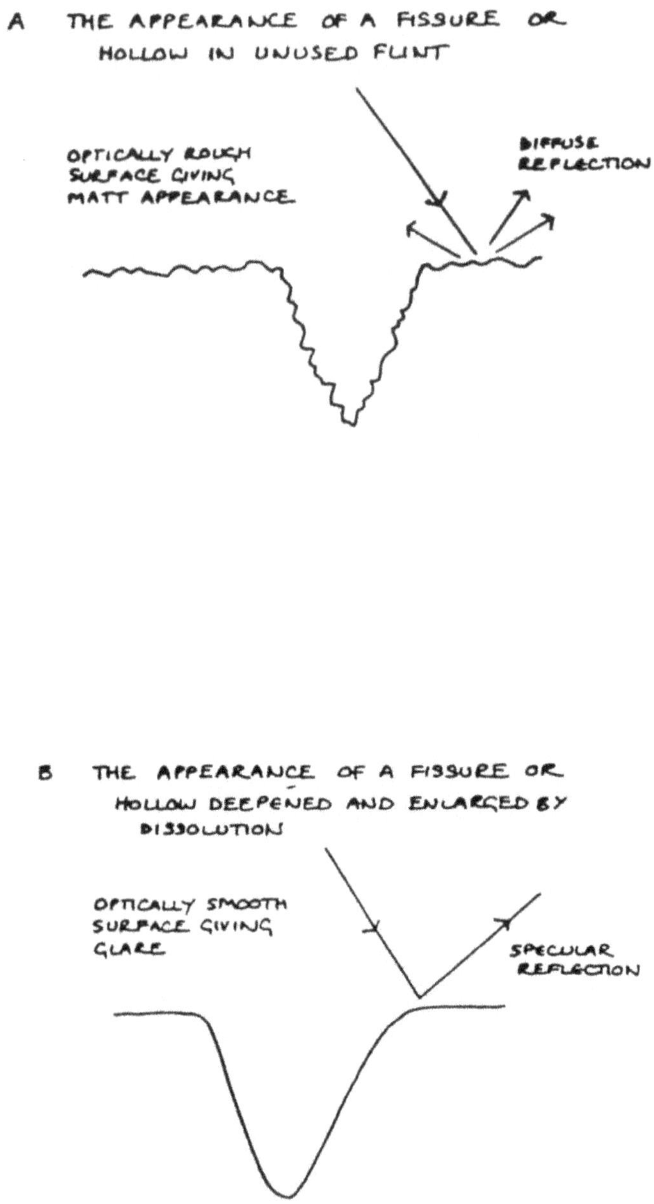

Figure 167: Glare on flint surfaces affected by dissolution.

Figure 168: Hollows appear deeper than in figure 165. Note white crystal.
EXP/105 x100

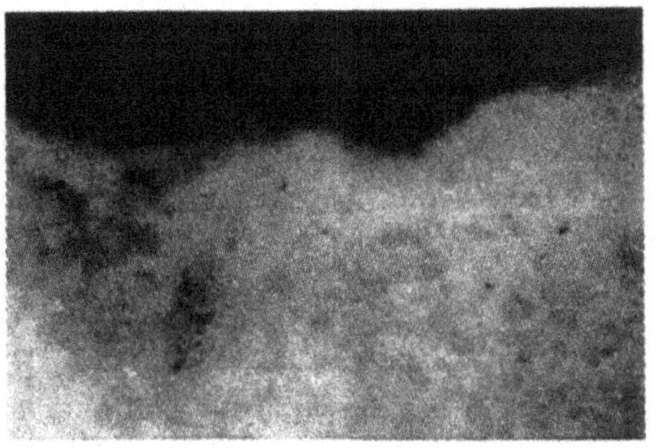

Figure 169: Completely patinated artefact with pockmarked surface.
EXP/52 x100

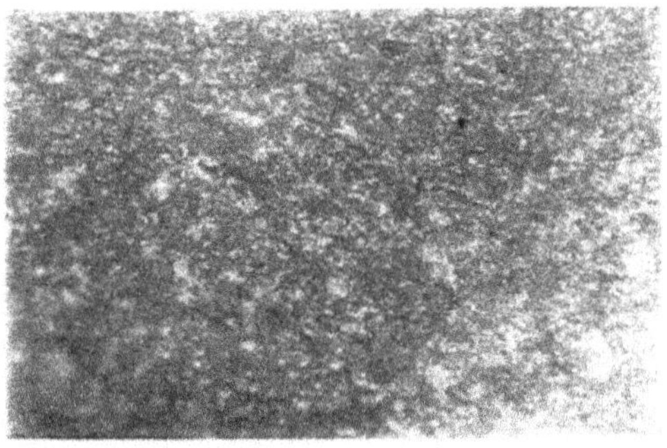

Figure 170: Surface before experiment.
EXP/100 x200

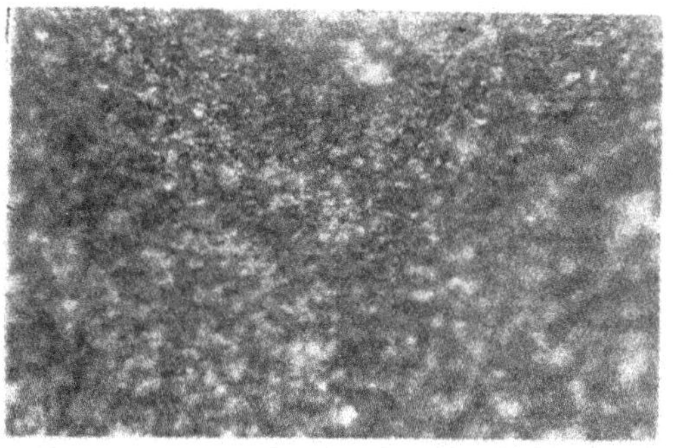

Figure 171: Surface below patina as it appeared after removing a small flake from the edge.
EXP/100 x200

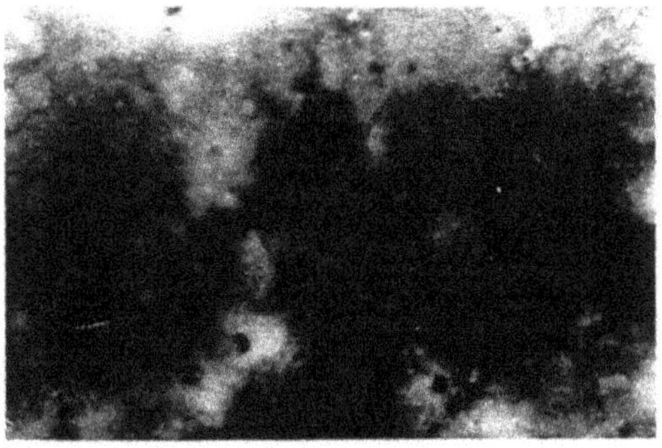

Figure 172: Several degrees of patination on one artefact.
EXP/100 x50

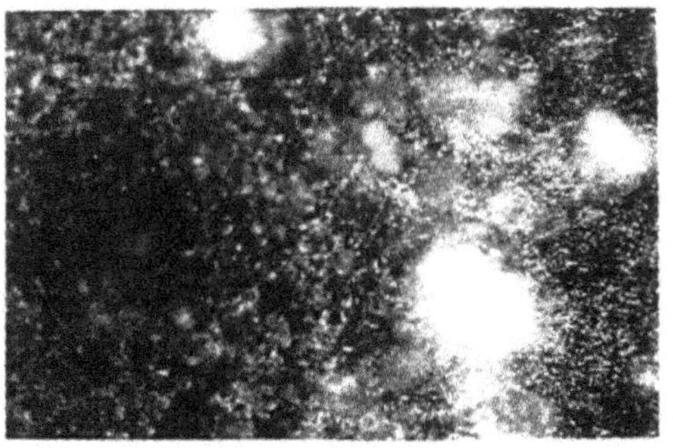

Figure 173: White crystals on surface after 24 hours in NaOH solution.
EXP/150 x100

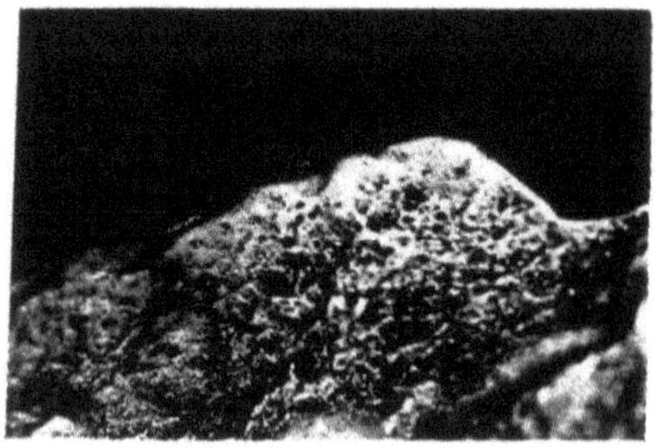

Figure 174: Polished edge before experiment.
EXP/150 x200

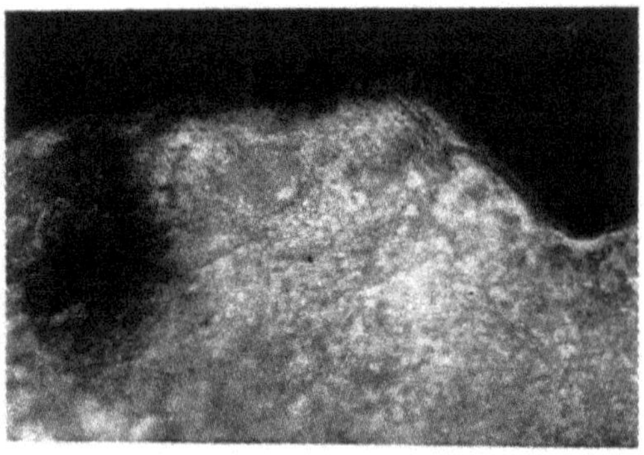

Figure 175: Polished edge after patination. Polish barely visible after 96hours.
EXP/150 x200

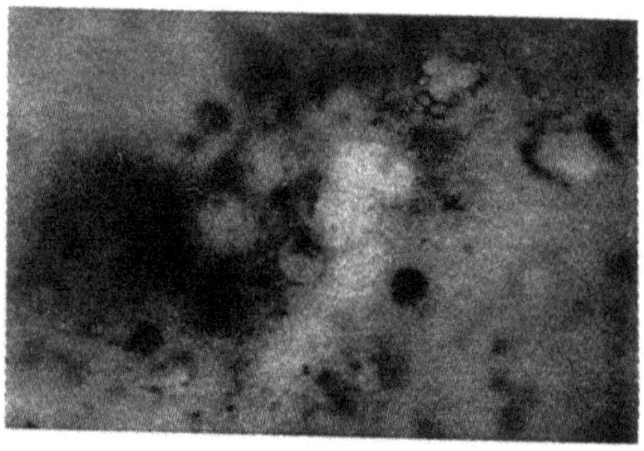

Figure 176: Pitted and totally patinated surface after 11 days.
EXP/150 x100

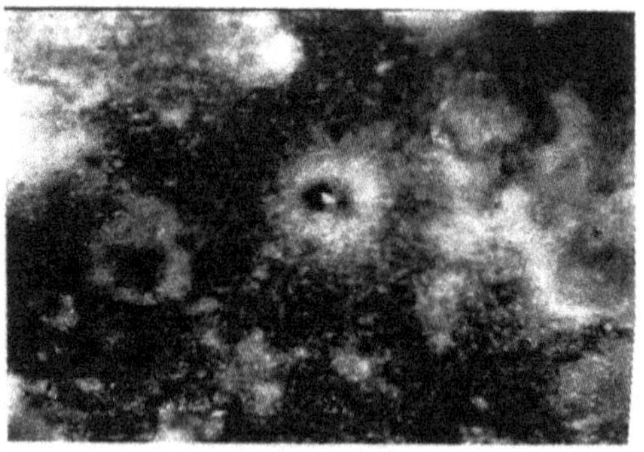

Figure177: Hollows become deeper after 8 days in NaOH solution.
EXP/152 x100

Figure 178: Complete patina after 11 days altered edge, cf. figure 179.
EXP/152 x100

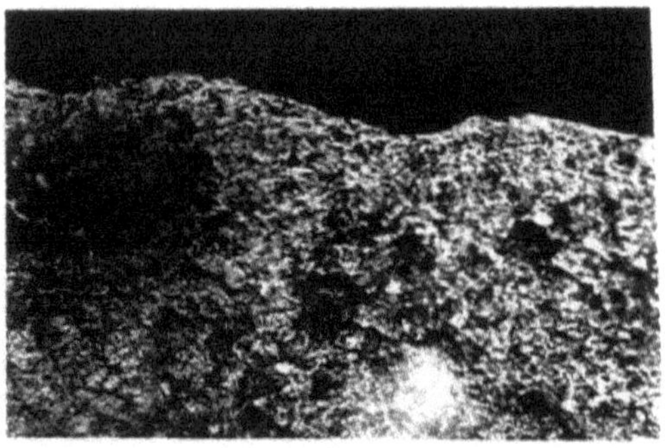

Figure 179: Edge before experiment as in figure 178.
EXP/152 x200

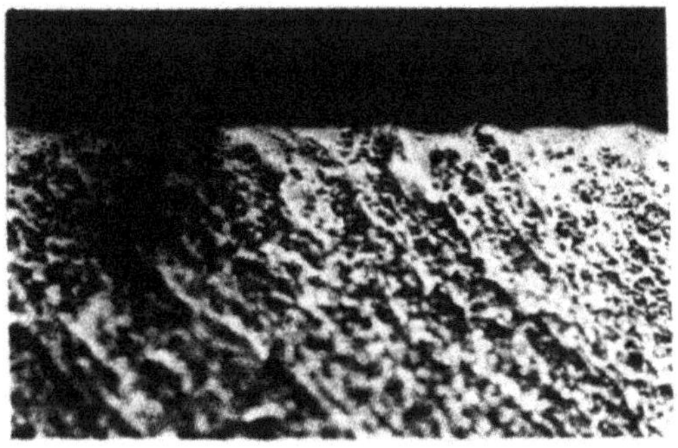

Figure 180: Polished edge before experiment.
EXP/147 x200

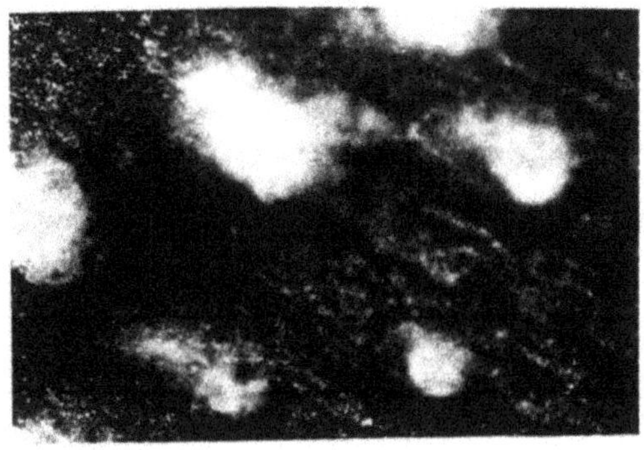

Figure 181: White crystals and glare after 96 hours.
EXP/147 x100

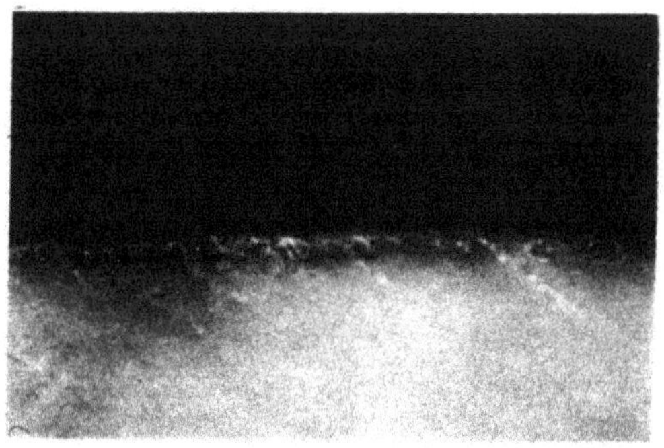

Figure 182: Edge as in figure 180 after patination.
EXP/147 x100

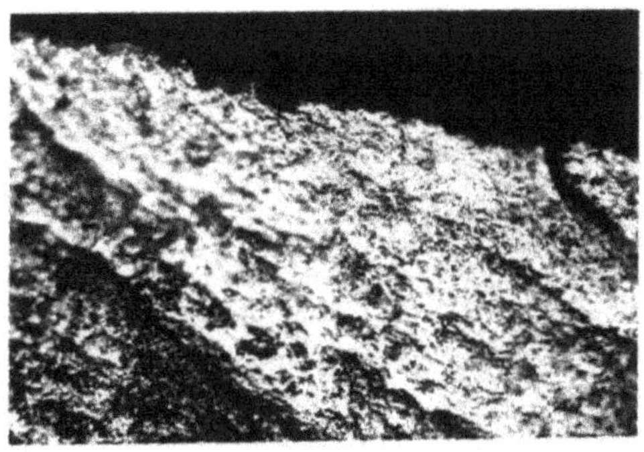

Figure 183: Microscopic sheen mimicking use-wear polish. 50 days in solution.
EXP/151 x200

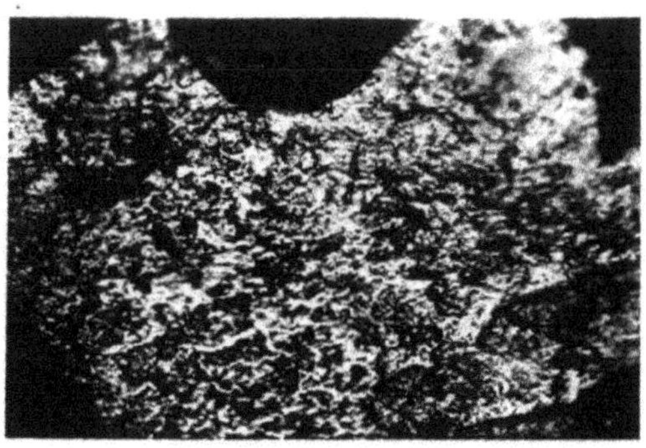

Figure 184: Microscopic sheen mimicking PDSM after 50 days in solution.
EXP/151 x200

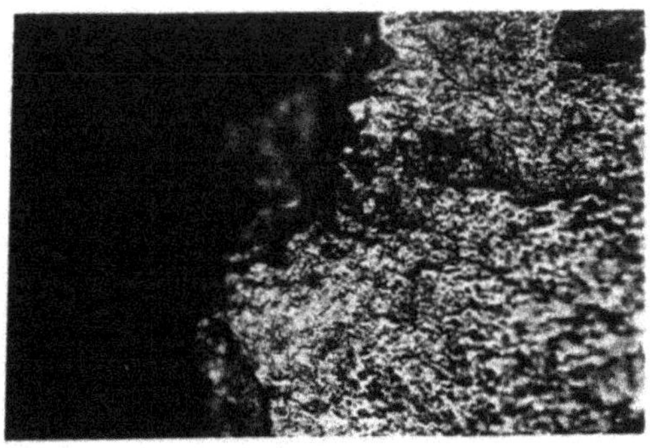

Figure 185: PDSM on artefact from West Heath similar to figure 184 x200

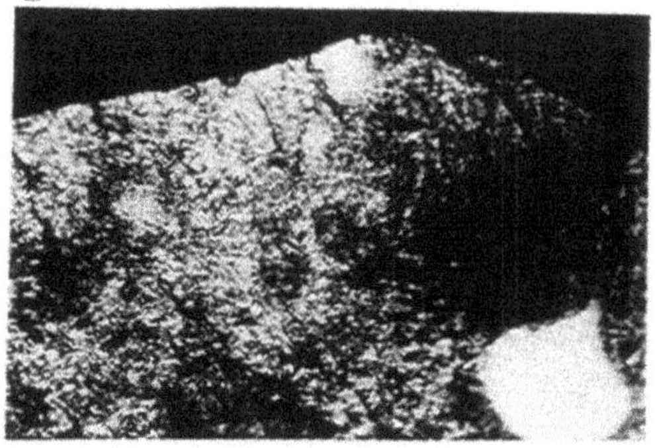

Figure 186: Surface of artefact before experiment.
EXP/153 x200

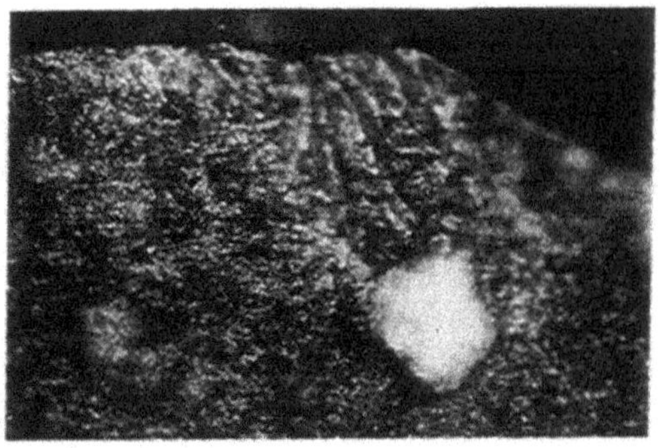

Figure 187: Smoother surface and new crystal to left of large one. 2 months in solution.
EXP/153 x200

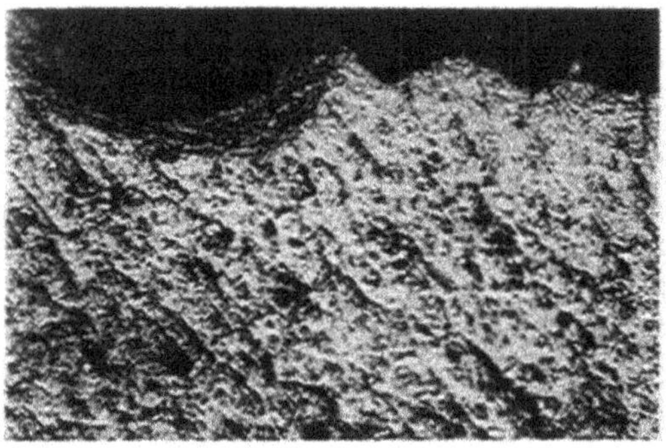

Figure 188: Sheen very similar to PDSM on archaeological artefacts.
EXP/149 x200

Figure189: Surface polished by damp hide altered after 12 weeks.
EXP/149 x200

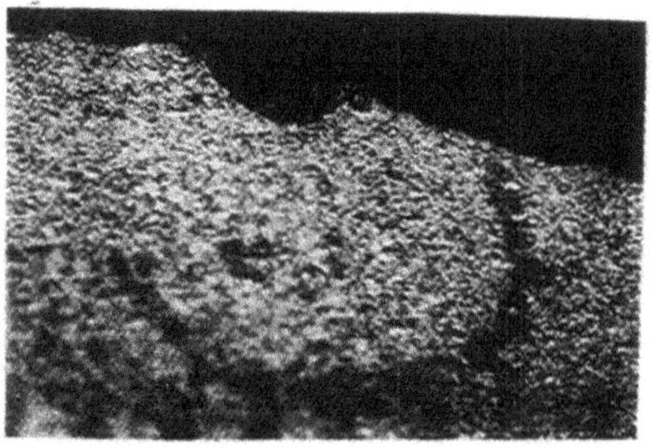

Figure 190: Glare and altered polish, see figure 189
EXP/149 x100

Figure 191: Archaeological PDSM like figure 188
x200

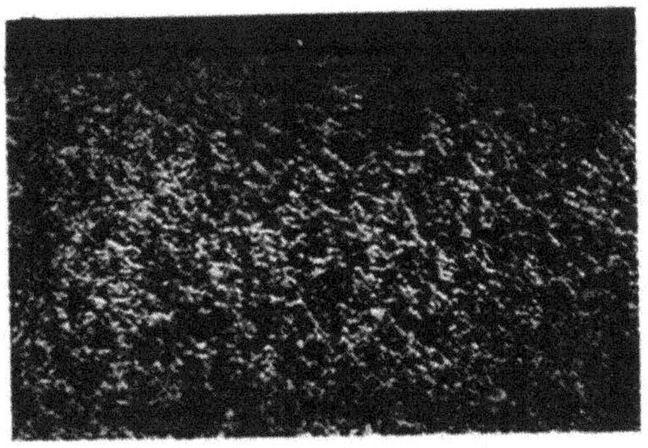

Figure 192: Polish on edge(dorsal) decreasing inland after 1 month.
EXP/29 x200

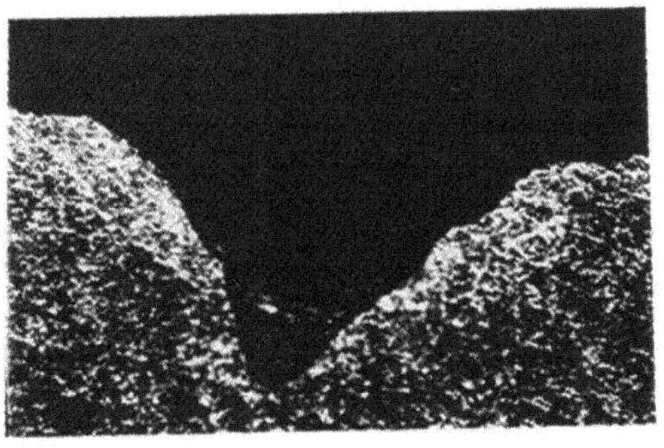

Figure 193: Polish on edge ventral aspect decreasing
EXP/29 x100

Figure 194: Macroscopic polish on edge (ventral) similar to PDSM after 5 weeks
EXP/29 x200

Figure 195: Traces on artefact from Kebara
x100

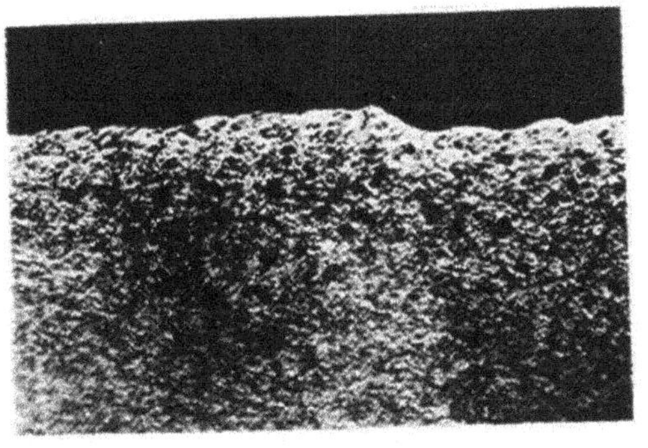

Figure 196: Polish on the edge before experiment
EXP/154 x100

Figure 197: Polish on edge after 2 months, intact but with glare
EXP/154 x100

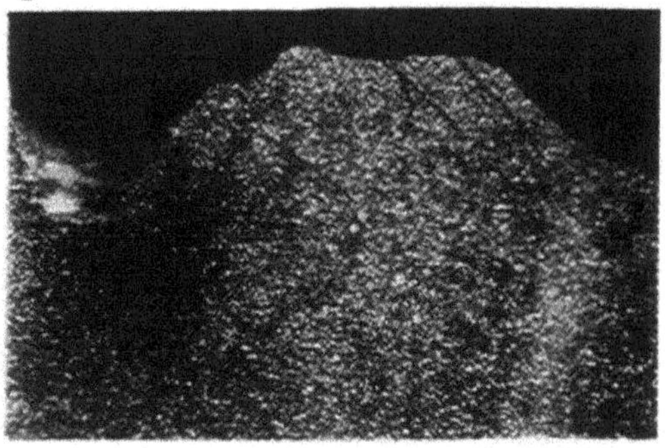

Figure 198: Edge before experiment.
EXP/29 x100

Figure 199: Polish on same edge as in figure198 after 1 month
EXP/29 x100

Figure 200: Macroscopic sheen better developed near the edge decreasing inland. After 50 days.

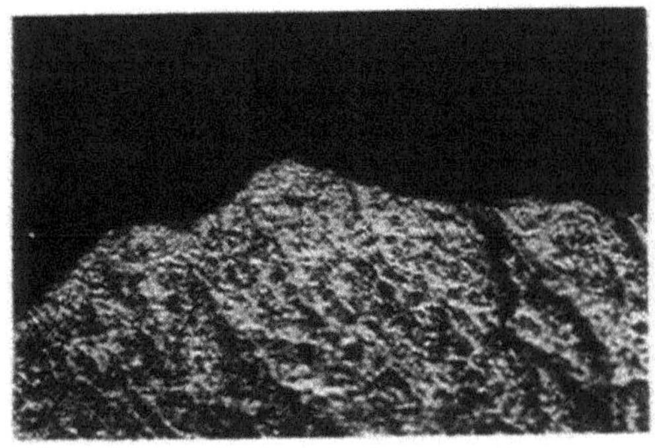

Figure 201: Magnified view of edge as in figure 200.
EXP/29 x200

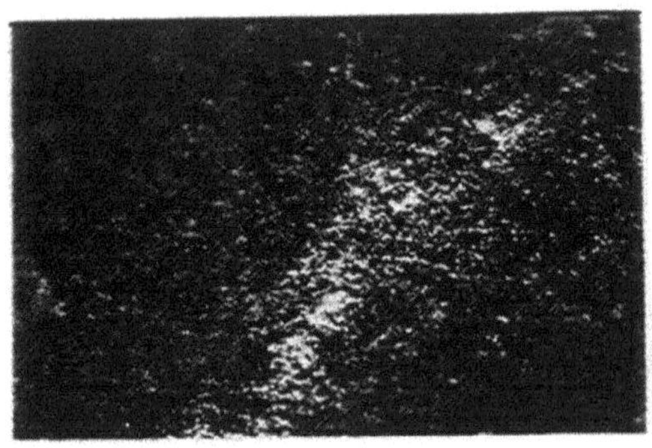

Figure 202: Macroscopic and microscopic bright spots after 25 days.
EXP/155 x100

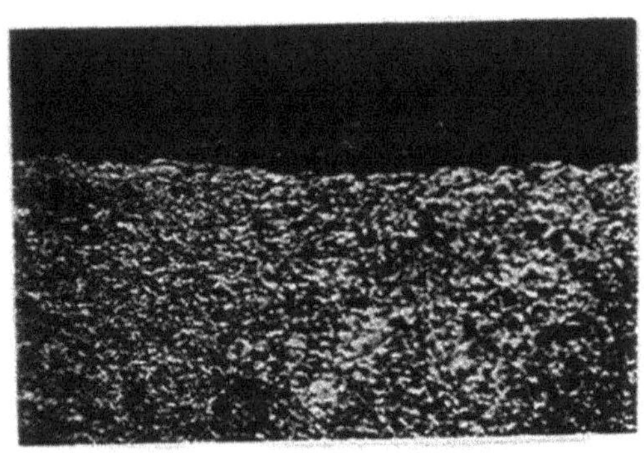

Figure 203: Another cluster of bright spots distributed like use-wear near the edge.
EXP/155 x100

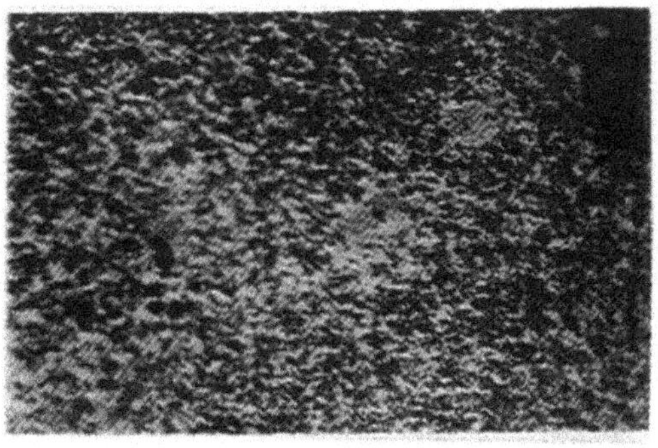

Figure 204: Flat patches of bright spots inland.
EXP/155 x200

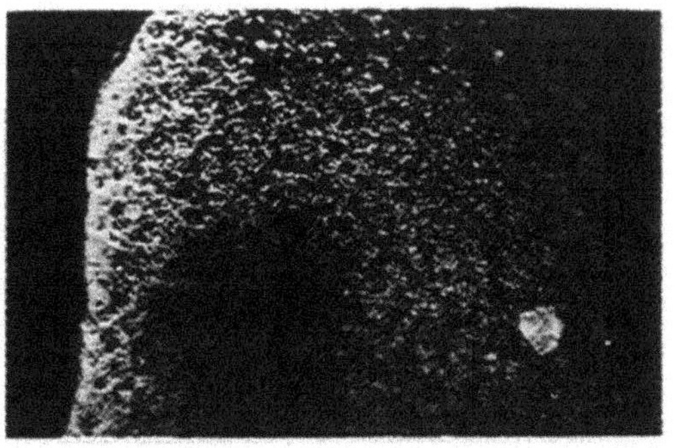

Figure 205: Polish produced by hide seems intact after 25 days.
EXP/155 x200

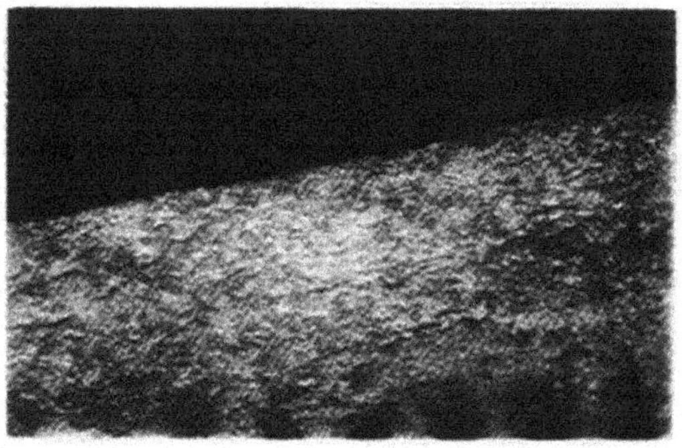

Figure 206: Initial development of polish after 2 months.
EXP/154 x100

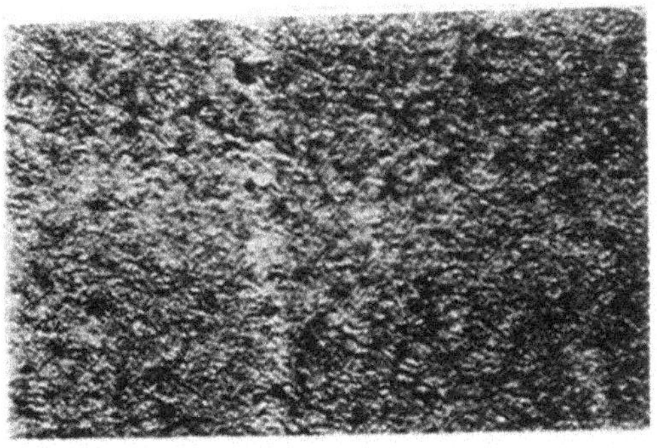

Figure 207: Smear from dry contact of flint with flint.
EXP/103 x100

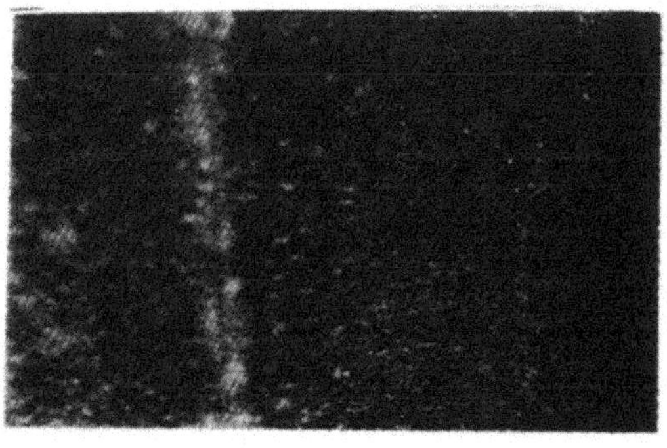

Figure 208: Groove in flint scratched with wet flint.
EXP/103 x100

Figure 209: Groove in figure 208 polished during tumbling.
EXP/103 x100

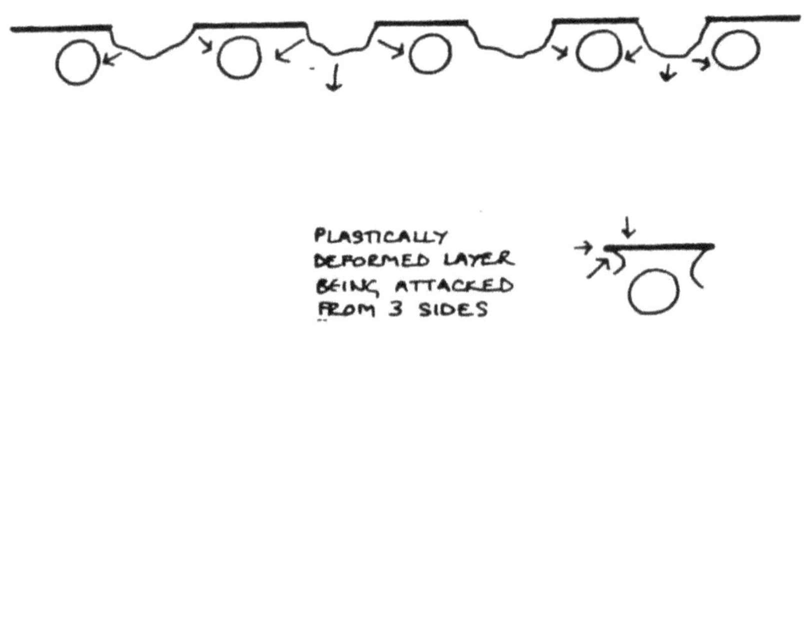

Figure 210: Flint surface polished to stage B offering many inroads for solvents to attack.

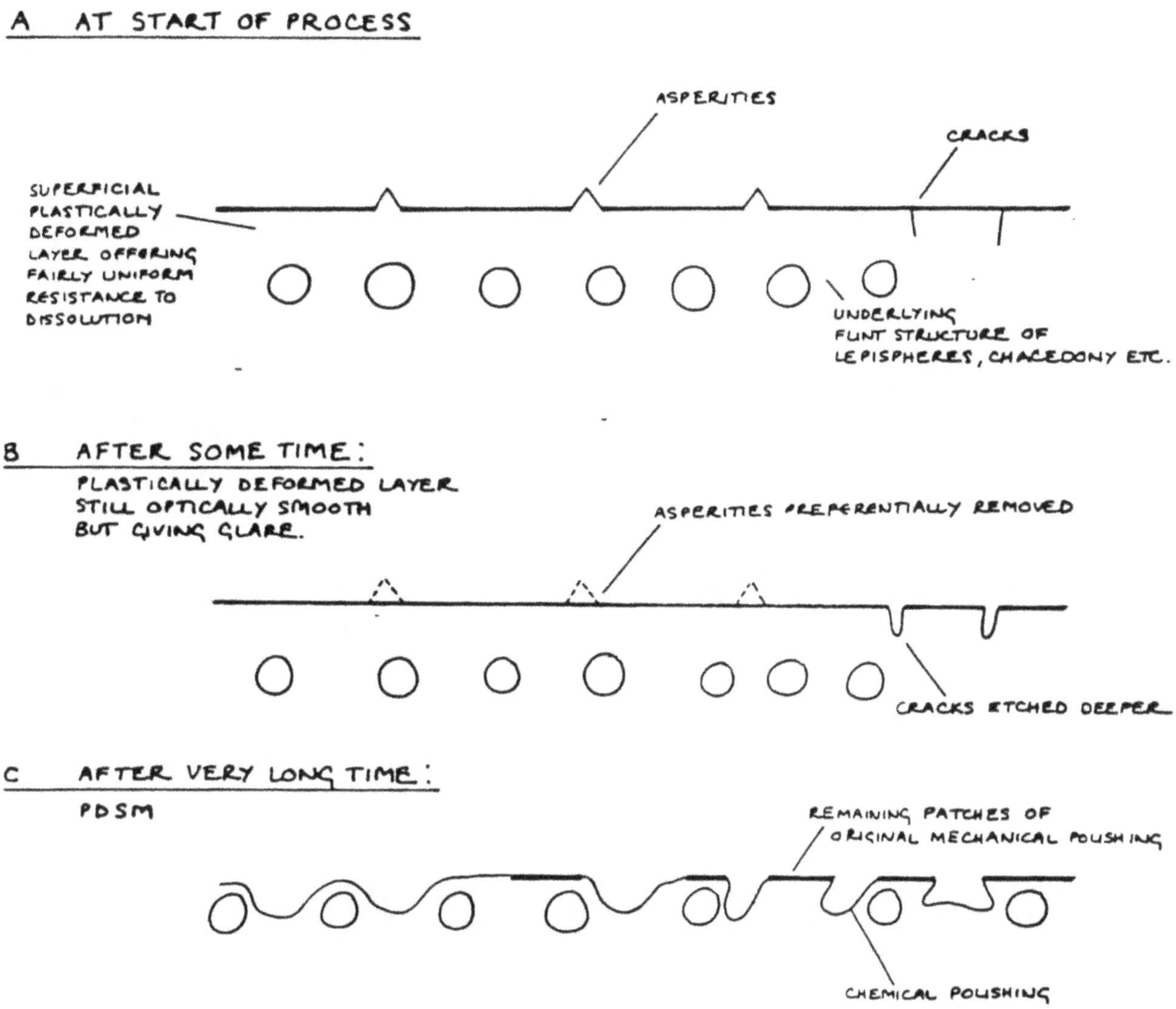

Figure 211: Flint surface polished to stage C+ offering rare inroads for solvents to attack.

Figure 212: Dry flint scratch on highly polished flint
EXP/103 x100

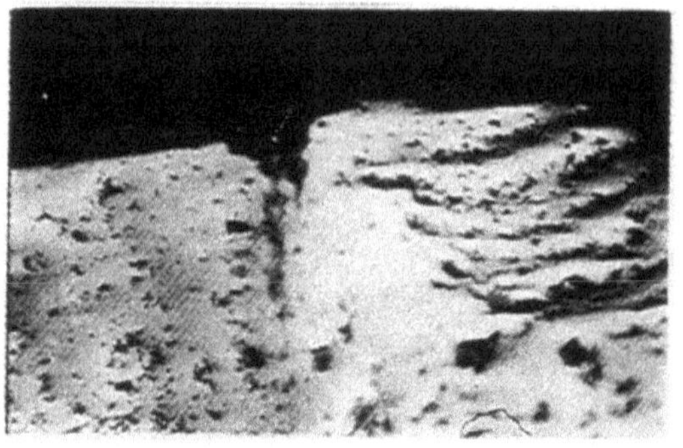

Figure 213: Scratch widened after tumbling
EXP/103 x100

Figure 214: Depression from wet flint on highly polished flint.
EXP/103 x100

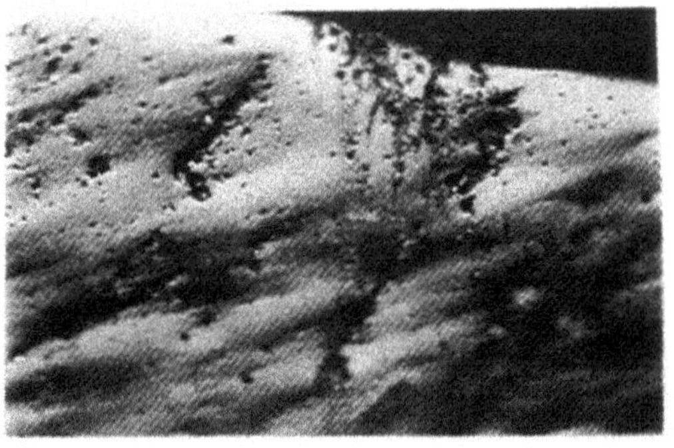

Figure 215: Scratch revealing polish as very thin layer, cf. figure 220.
EXP/103 x100

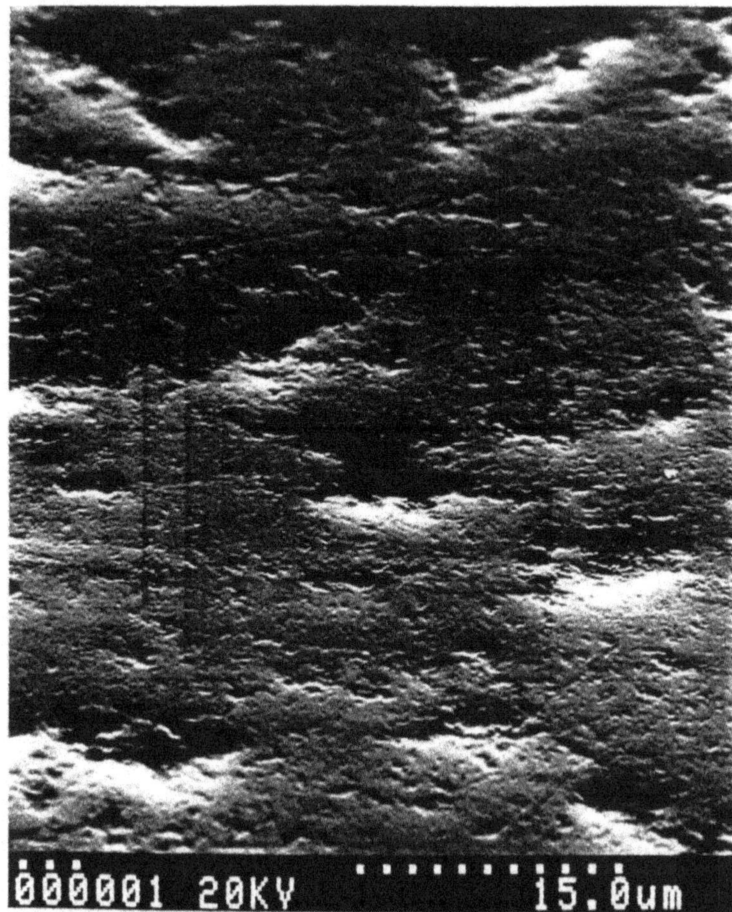

Figure 216: Smooth surface on Neolithic blade. Note elongated grooves parallel to edgeSEM

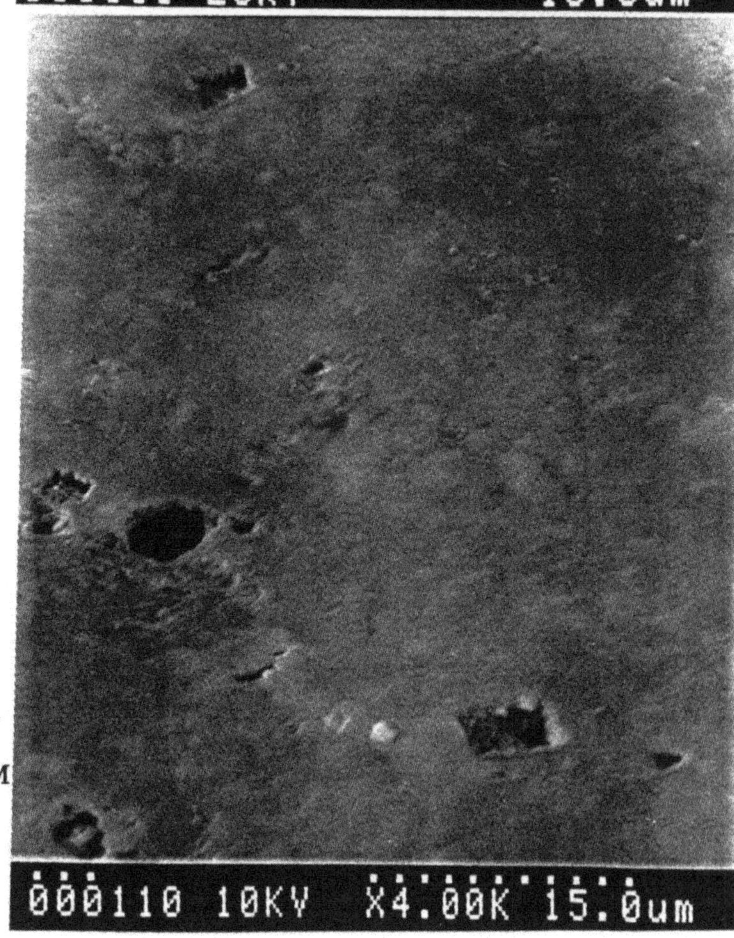

Figure 217: Very smooth surface from polishing with ¼μ grit. Note smooth-edged pits.
EXP/103 SEM

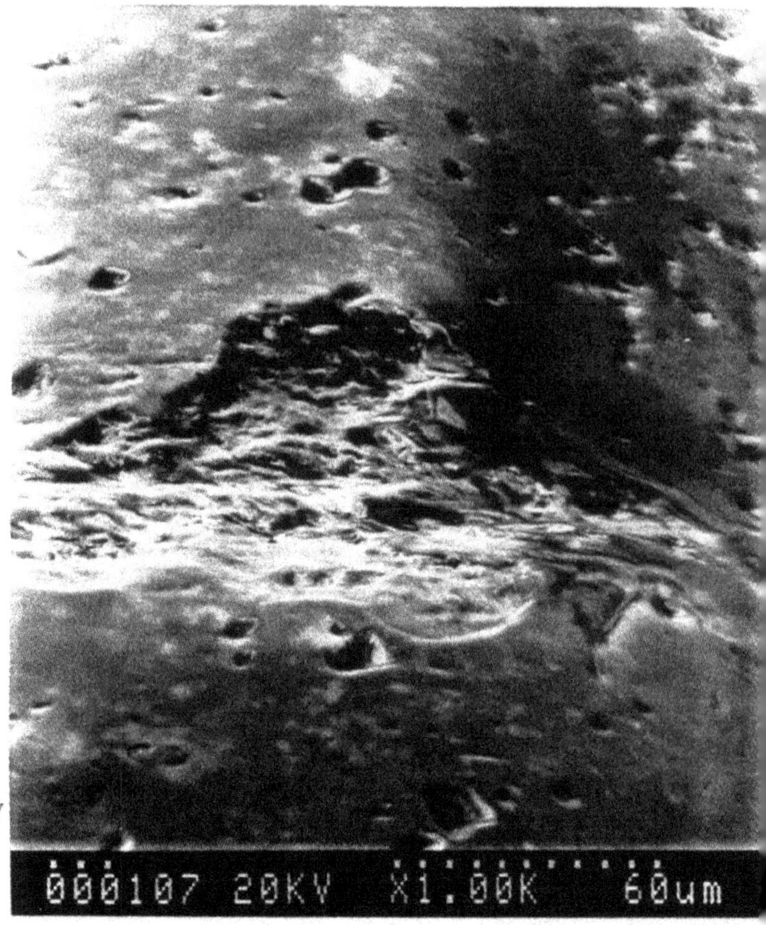

Figure 218: Lateral crack produced by scratching highly polished flint withdry flint.
EXP/103 SEM

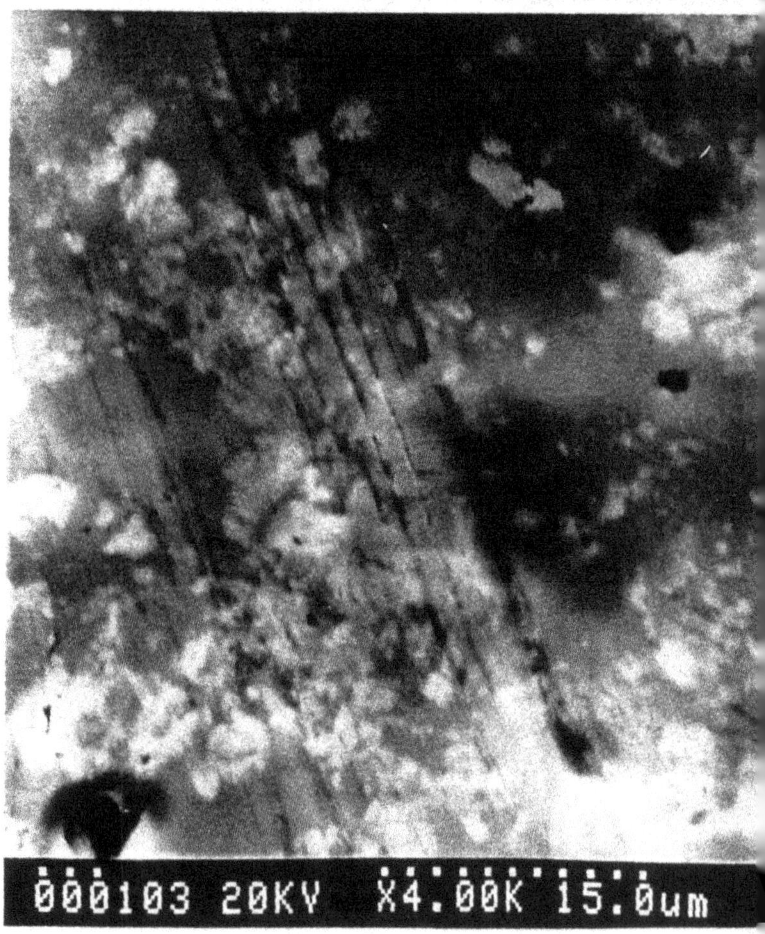

Figure 219: Depressions left on highly polished flint by wet flint. Note surface unbrokencontrary to figures 218, 68 and 69
EXP/103 SEM

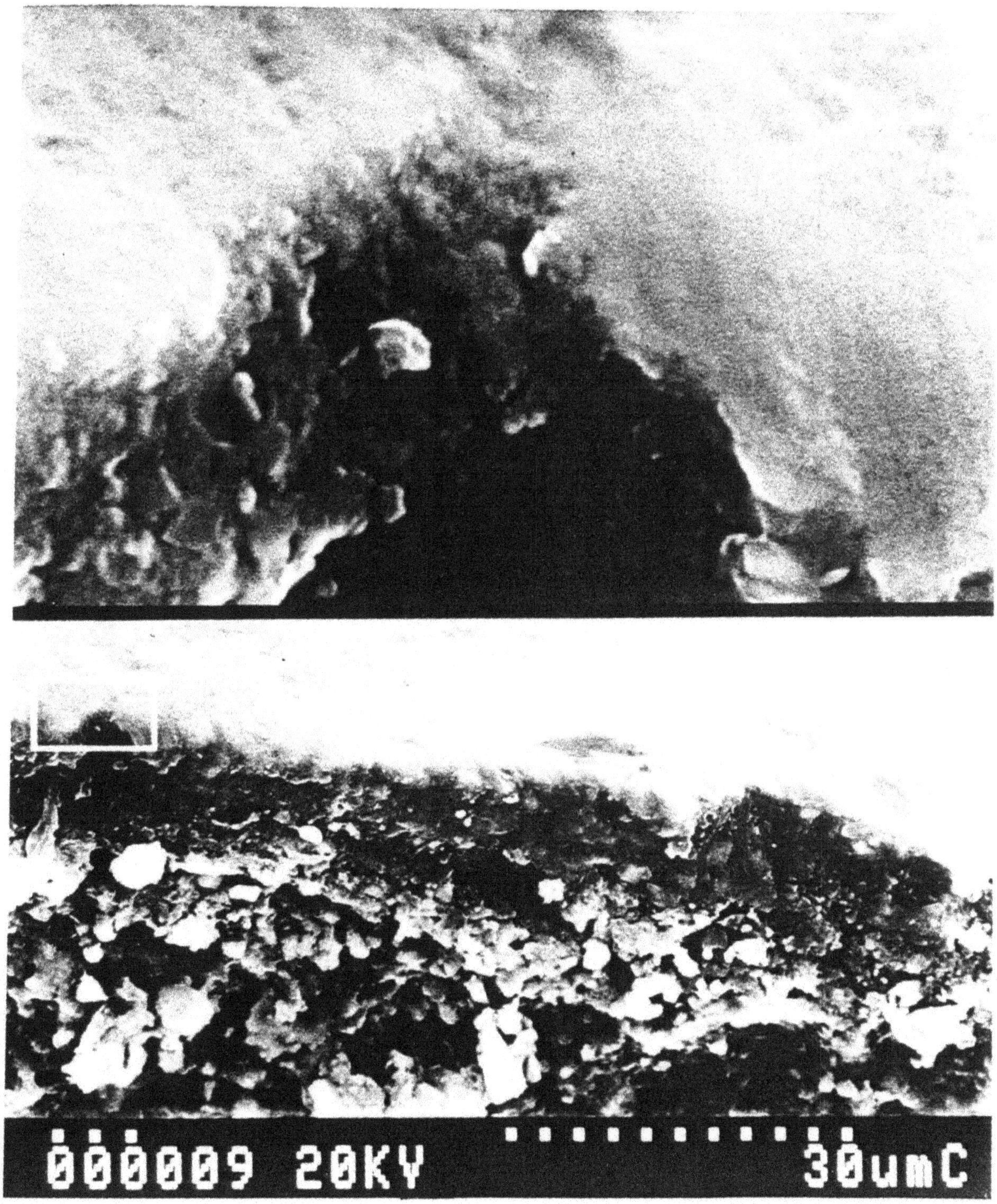

Figure220: Note area highlighted by white rectangle showing the deformed layer on a Neolithic "sickle blade" from Far'ah to be a fraction of a micron. Top picture is magnified 10 times more than the bottom.

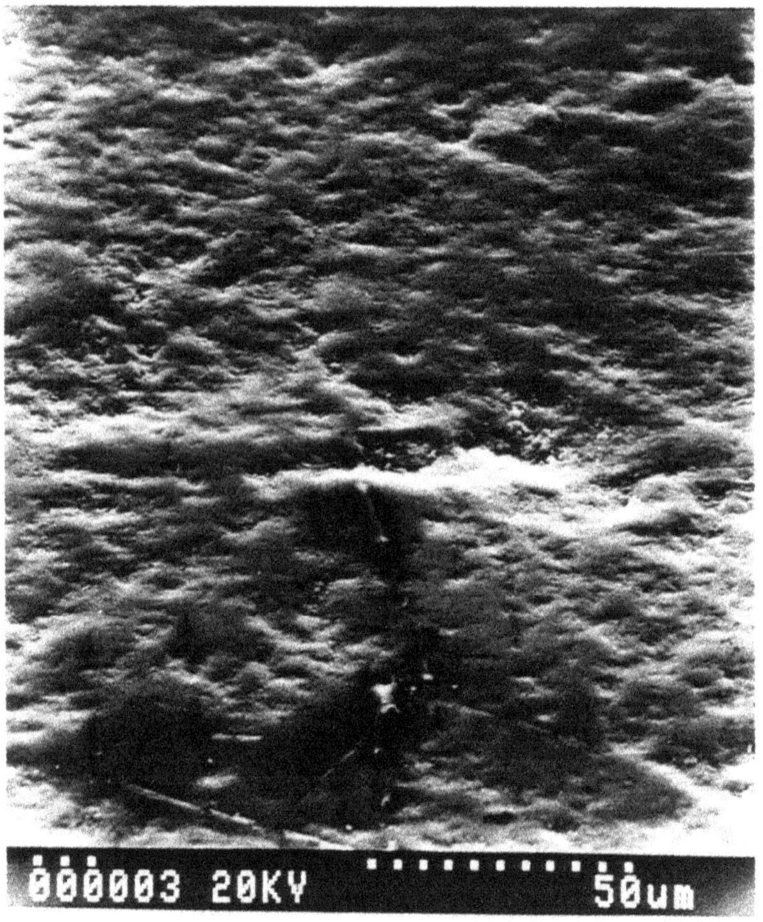

Figure 221: Elongated striation on Neolithic sickle blade.

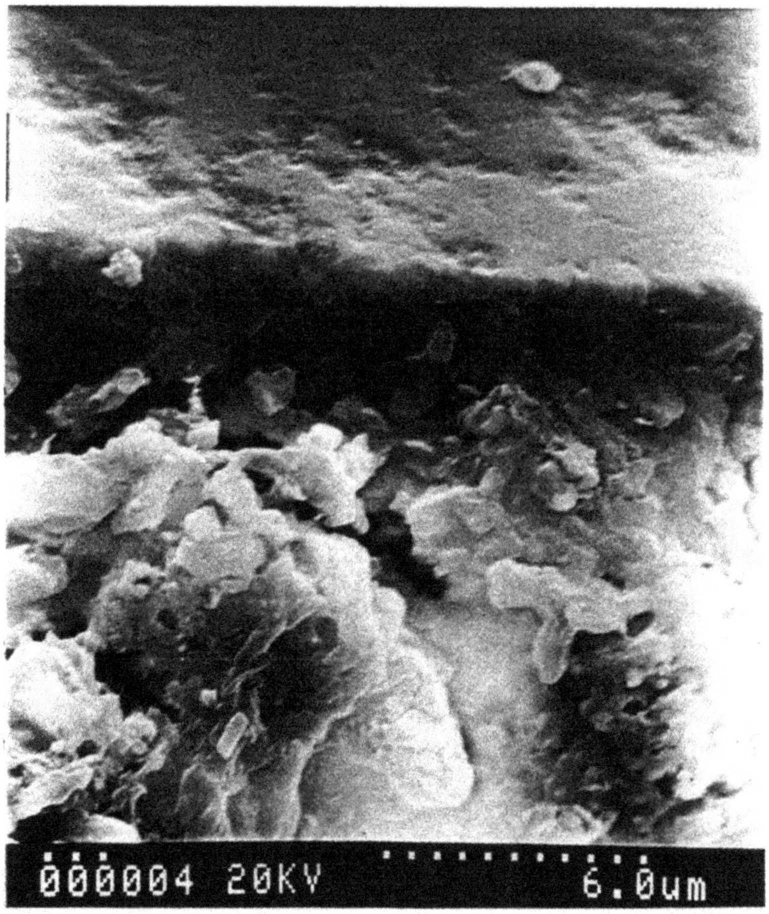

Figure 222: View of deformed layer on Neolithic sickle blade.

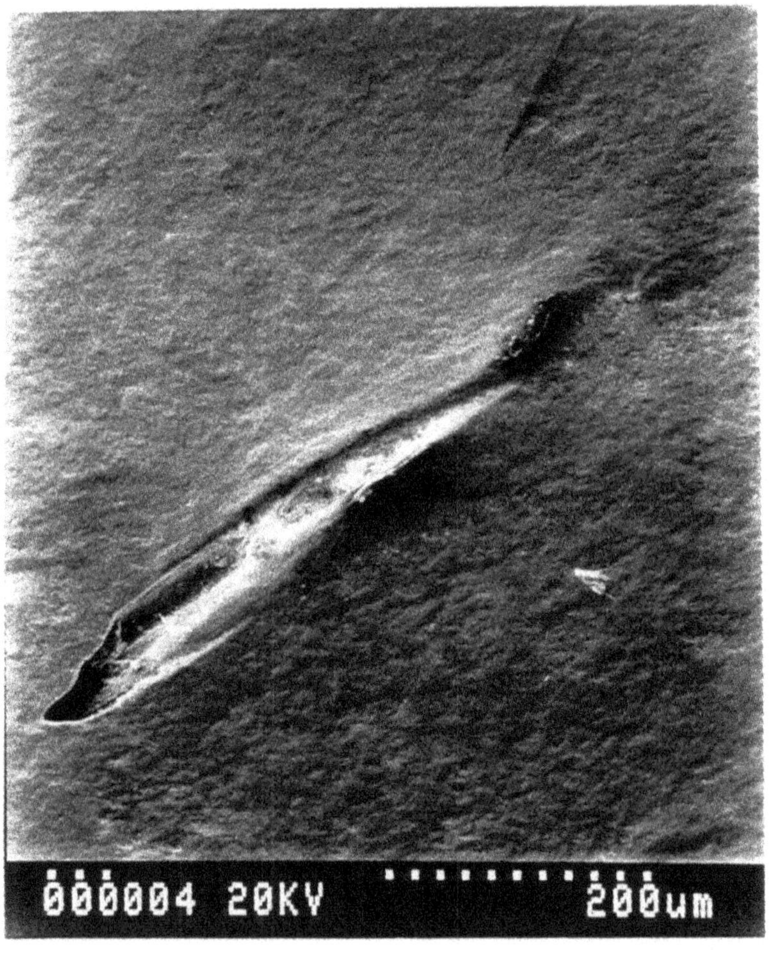

Figure 223: Cigar-shaped hollow on Neolithic sickle blade.

Figure 224: Enlarged view of figure 223.

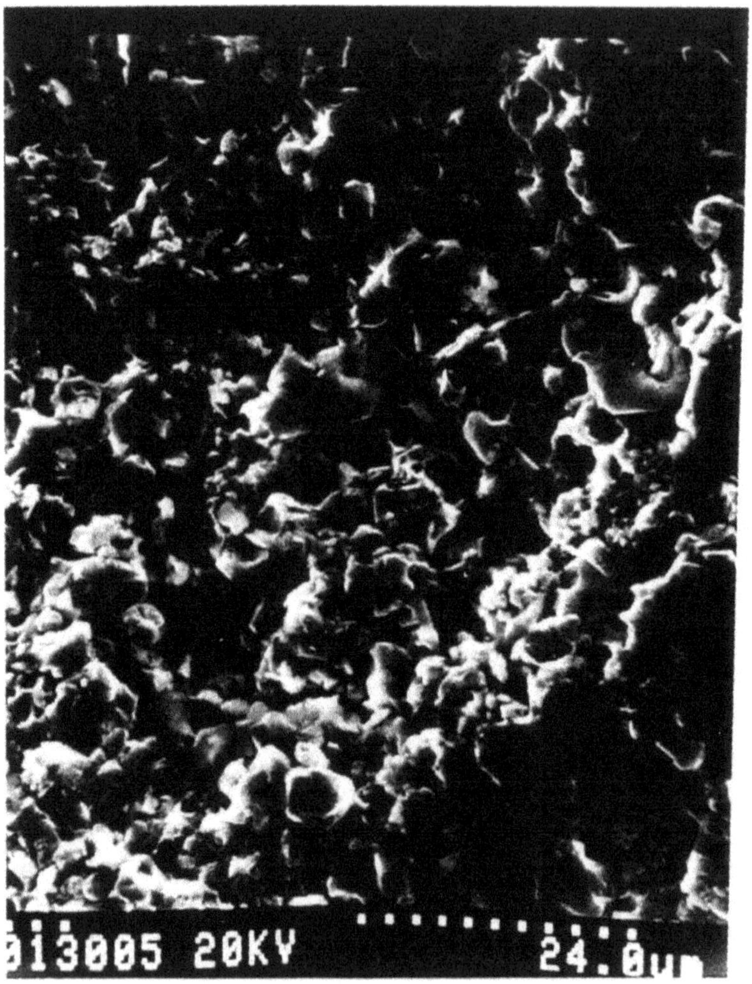

Figure 225: Unused Potters Bar flint.

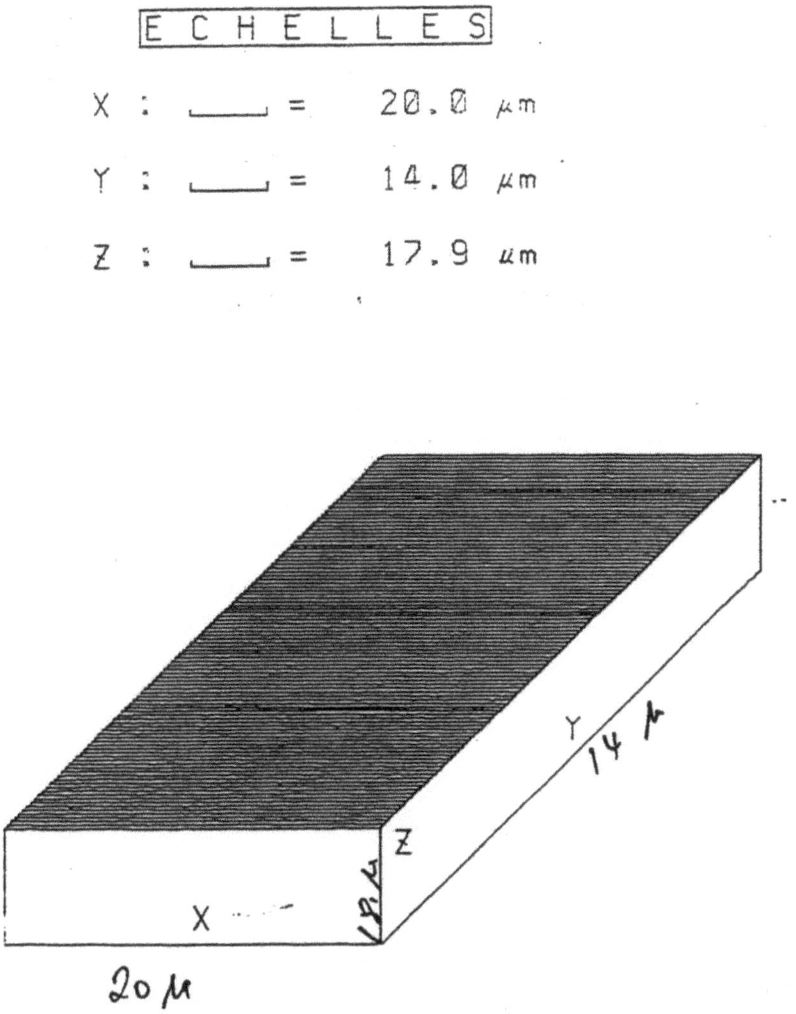

Figure 226: Profilometer rendering of EXP/101, Figure 119.

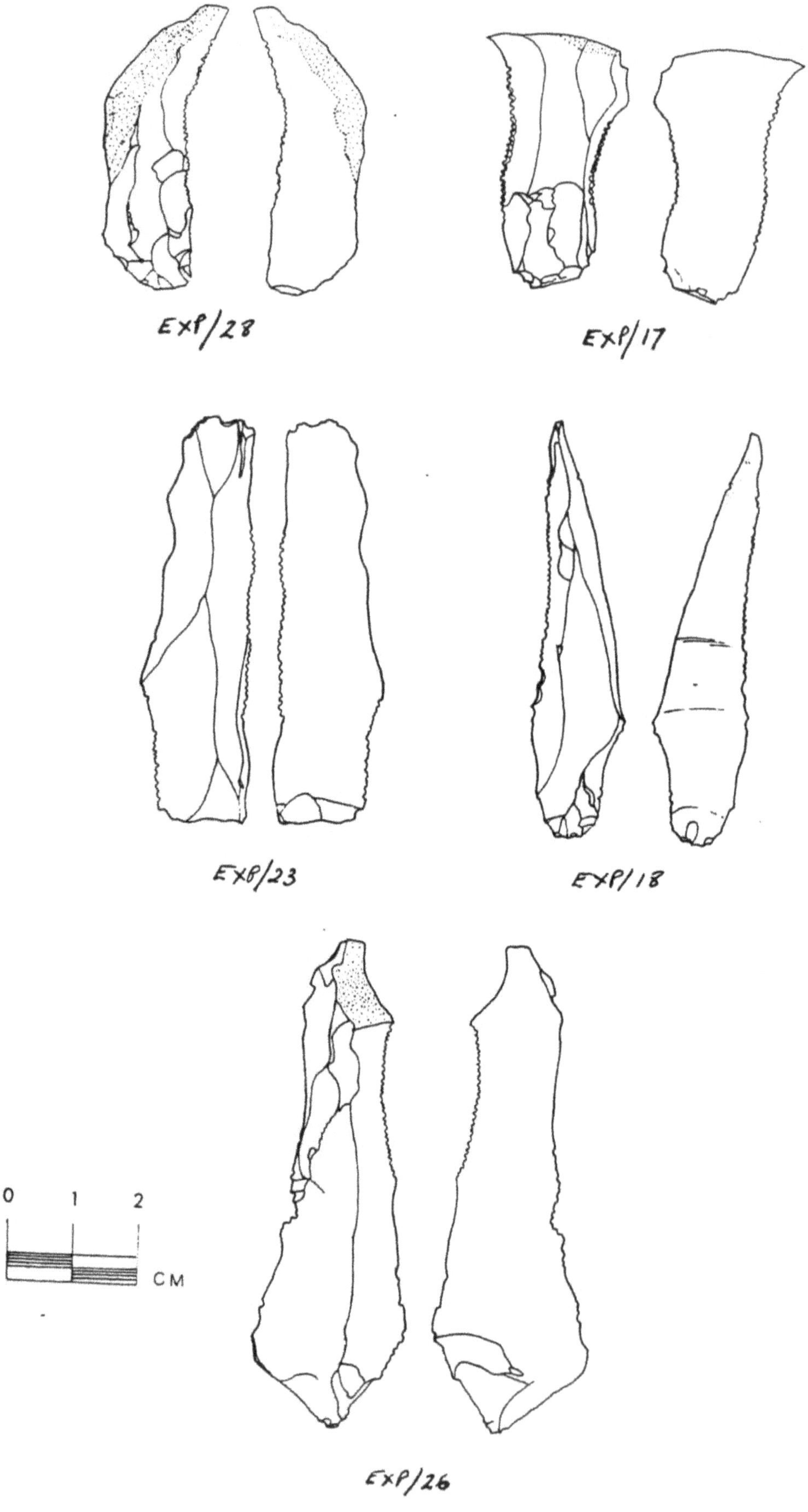

Figure 227: Artefacts used in use-wear experiments Section 6.4 and some also used in PDSM experiments Section 6.5.1.

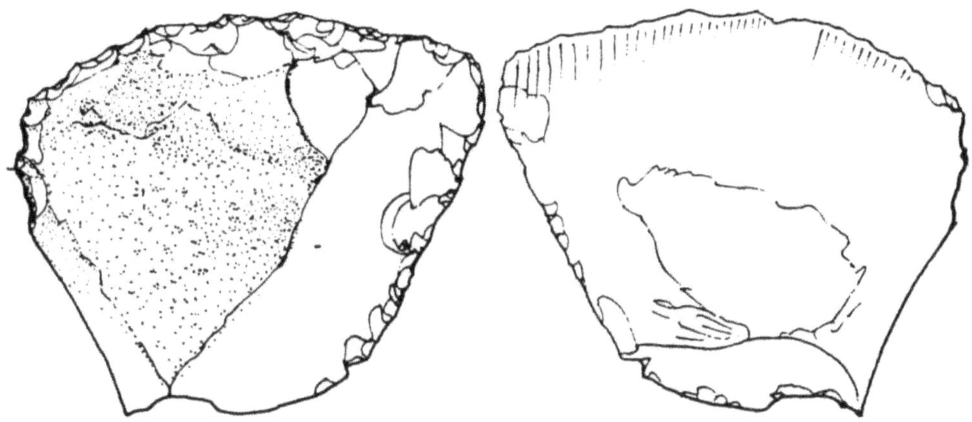

EXP/83

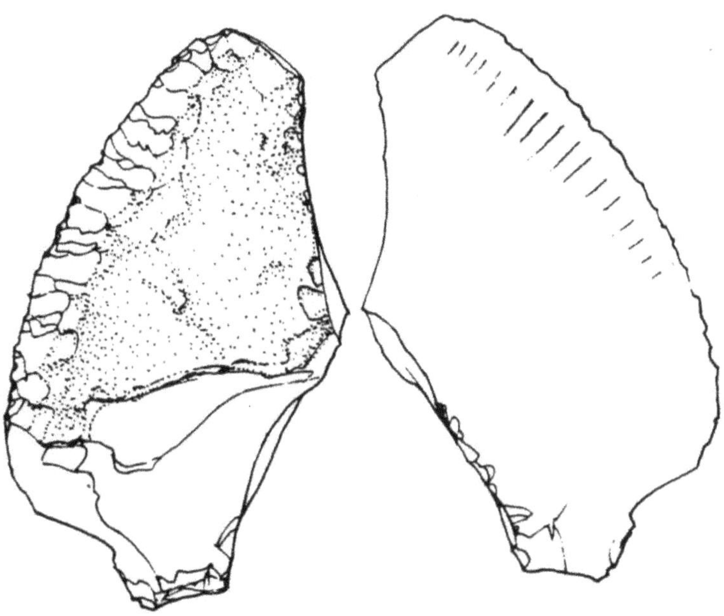

EXP/78

Figure 228: Artefacts used in use-wear experiments. Hachuring indicates macroscopic gloss removed by cleaning.

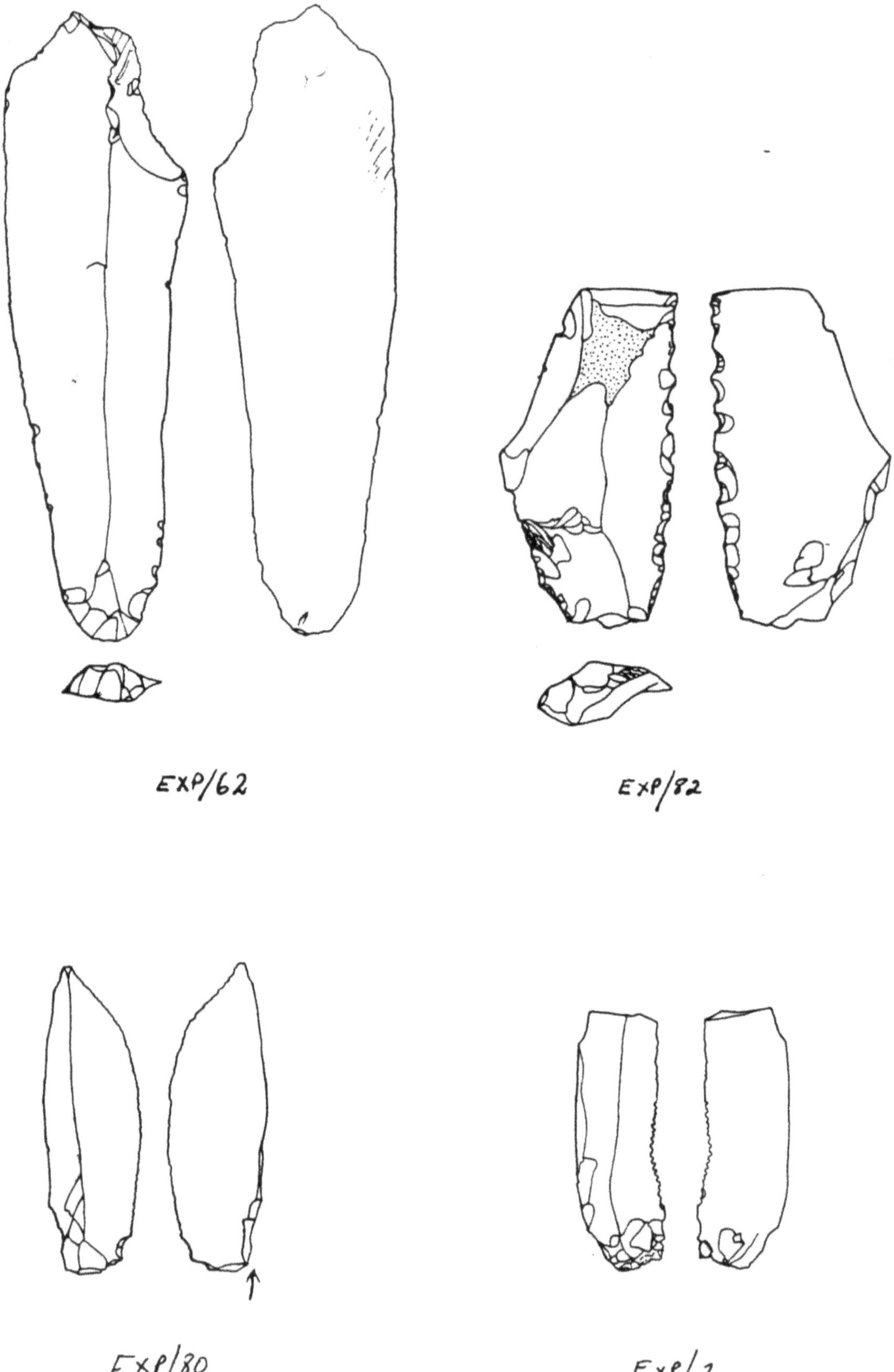

Figure 229: Artefacts used in use-wear experiments.

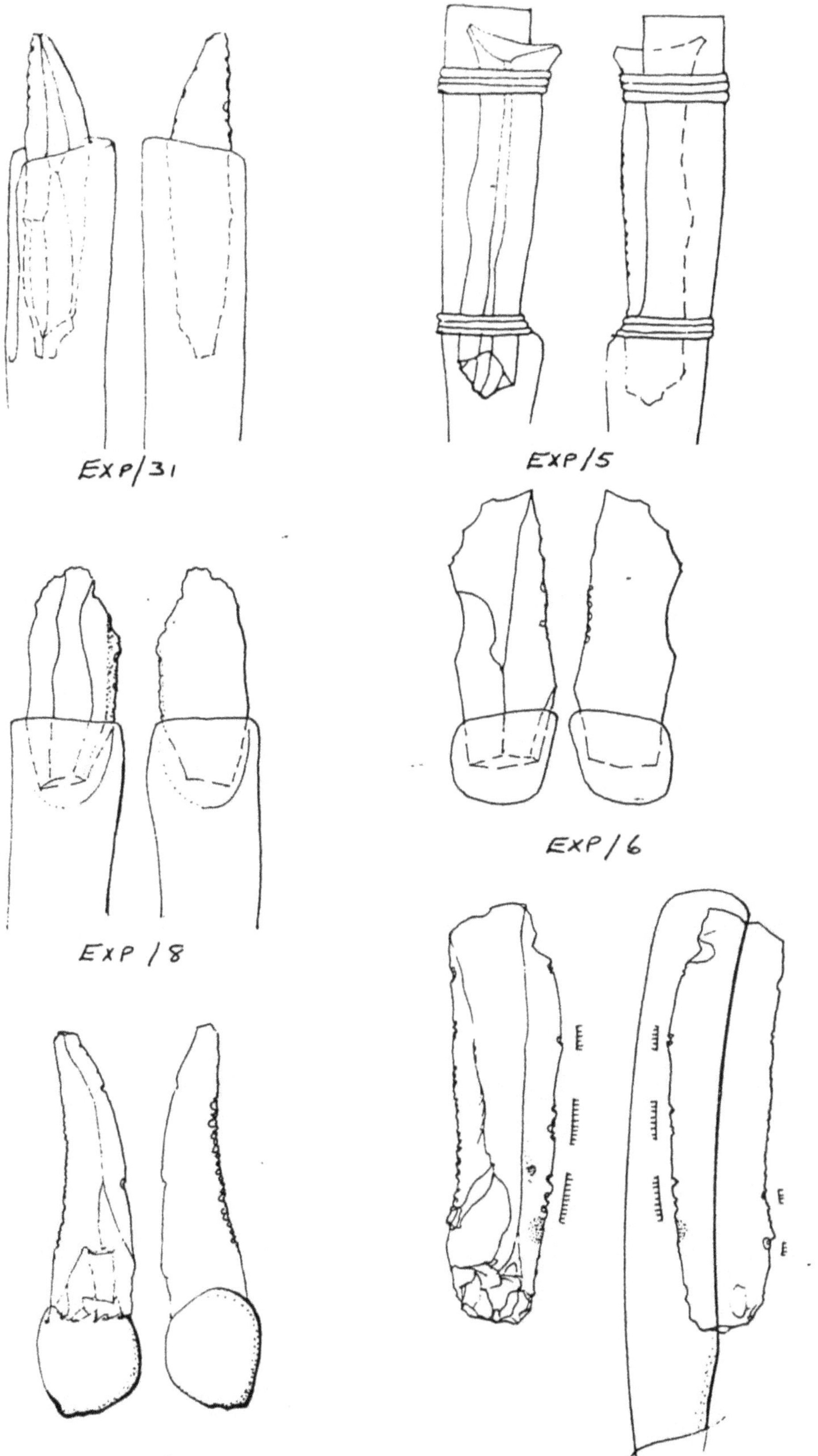

Figure 230: Examples of hafts used on microdenticulates.

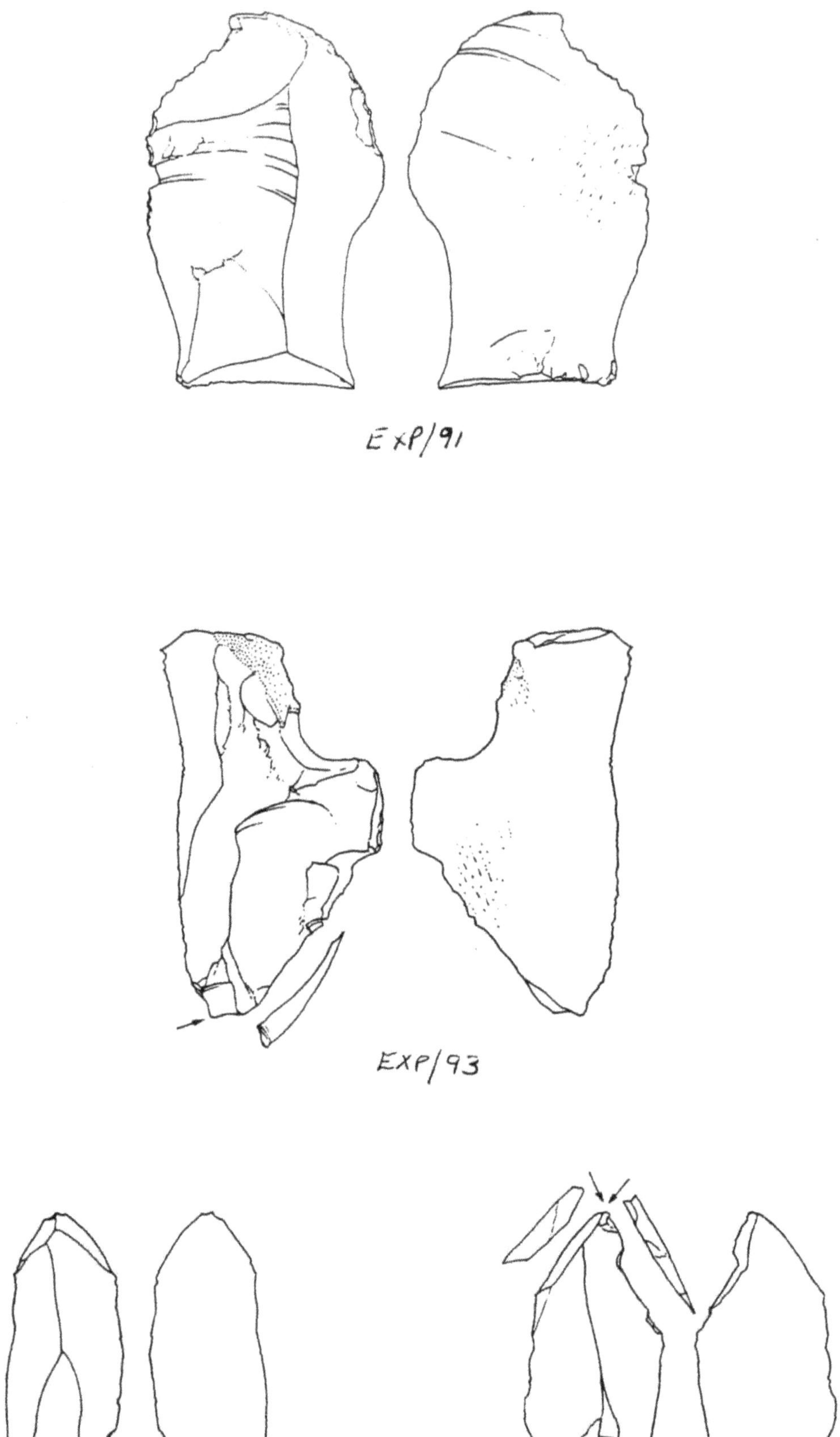

Figure 231: Artefacts used in wood grooving experiments. See Section 6.4.6.

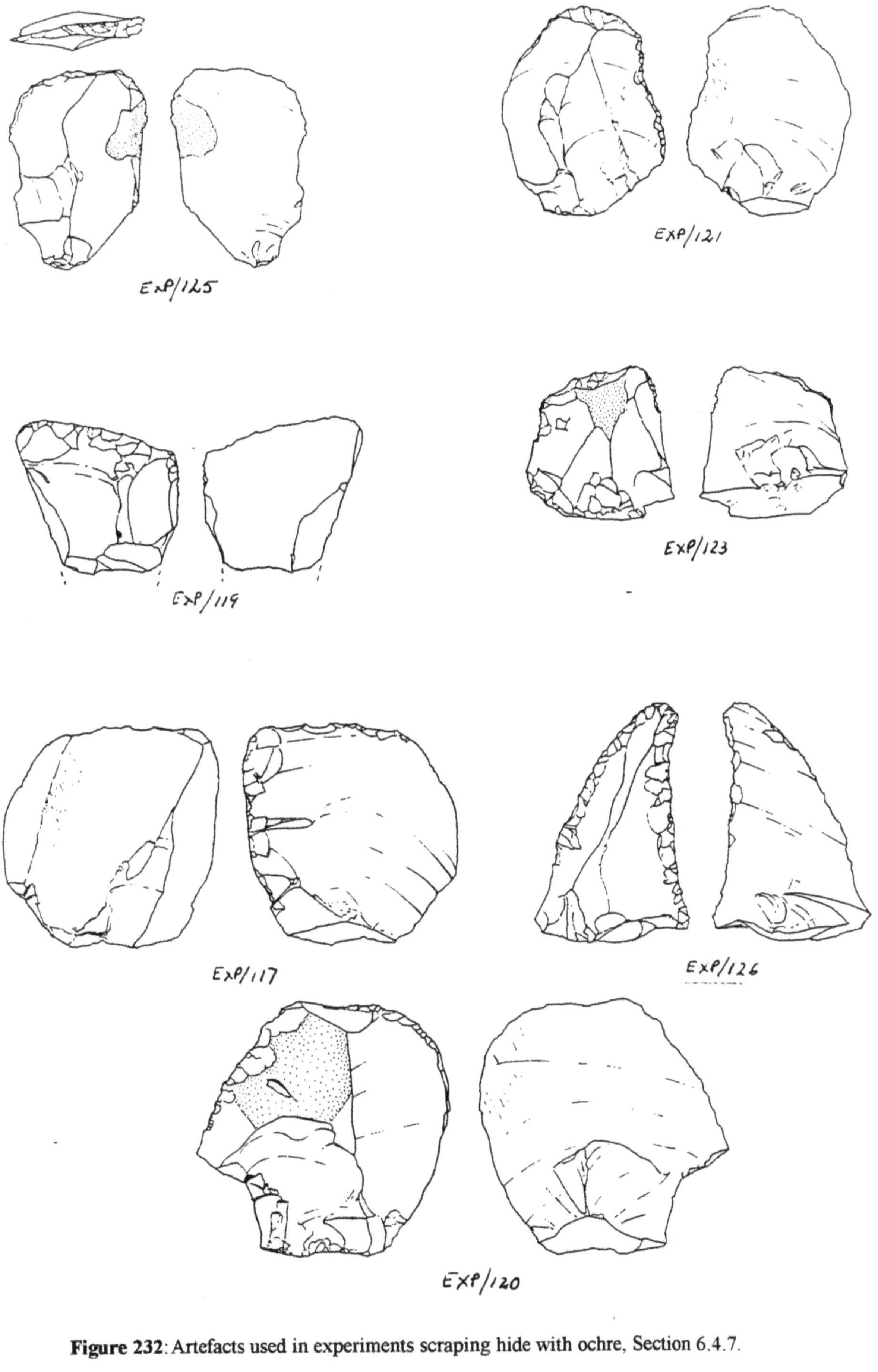

Figure 232: Artefacts used in experiments scraping hide with ochre, Section 6.4.7.

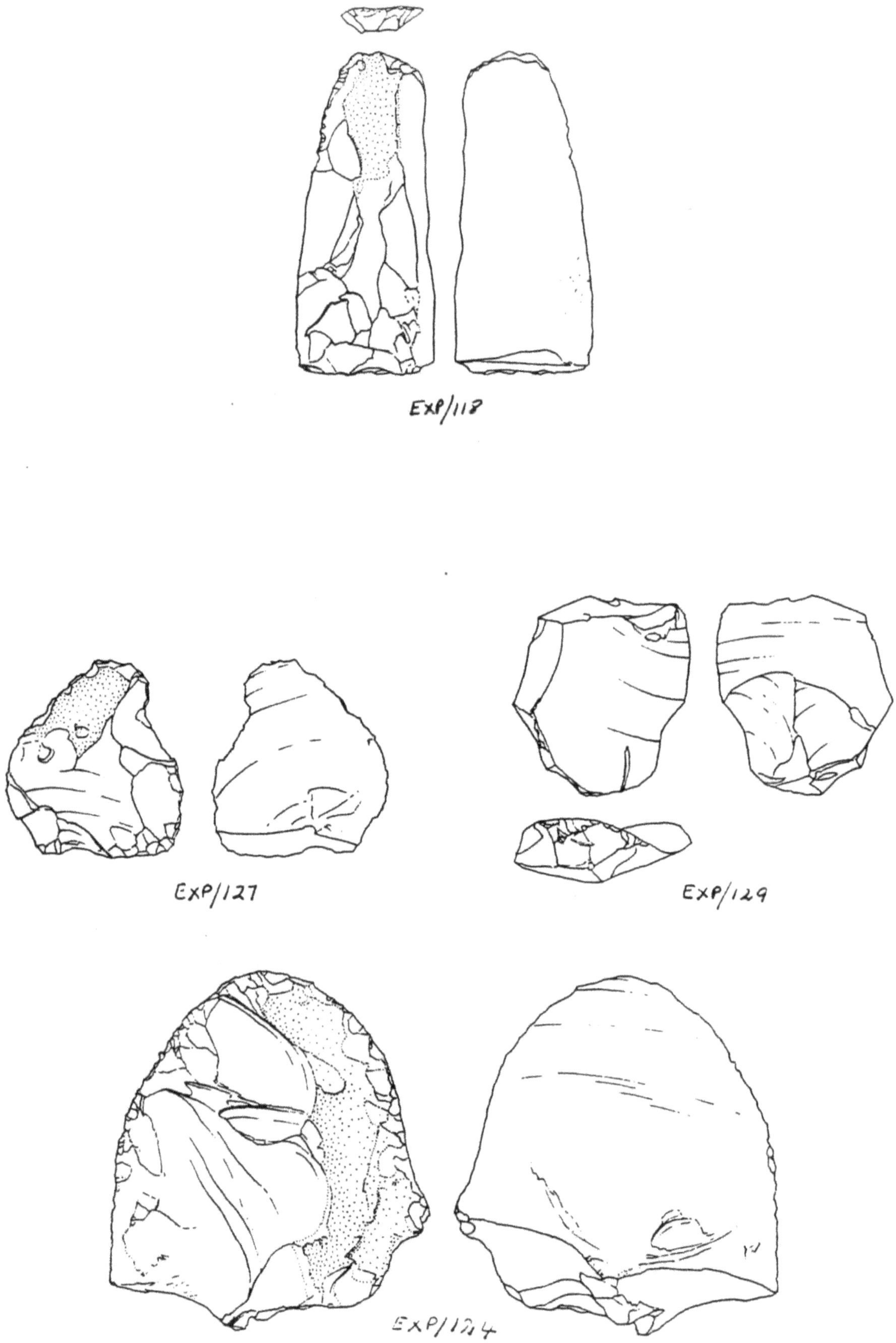

Figure 233: Artefacts used in experiments scraping hide with ochre, Section 6.4.7.

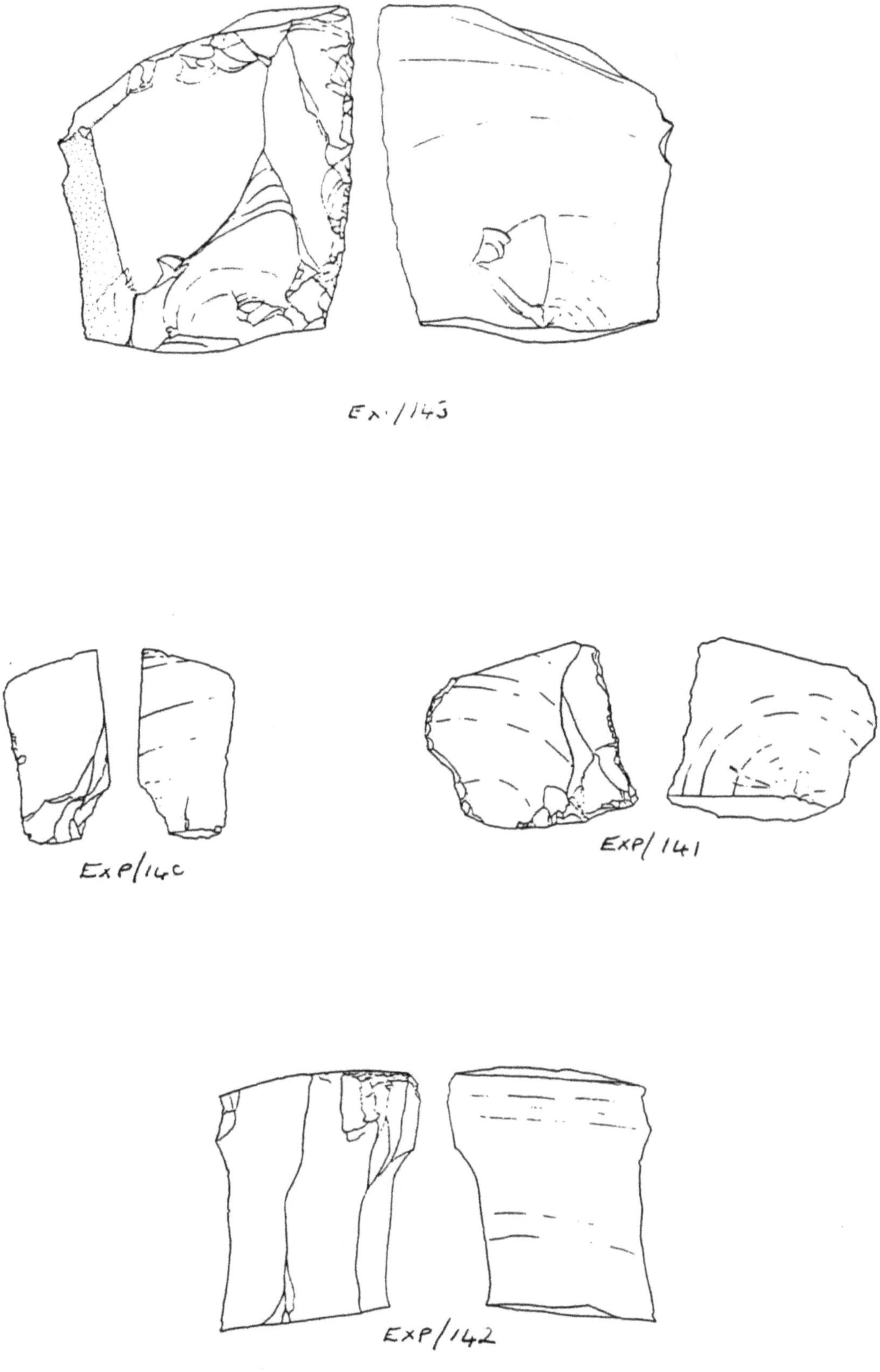

Figure 234: Artefacts used in polish formation experiments, Sections 6.4.10 and 6.4.12.

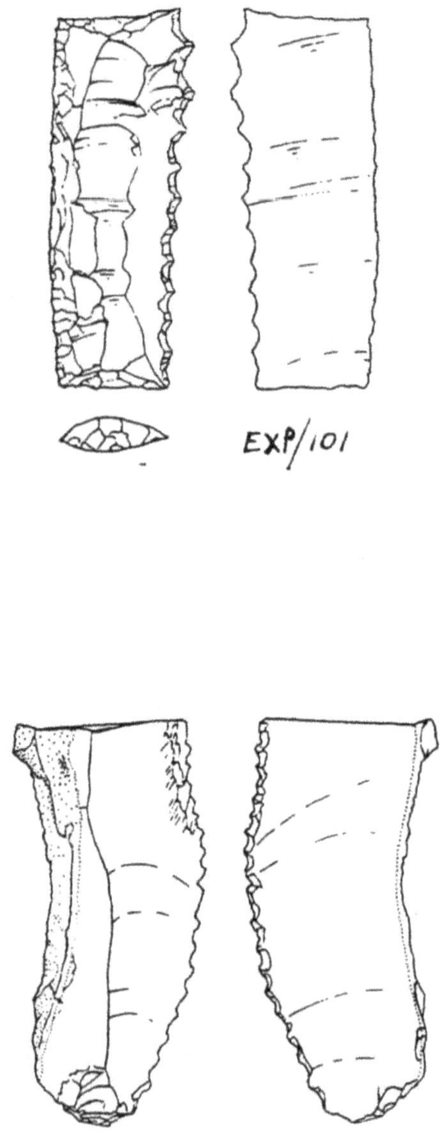

Figure 235: Artefacts used in polish formation experiments, Section 6.4.8 and in PDSM experiments 6.5.13.

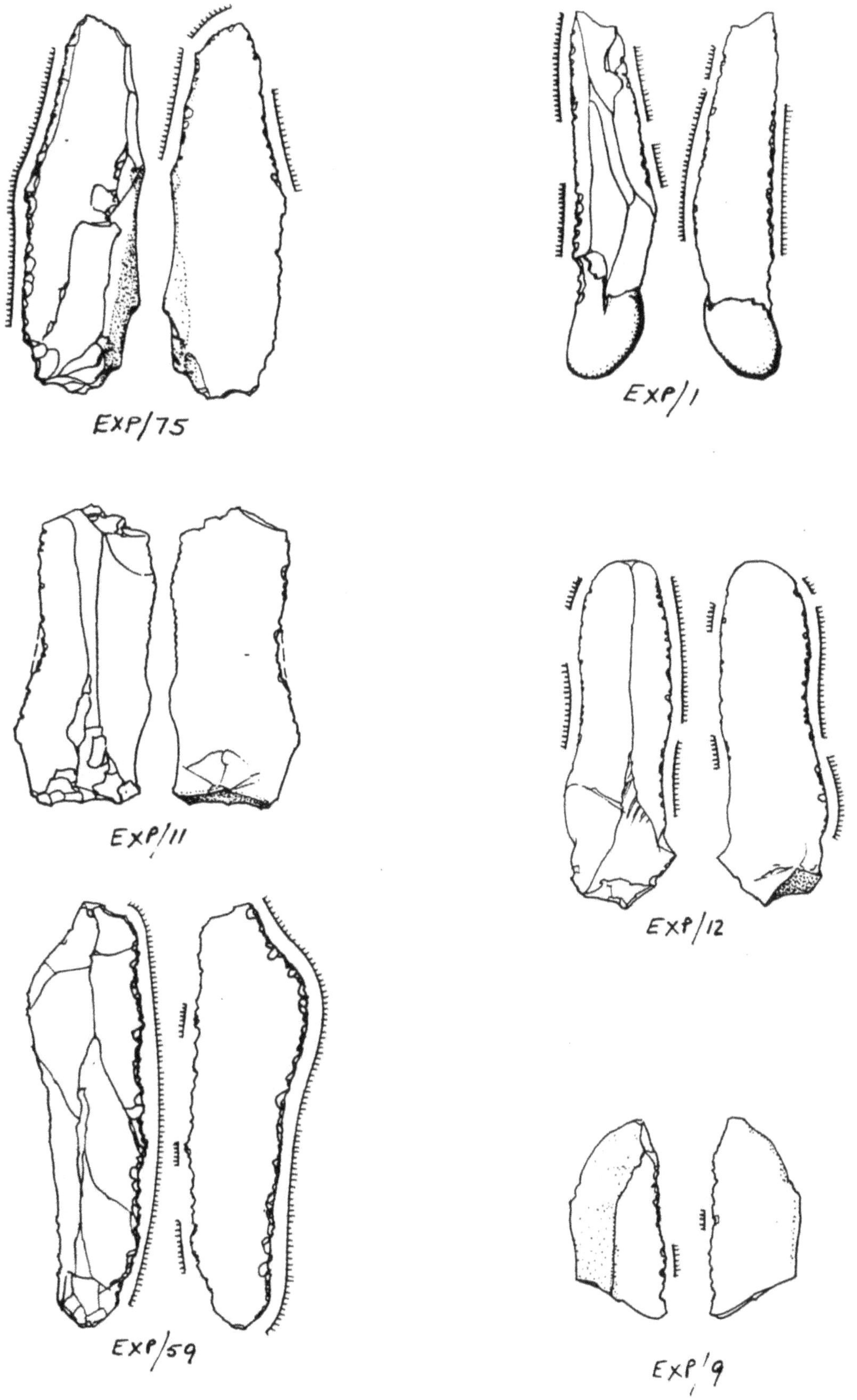

Figure 236: Artefacts used in use-wear and PDSM experiments, Sections 6.5.1 and 6.5.2. [scale-bar symbol] indicates post-depositional edge damage.

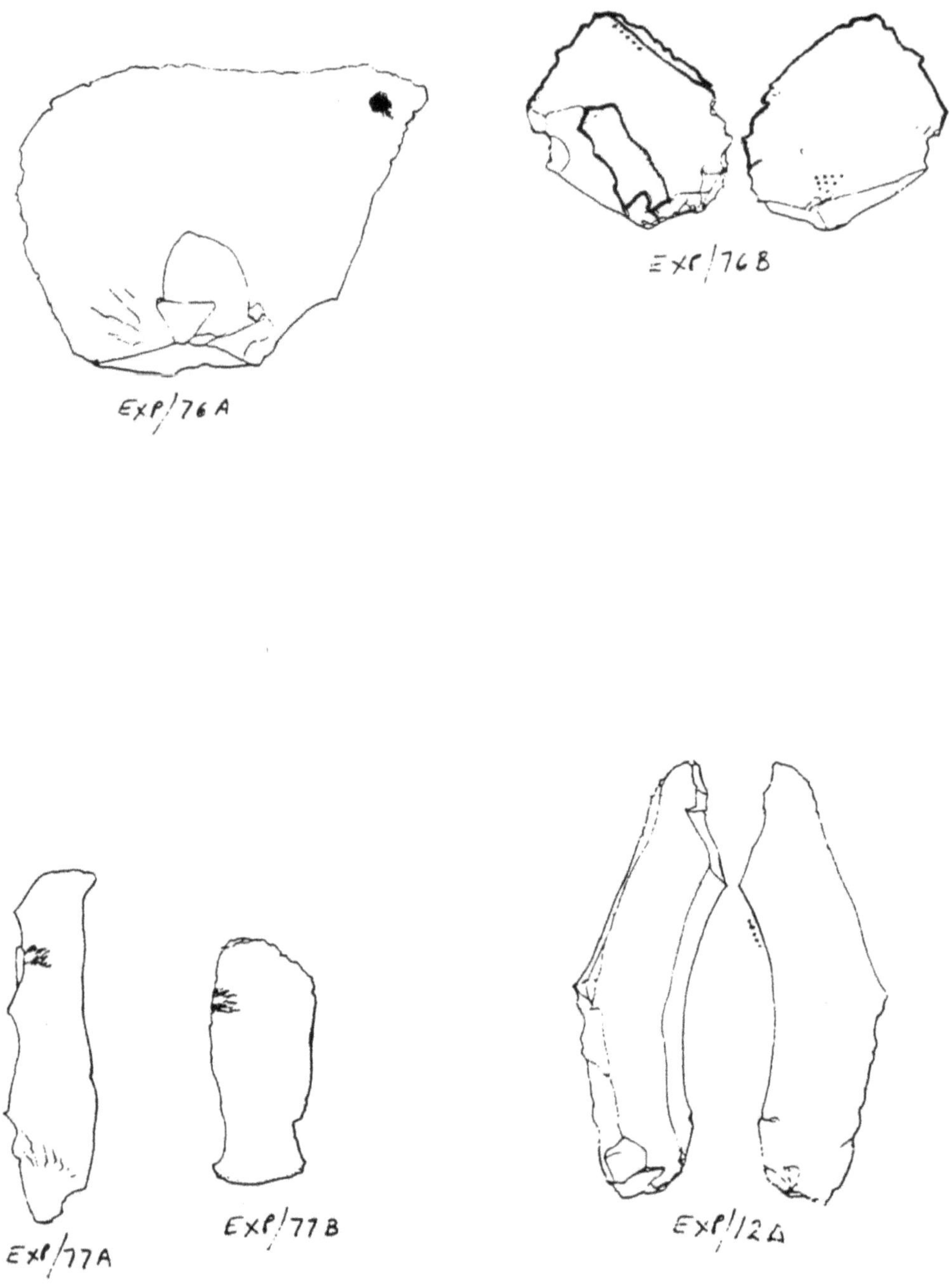

Figure 237: Artefacts used in PDSM experiments, Sections 6.5.1 and 6.5.3. Dots and shading indicate bright spots.

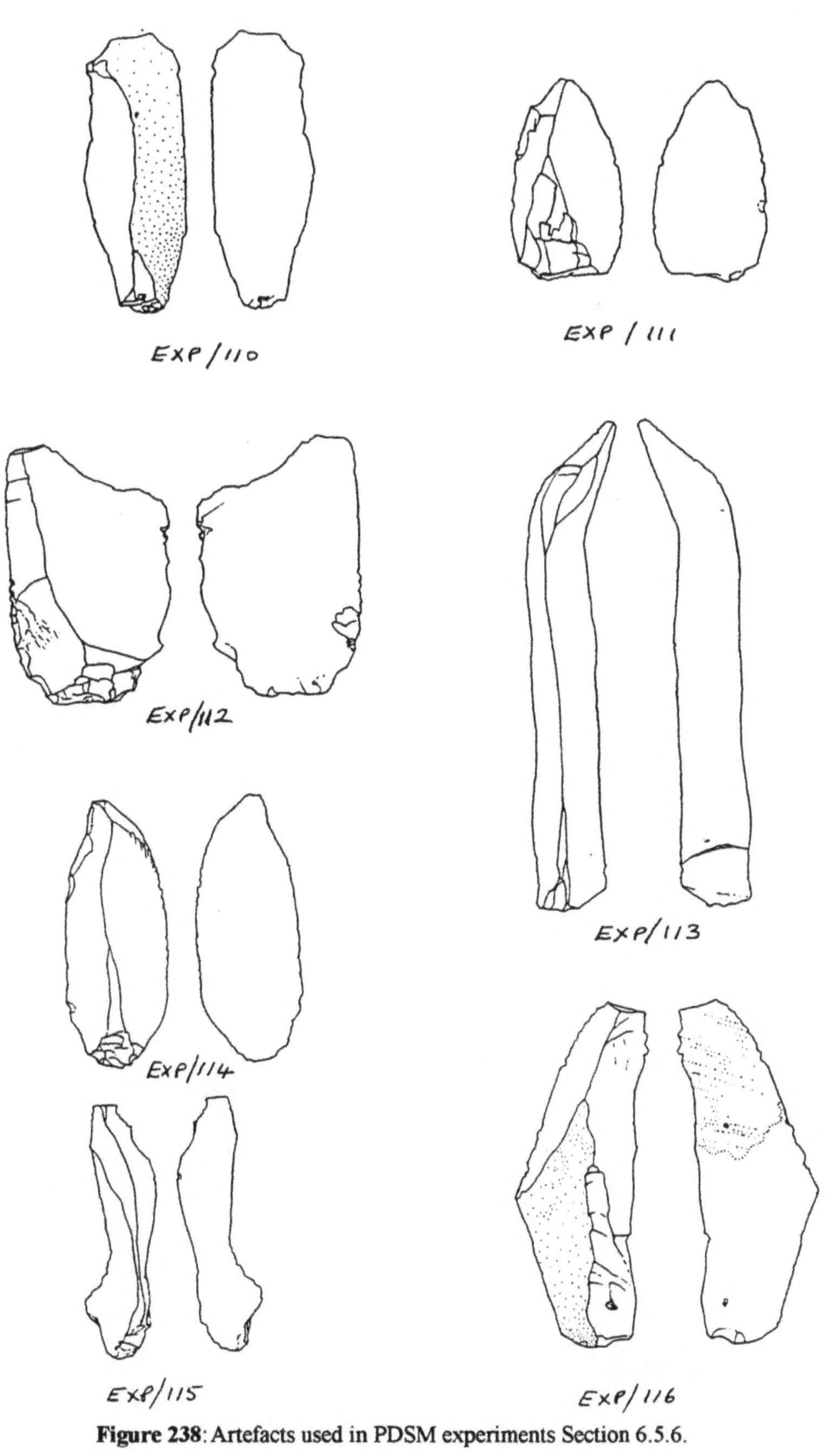

Figure 238: Artefacts used in PDSM experiments Section 6.5.6.

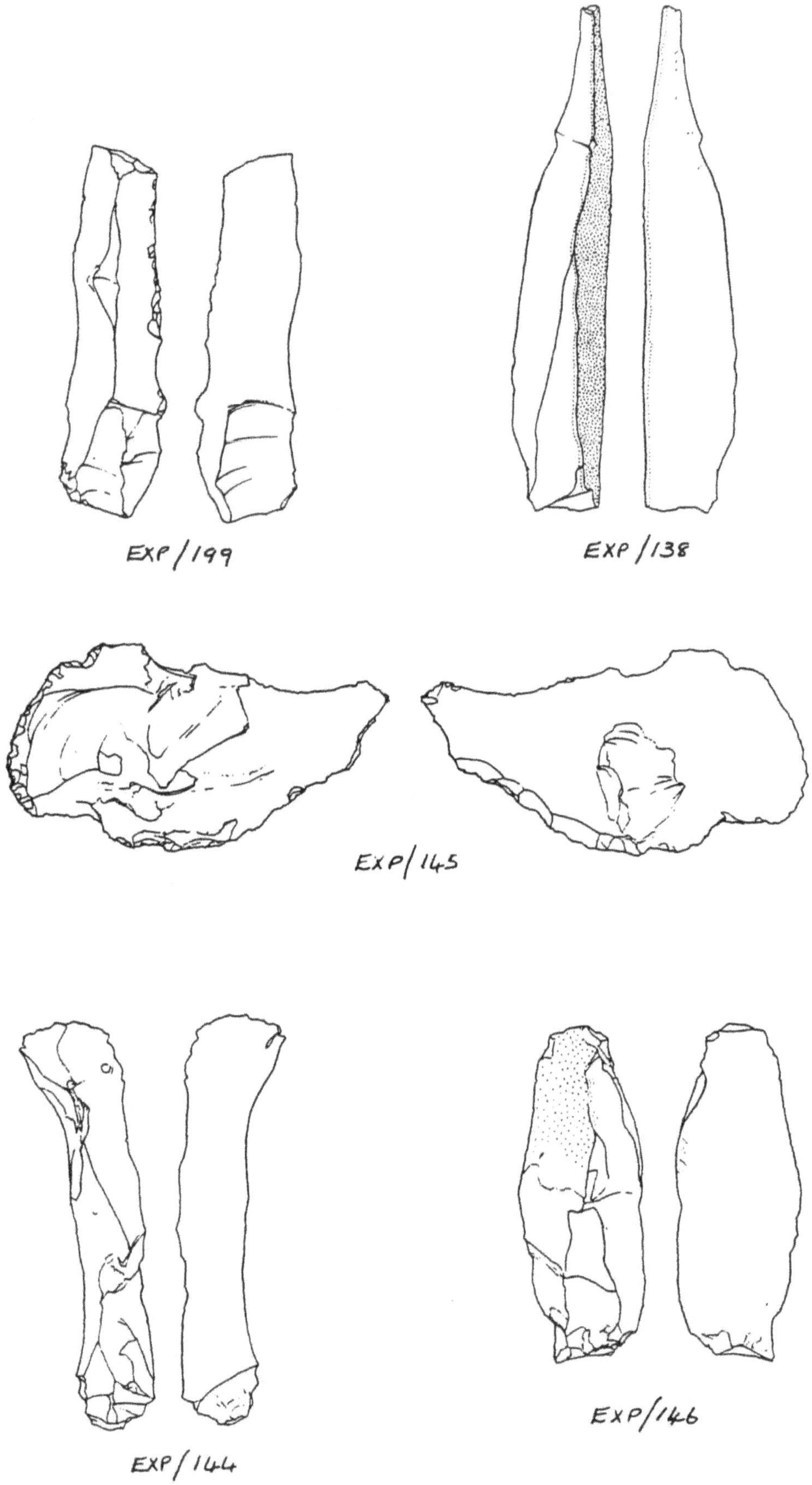

Figure 239: Artefacts used in PDSM experiments Section 6.5.7.

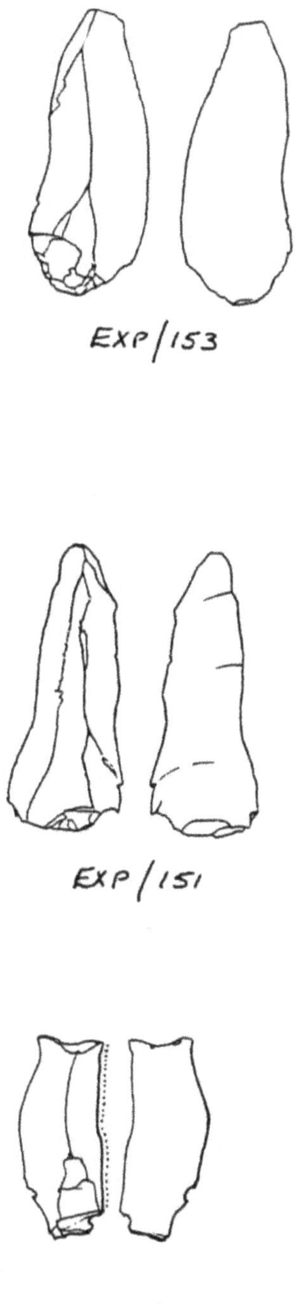

Figure 240: Artefacts used in PDSM experiments. Section 6.5.11(b).

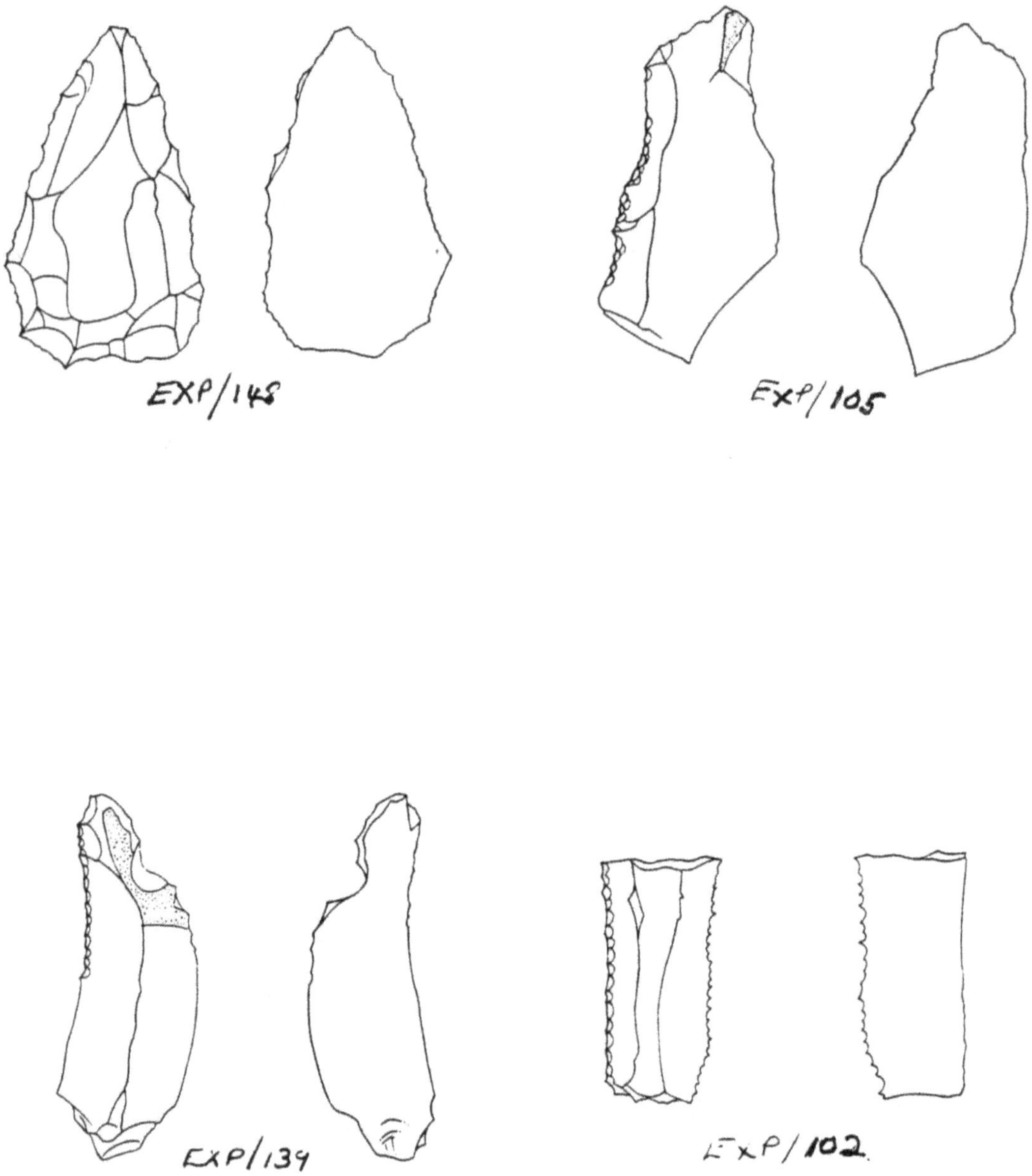

Figure 241: Artefacts used in PDSM experiments Section 6.5.8.

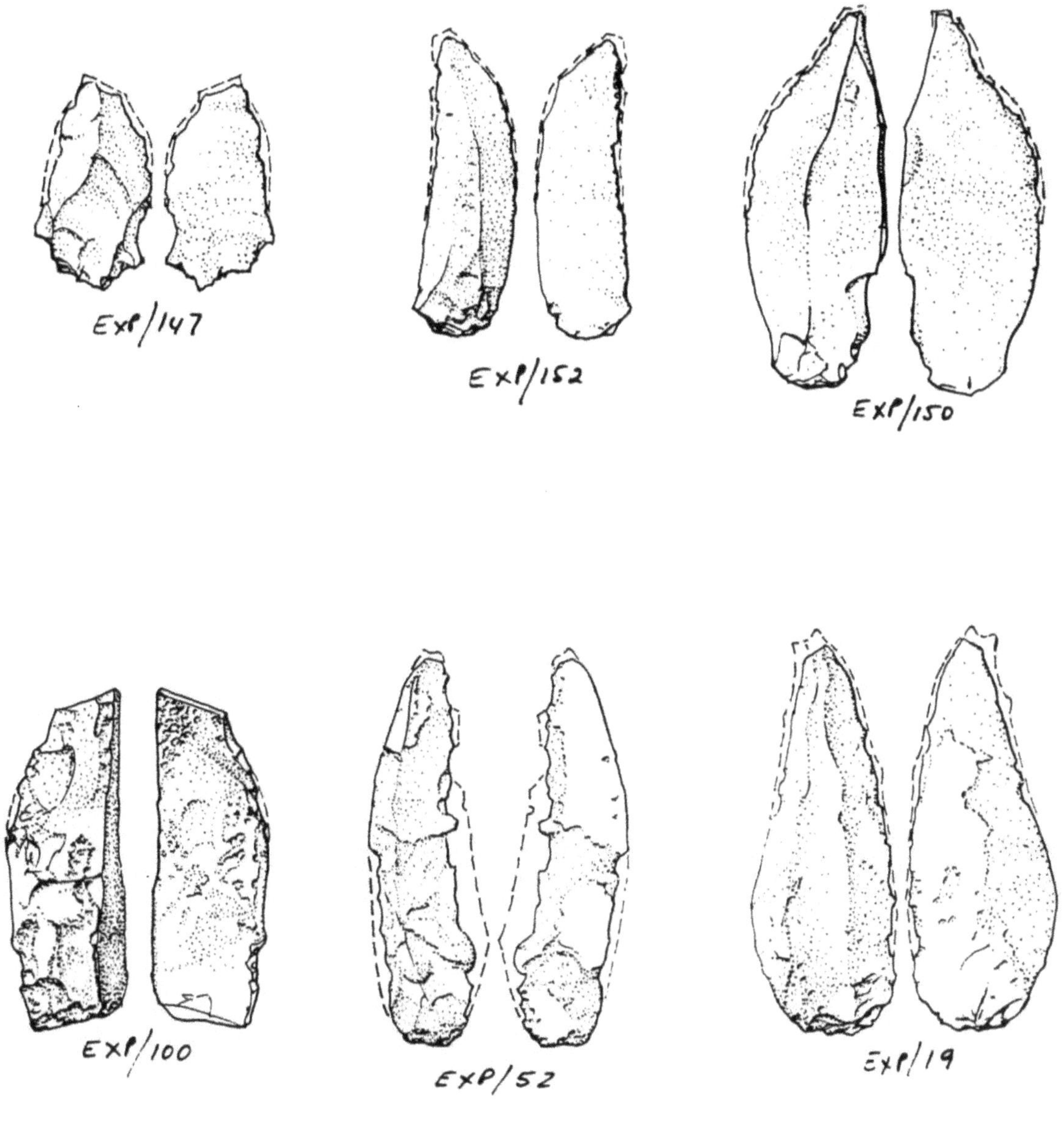

Figure 242: Artefacts used in patination experiments, Sections 6.5.10 and 6.5.11(a). - - - - indicates original outline of artefact. Stipling indicates patination.

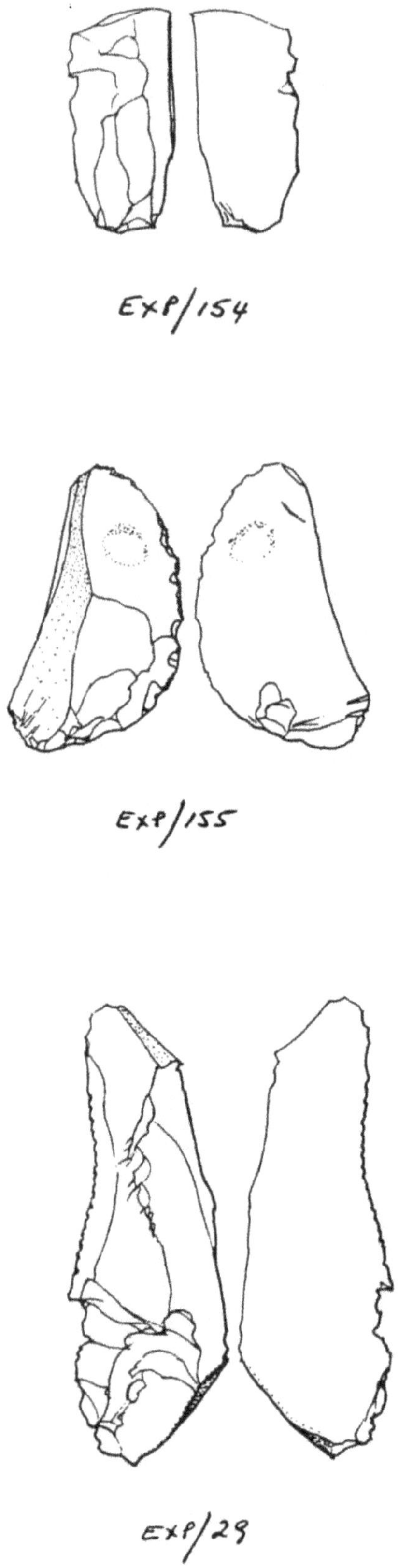

Figure 243: Artefacts used in soil sheen experiments, Section 6.5.12. Stipling indicates blue patina.

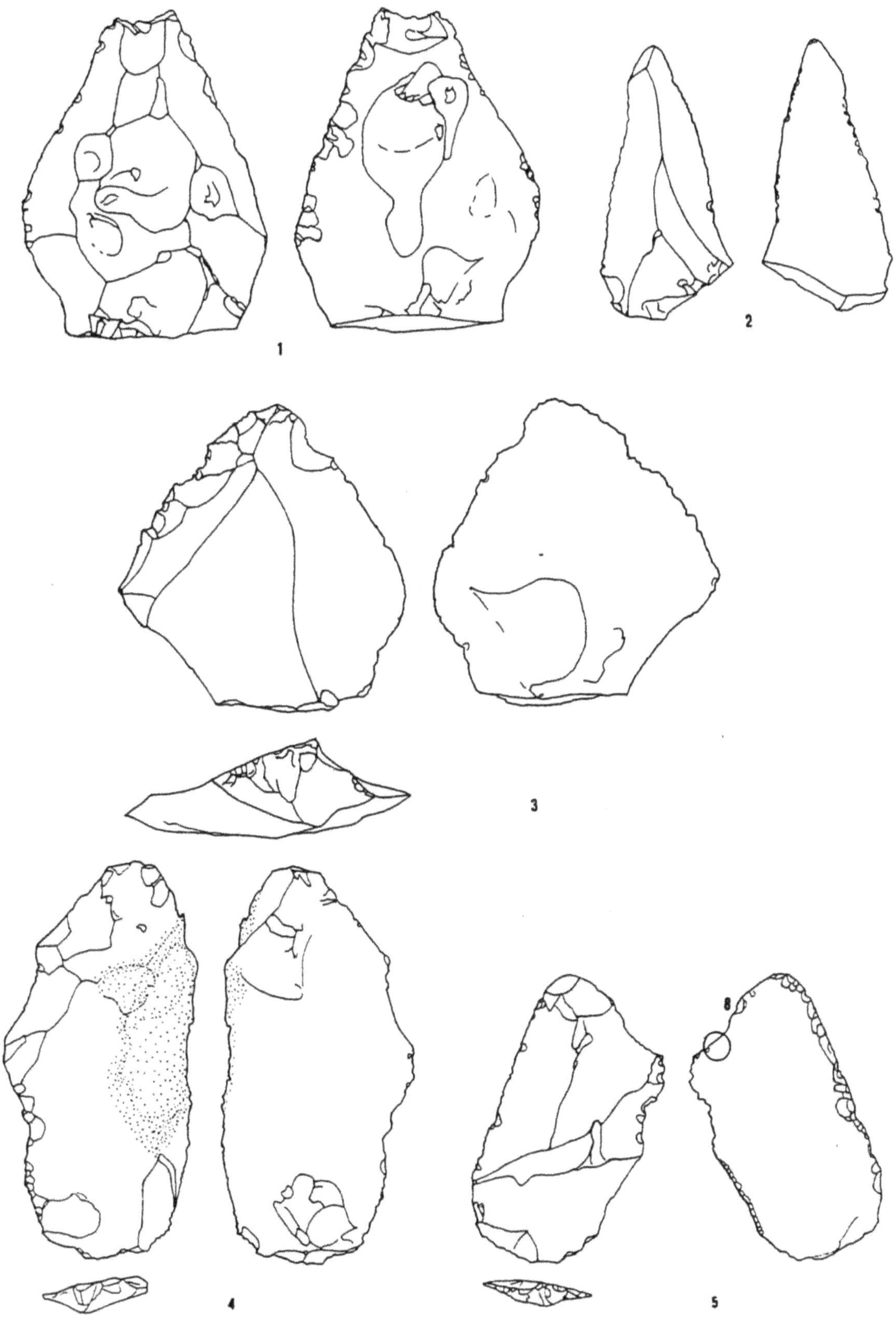

Figure 244: Artefacts from the Mousterian of Kebara cave.

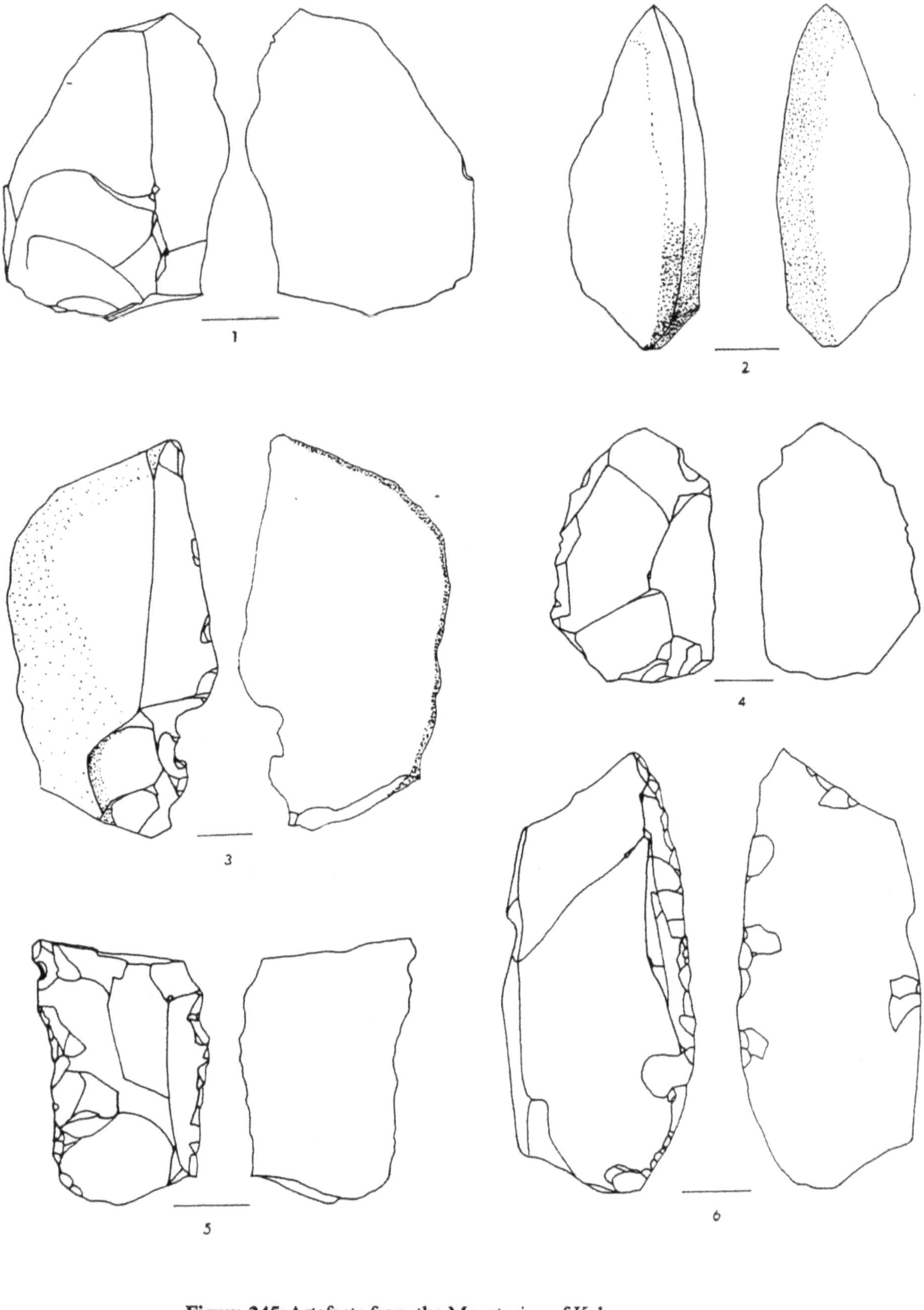

Figure 245: Artefacts from the Mousterian of Kebara cave.

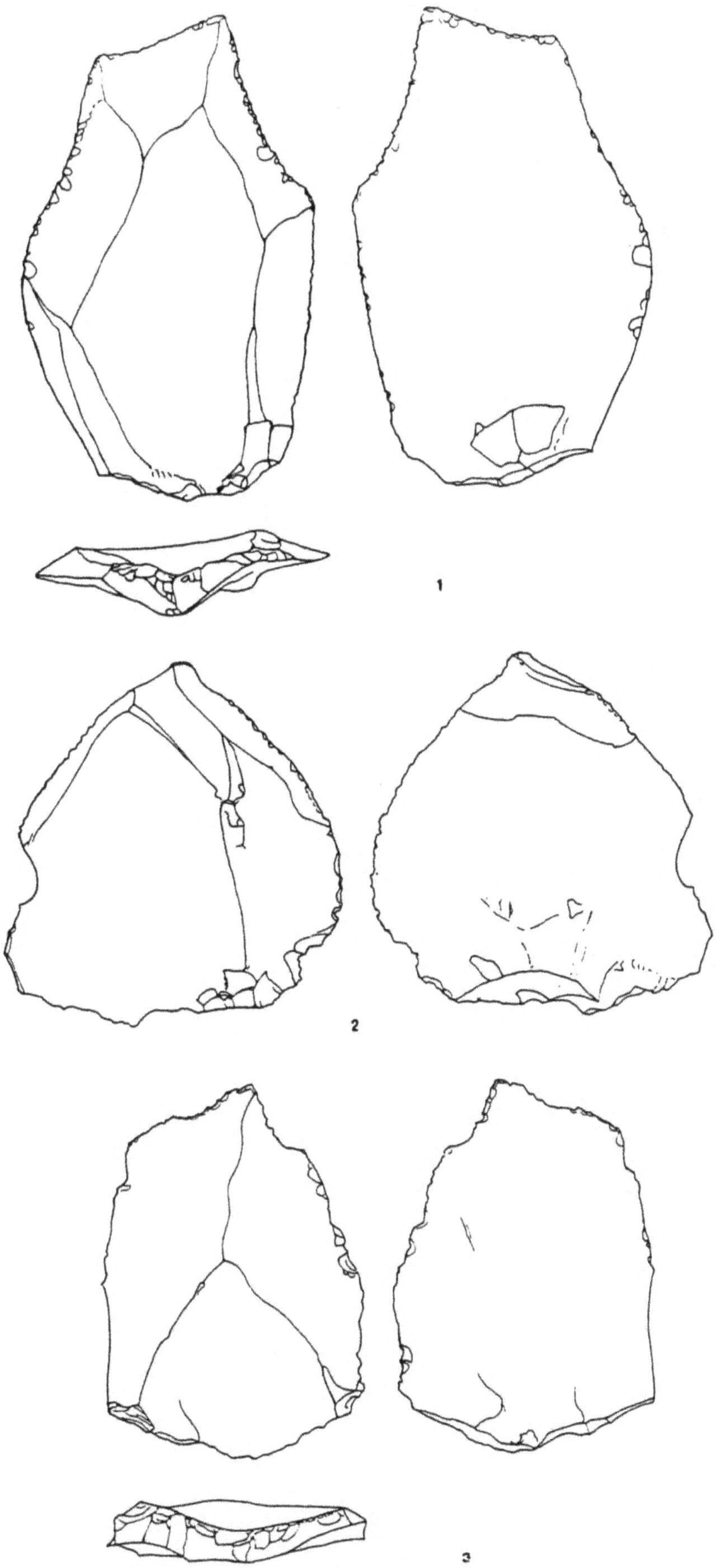

Figure 246: Artefacts from the Mousterian of Kebara cave.

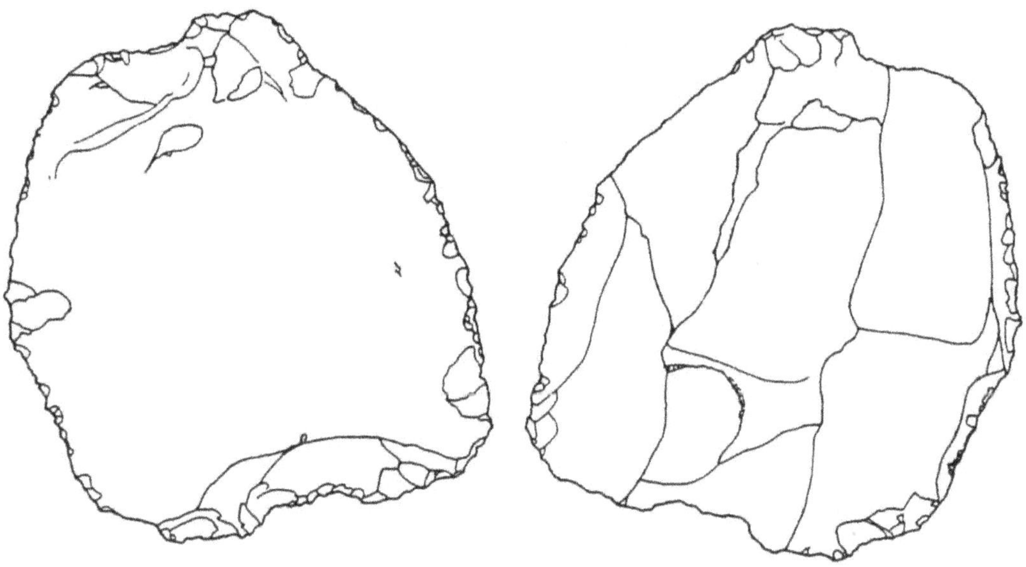

1

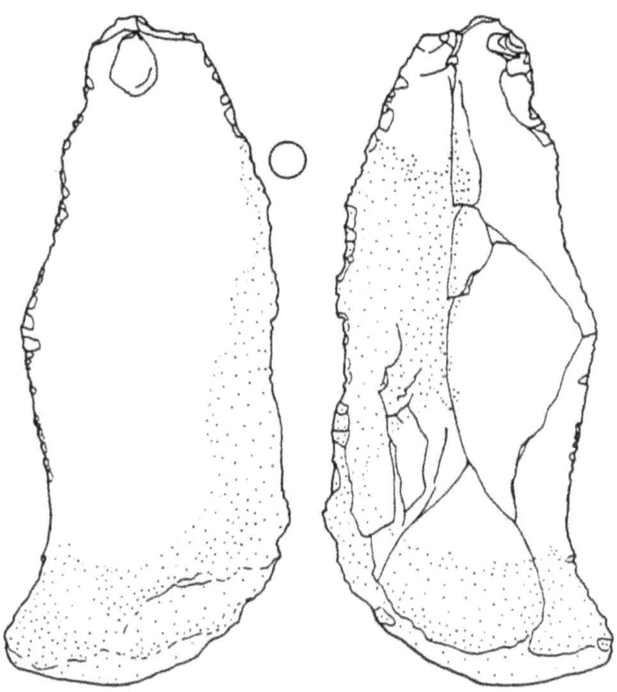

2

Figure 247: Artefacts from the Mousterian of Kebara cave.

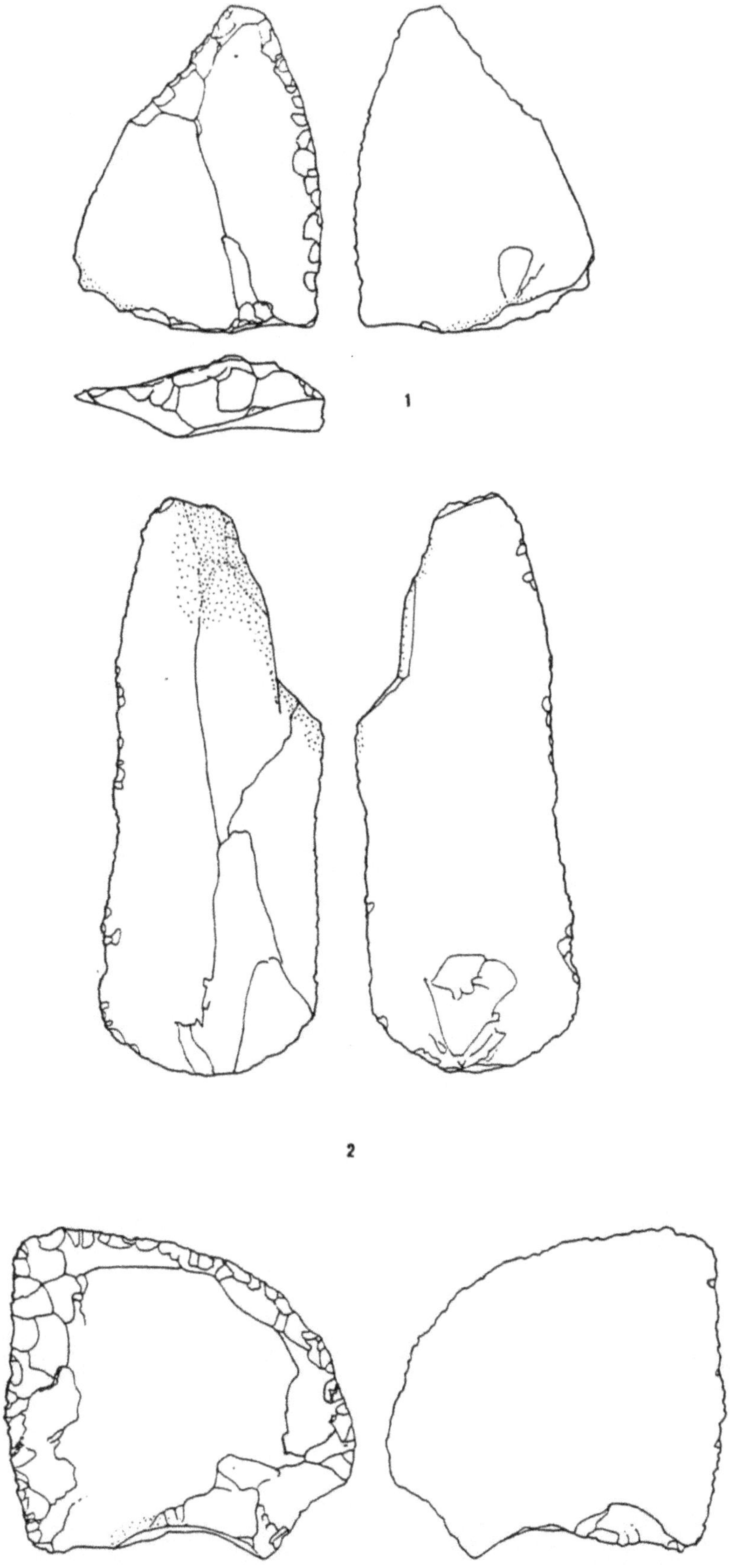

Figure 248: Artefacts from the Mousterian of Kebara cave.

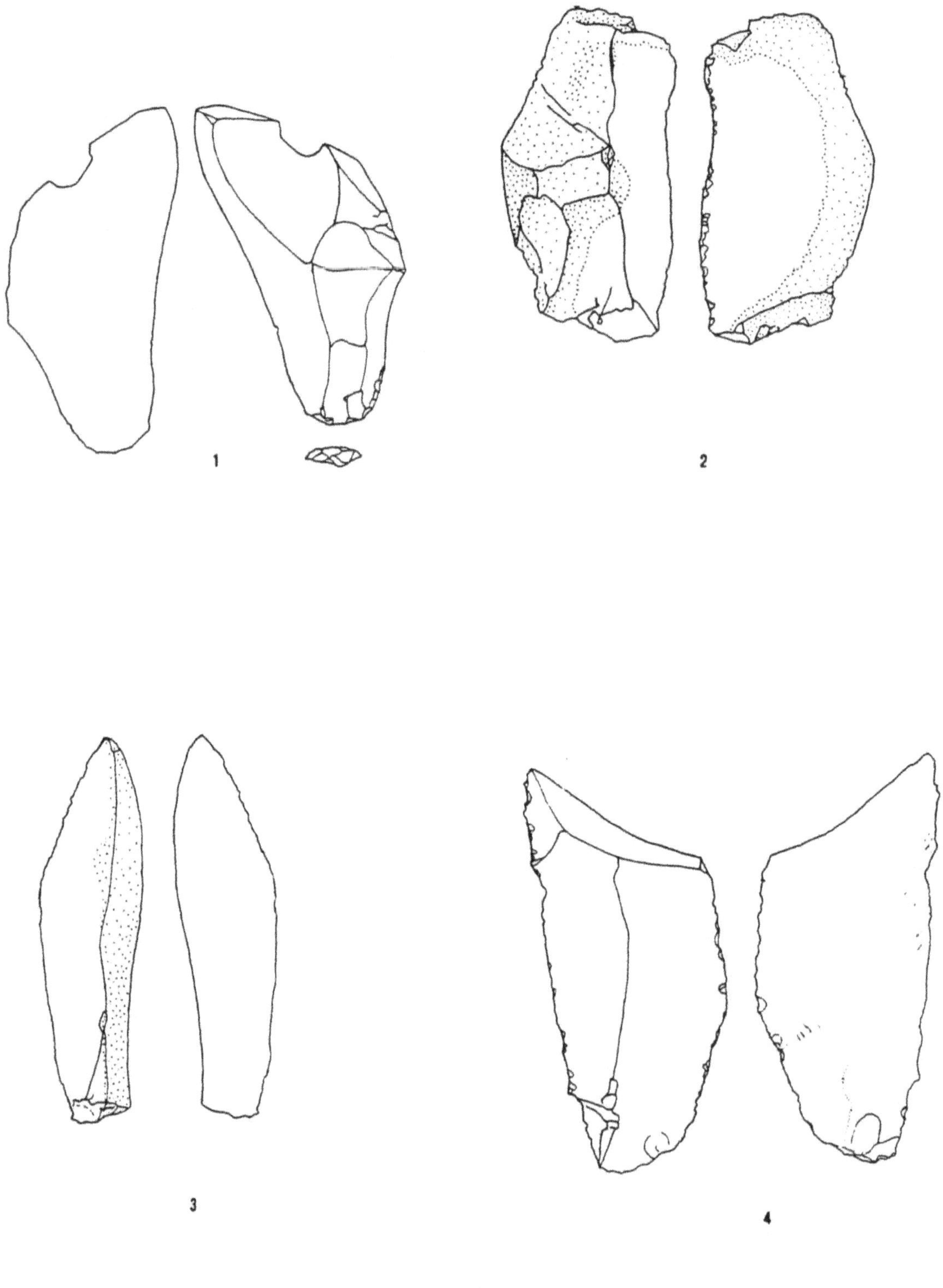

Figure 249: Artefacts from the Mousterian of Kebara cave.

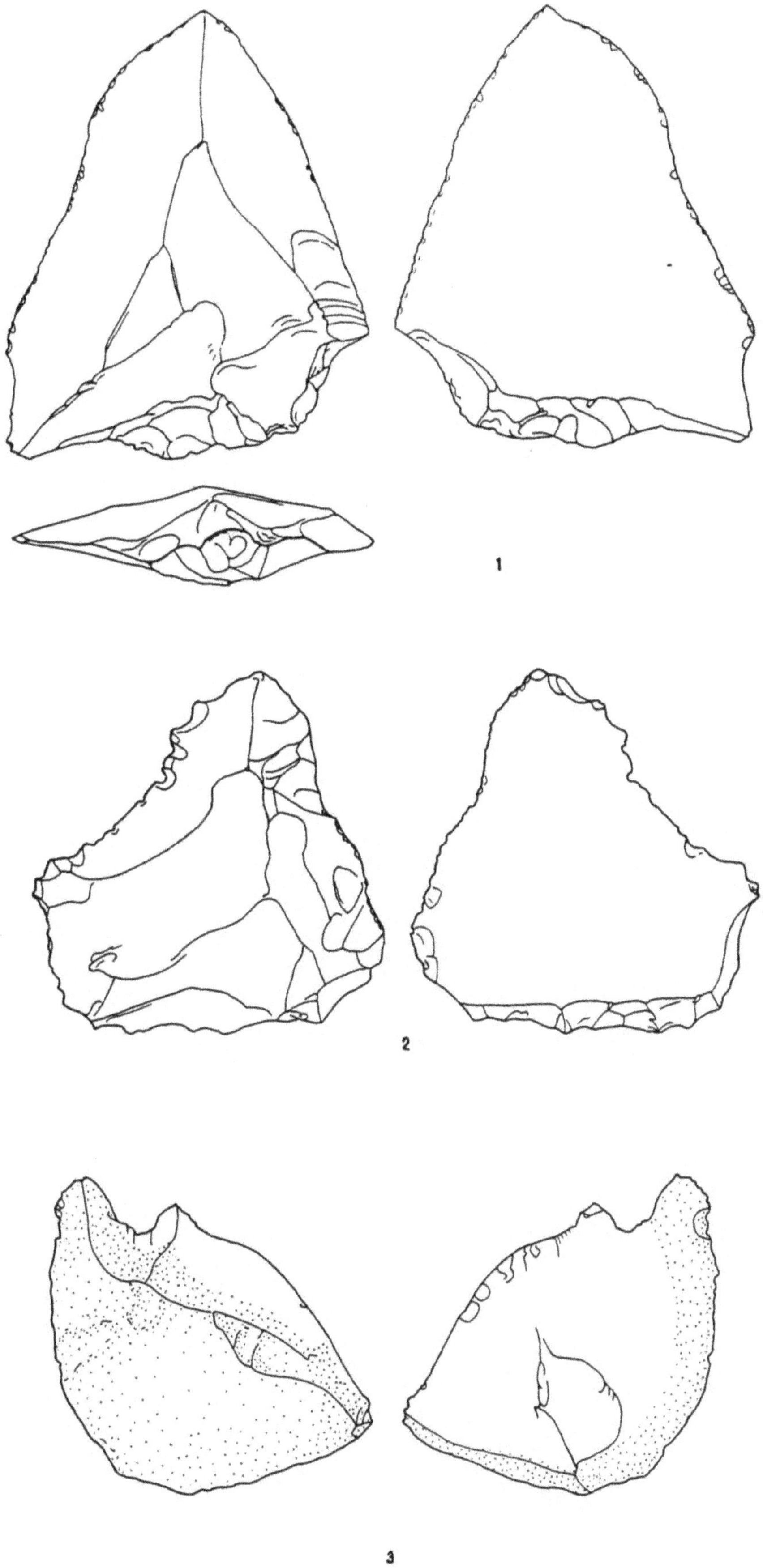

Figure 250: Artefacts from the Mousterian of Kebara cave.

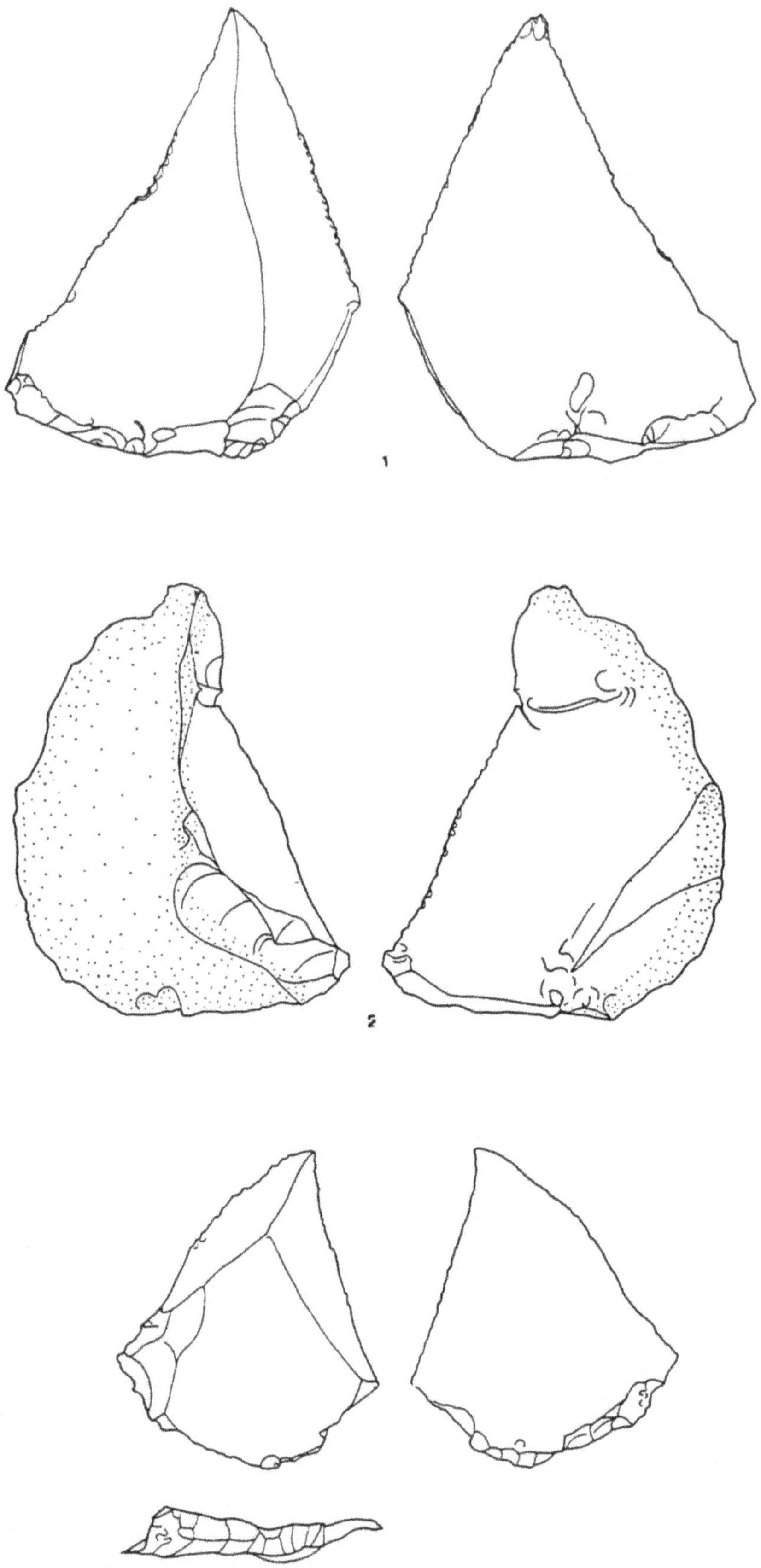

Figure 251: Artefacts from the Mousterian of Kebara cave.

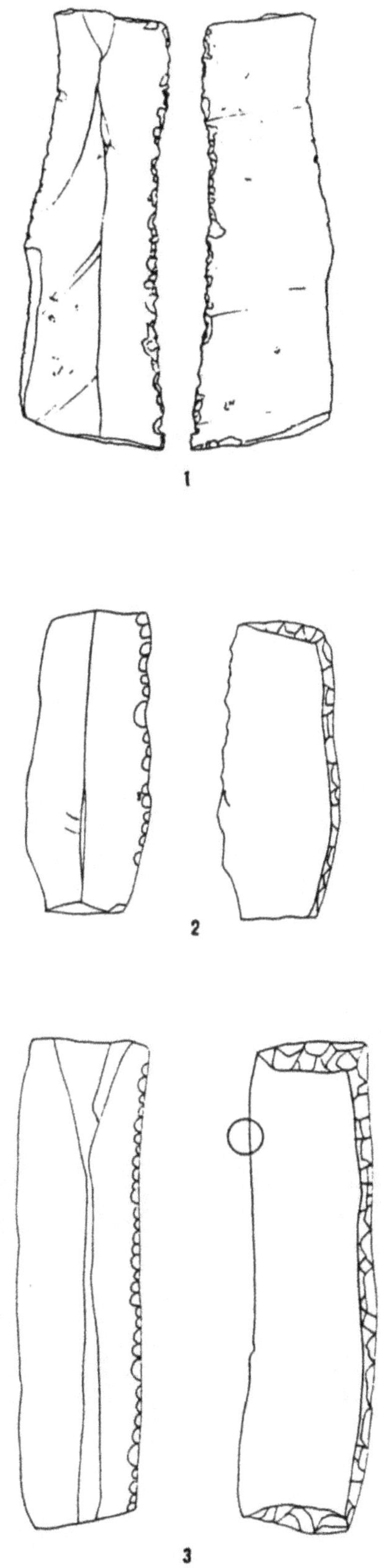

Figure 252: Sickle blades with gloss from Far'ah.

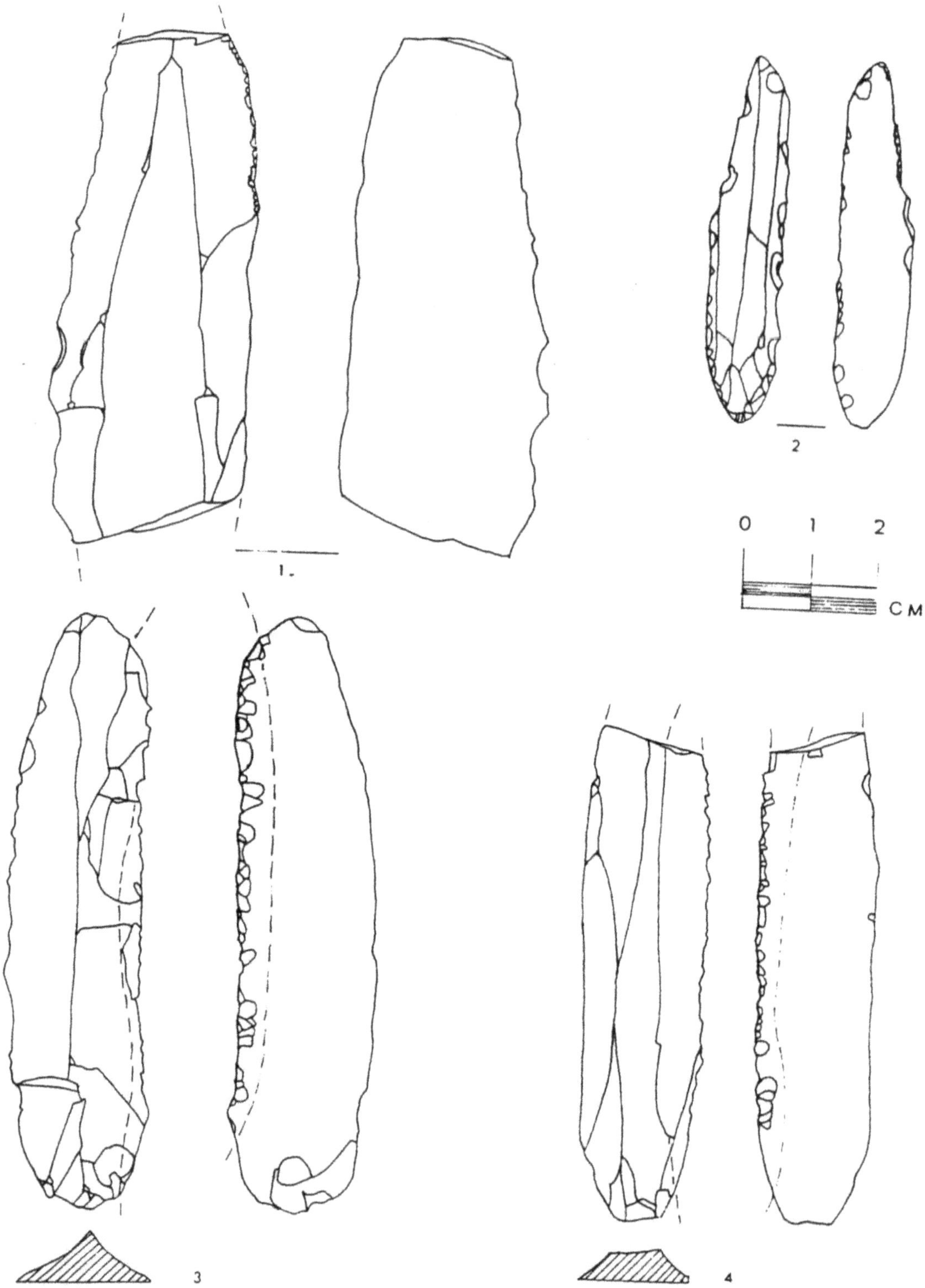

Figure 253: Artefacts from Nahal Heimar cave.

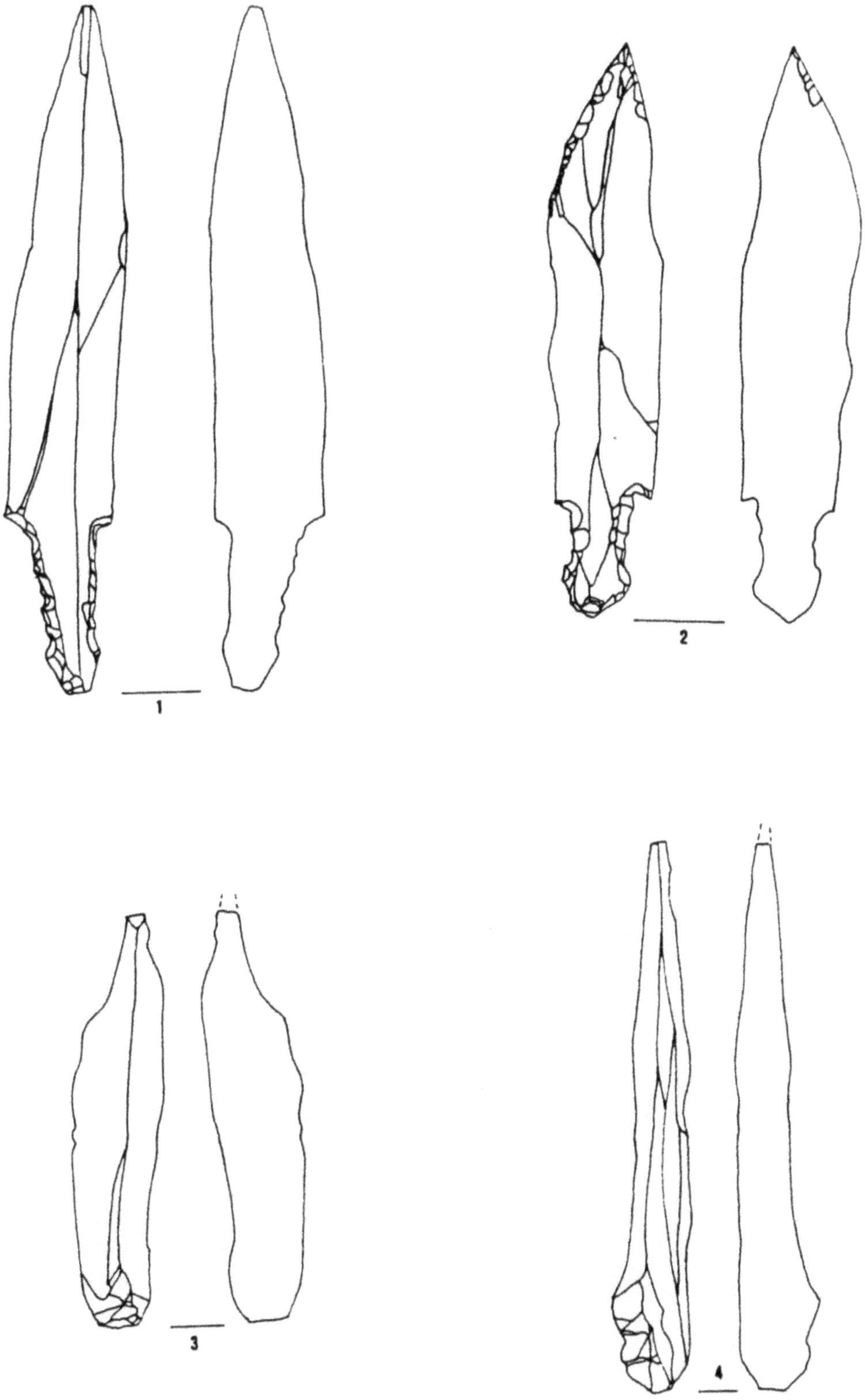

Figure 254: Artefacts from Nahal Heimar cave.

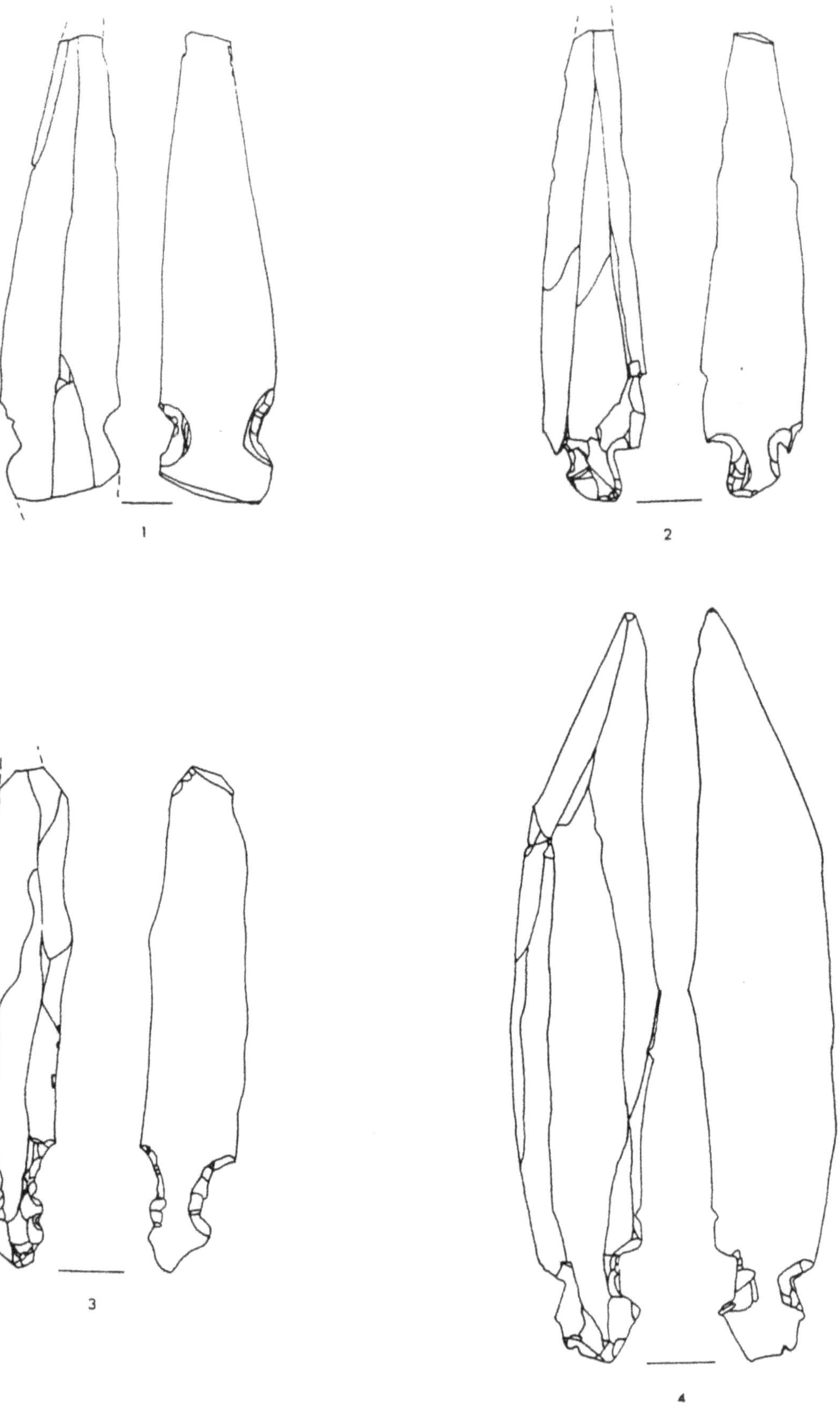

Figure 255: Artefacts from Nahal Heimar cave.

www.ingramcontent.com/pod-product-compliance
Lightning Source LLC
LaVergne TN
LVHW071210210926
848482LV00015B/772